Aboriginal Peoples and Birds in Australia

To all Australians who love birds.

Aboriginal Peoples and Birds in Australia

HISTORICAL AND CULTURAL RELATIONSHIPS

PHILIP A. CLARKE

CSIRO

PUBLISHING

A catalogue record for this book is available from the National Library of Australia.

ISBN: 9781486315970 (pbk)
ISBN: 9781486315987 (epdf)
ISBN: 9781486315994 (epub)

How to cite:
Clarke PA (2023) *Aboriginal Peoples and Birds in Australia: Historical and Cultural Relationships.* CSIRO Publishing, Melbourne.

Published by:

CSIRO Publishing
36 Gardiner Road, Clayton VIC 3168
Private Bag 10, Clayton South VIC 3169
Australia

Telephone: +61 3 9545 8400
Email: publishing.sales@csiro.au
Website: www.publish.csiro.au
Sign up to our email alerts: publish.csiro.au/ earlyalert

Front cover: Forehead ornament, made from Major Mitchell cockatoo feathers. Mary Laughren collection, Yuendumu, Northern Territory, 1970s.

Edited by Adrienne de Kretser, Righting Writing
Cover design by Cath Pirret
Typeset by Envisage Information Technology
Printed by Ingram Lightning Source

CSIRO Publishing publishes and distributes scientific, technical and health science books, magazines and journals from Australia to a worldwide audience and conducts these activities autonomously from the research activities of the Commonwealth Scientific and Industrial Research Organisation (CSIRO). The views expressed in this publication are those of the author(s) and do not necessarily represent those of, and should not be attributed to, the publisher or CSIRO. The copyright owner shall not be liable for technical or other errors or omissions contained herein. The reader/user accepts all risks and responsibility for losses, damages, costs and other consequences resulting directly or indirectly from using this information.

CSIRO acknowledges the Traditional Owners of the lands that we live and work on across Australia and pays its respect to Elders past and present. CSIRO recognises that Aboriginal and Torres Strait Islander peoples have made and will continue to make extraordinary contributions to all aspects of Australian life including culture, economy and science. CSIRO is committed to reconciliation and demonstrating respect for Indigenous knowledge and science. The use of Western science in this publication should not be interpreted as diminishing the knowledge of plants, animals and environment from Indigenous ecological knowledge systems.

Foreword

When one looks at any map of Australia with some sense of awareness it is possible to understand perhaps two things; one is the vast size of the country and the second is the many various ecologies that exist, ecologies that range from high alpine pastures through to arid deserts and tropical rainforests. The next layer that could be imposed upon this understanding are the Indigenous groups that also covered this continent and found homes in every ecological zone that we might wish to consider. A direct result of this is a remarkable variety of ways of knowing the land, the sea, the plants, fish, animals, birds and natural phenomena. From the outset of colonisation documentation began regarding what knowledge Indigenous peoples might have had about their country and the flora and fauna that existed in it. Such documentation has continued up until the present day though radically transformed in scope and intent, and perhaps also with a certain intensity as many of Australia's Indigenous languages are critically endangered. In the past very few of the men and women who documented such knowledge spoken any of the Indigenous languages, they had little experience working with oral traditions and they knew little about the Indigenous political and social organisations that held the country, and often the intent of the documentation was to preserve knowledge before it all 'passed away'. There were of course men and women of more serious intent who really wanted to know about Australia's flora and fauna and they were assisted and often lead by Indigenous specialists, such work is documented in *Australia's First Naturalists* by Olsen and Russell (2019).

To undertake any research into the history of this kind of documentation is hard work; there is not only the archive to contend with and the various ways that many researchers and documenters have chosen to represent Indigenous knowledges, it is also having to sometimes contend with fragments of knowledge, often seemingly contradictory, or of species that now may be extinct. Yet carefully done, such research opens up ways to reengage and understand the place of Indigenous knowledges in Australia. Therefore, such books as this one in regard to birds is an important contribution to such knowledge, and given what has been said above, the hours of work researching and creating a ways of documenting this knowledge has been well thought through and thorough in terms of content and specificity.

It is important to recognise that such texts are not just for a non-indigenous audience. Increasingly, Australia wide, Indigenous peoples are reading such texts as this in an attempt to revivify cultural practice and knowledge as well as research for language and information relating to various species and country. Such knowledge too increasingly finds its way in Native Title conversations and more general discussions about the knowledge of ancestors. As such, documentation of the kind found in this book needs to be cognizant of such uses and employ ways of writing and documenting that breaks the colonial nexus that is all too often apparent in the archives and academic writing. This is sometimes a difficult task, because books such as the one under discussion here document knowledge that is still alive and important to various Indigenous groups and yet on the other hand there is knowledge recorded that may not have been seen by the descendants of other Indigenous groups who are in the process of reclaiming knowledge. This book, whilst not having all the answers, does at least acknowledge that there is a continued and increasingly problematic use of terms, common in the Western academy in regards to Indigenous knowledge of fragmenting knowledge into mirror sets of ethno-disciplines, for example ethno-botany, ethno-musicology, ethno-astronomy, and in regards to this book ethno-ornithology. These ethno-disciplines reflect specific scientific disciplines and divisions and serve to add significance to certain aspects of Indigenous knowledges while simultaneously disregarding and excluding other forms and possibilities of knowledge which do not fall within the criteria of western knowledge, it is a way of fragmenting knowledge and subverting a knowledge based upon inter-relationships in which humans are just one part. Similarly, words such as bio-cultural knowledge or the often-used TEK (Traditional Ecological Knowledge) and its variants do the same. Western knowledge too is an ethno knowledge, and given contemporary efforts at decolonial practice perhaps a term as simple as Indigenous knowledges is enough to use and when, the source of knowledge is actually known acknowledging the source of this knowledge to region, Country and language. This book does acknowledge that there are – and needs to continue to be – new ways of thinking about this knowledge and making it a space for Indigenous people to also access knowledge without having to confront the myriad ways the West has decided Indigenous knowledges should be spoken about. It is not being 'new age' or 'trendy' to remind readers what is still colonial about forms of western knowledge and the recording of Indigenous knowledges. Any act of decolonisation should unsettle certain complacencies. Decolonisation brings about the repatriation of Indigenous land and life and knowledges which is so important in a text like this.

I understand the term spirituality is not used in this book, but the point needs to be made that both Indigenous and non-indigenous people will use a book like this to explore such ideas. It is then perhaps worth noting that in the way it is now used, spirituality is a concept that is not susceptible to a dictionary-type definition. For many Indigenous people there are four main concepts – respect, complexity, creation and connection – that come together in spiritual practices. It is probably best to think of spirituality as action rather than a thing. In many instances spiritual practice is not confined to humans, the living world of sentient and what we might consider to be non-sentient beings also convey understandings of spirituality. There is if you like a moving, a concept of motion from inside the person to outside into the country. Books such as this one will be used for deeper exploration.

While there are complexities around Indigenous knowledge and its documentation, this book is an important contribution to our understanding of Indigenous peoples and their relationships with Australian birds.

Dr John J. Bradley
Associate Professor
Acting Head Monash Indigenous Studies Centre
Monash University, Victoria

Contents

Prologue

Since the dawn of time our people have had a spiritual, environmental and cultural connection to all birds and their habitats.

Whether it be through our Ngartji [*ngaitji*] system which defines our roles and responsibilities of caring for Country and all in it, our storylines and songlines, as a food source, as a material source or as a transmitter of messages – birds are and have always been a part of our daily rituals and wellbeing.

This book encompasses all that is our connection to Country and self.

Mark Koolmatrie
Ngarrindjeri Elder

Acknowledgments

The following people assisted me with the writing of this book. Kim Akerman, Robert Graham, David Jones, Peter Sutton and Philip Weinstein generously made time to provide detailed comments on drafts. Alexis Ee-Khem Aw, Cameron Clarke, Daryl Clarke, Kyle Clarke, Li Fen and Li Xiaoli assisted with photography. Lea Gardam at the South Australian Museum Archives arranged for permission to reproduce the William Barak painting. Bereline Loogatha selected photographic images from the archives at Miart, Mornington Island. Additional images were provided by Kim Akerman, Valerie Boll, Mark Crocombe, Robert Graham, Jeff Hardwick, Mary Laughren, Jamie Robertson, Peter Sutton and the National Library of Australia. As background for the research material covered in this volume, many Aboriginal peoples on their own Country have enthusiastically discussed their perception and use of the environment with me. Working as an anthropologist from 1982 to 2011 at the South Australian Museum, and since then at Griffith University, Federation University and in private practice, has given me many exciting opportunities to explore Aboriginal culture.

Cultural sensitivity warning

Readers are warned that there may be words, descriptions and terms used in this book that are culturally sensitive, and which might not normally be used in certain public or community contexts. While this information may not reflect current understanding, it is provided by the author in a historical context.

This publication may also contain quotations, terms and annotations that reflect the historical attitude of the original author or that of the period in which the item was written, and may be considered inappropriate today.

Aboriginal and Torres Strait Islander peoples are advised that this publication may contain the names and images of people who have passed away.

1
Introduction

Throughout the world, birds have a prominent place in the cultural traditions of human societies, being celebrated in painting, craftwork, song, ceremony, dance, story and literature. In many Indigenous cultures they are perceived as having been involved with the original creation of the world and as spirit beings to have since then maintained their supernatural powers that can transform the day-to-day existence of people. Birds have a symbolic significance relating to both life and death. They are routinely seen as portents of impending calamity and even death, are considered carriers of the spirits of dead people, and often embody those same spirits themselves. Conversely, birds are also associated with life and fertility. Birds are significant as sources of food and medicine. In the material culture, which is the totality of objects that people make and use, birds appear symbolically in art, while bird parts are routinely used in the construction of artefacts. Birds have complex behaviours, some of which are analogous to those of humans.

When the British colonists first arrived at Sydney in 1788, Aboriginal Australia was culturally complex, with over 200 language groups and many more clans spread across an ecologically highly varied landscape.[1] In spite of major regional cultural differences, the importance of birds for Aboriginal peoples was universal.[2] Over the millennia of their occupation of this continent, the Aboriginal custodians of the environment had developed deep understandings of the immense diversity, seasonality and spatial distribution of the organisms within their foraging territories. This Indigenous-held knowledge was so detailed and useful to the European colonists that the first collectors of natural history specimens routinely employed local Aboriginal peoples as guides and assistants.[3] This relationship between Aboriginal peoples and scientists persisted into the 20th century, and is perhaps best demonstrated by the work of biologist cum anthropologist Donald Thomson, who incorporated observations from his Aboriginal field assistants into his zoological writings.[4] He was not alone in appreciating Indigenous knowledge. Geologist and anthropologist Charles Chewings remarked that Aboriginal hunter-gatherers:

> … know the habits of every living thing around them, great or small. Captain S.A. White [who was a South Australian naturalist] has expressed the opinion that they are, in this respect, the most competent naturalists that ever were, or ever will be. With the sure knowledge that they can find food when they go forth to hunt tomorrow, they make the best and most of the present.[5]

This book focuses on the historical and cultural relationships of Aboriginal peoples and the Australian avifauna. While it provides an overview of Aboriginal relationships with birds across the whole continent, it accesses only a small part of the immense body of knowledge that Aboriginal peoples have held about a group of organisms that Europeans refer to as birds.

Much of the data utilised in this book was found in obscure historical sources, which belong to a period in which Aboriginal peoples were often harshly treated. Much of the language used by Europeans in the past to describe Aboriginal peoples is unsuitable for contemporary use, but in order to maintain the integrity of the original sources it is, in a few cases, retained in quotes. The book highlights what, to many Australians, are different ways of looking at a much-loved part of the continent's fauna. It aims to raise an awareness of the alternative bodies of ornithological knowledge that reside outside of Western science.

For Aboriginal readers, this book makes information about their culture that has lay hidden in often hard to find places more readily available. For those requiring deeper insights it plots a course through the myriad of colonial archival sources and academic writings. While the book looks at the historical ways that information about Aboriginal relationships with birds has been recorded, it also acknowledges that Aboriginal traditions concerning birds are not just from the past. For many Aboriginal peoples today, birds are a crucial element of their living, active knowledge about the world they live in.

Ethno-ornithology and environmental knowledge

The academic study of local-based knowledges concerning what Western Science defines as the physical world is conducted by scholars within the ethnoscience field. The aim of this field is to gain a more complete description of cultural knowledge by using the intellectual skills of an anthropologist as an ethnographer rather than a scientist.[6] The ethnographer engages in ethnography, which is the systematic and in-depth study of a single cultural entity, usually through fieldwork.[7] As a field of study, ethnoscience provides a way of discovering how people perceive their physical environment and relate to it, as reflected in their own words and actions. There are more specific fields within ethnoscience, such as ethnobotany which studies the cultural relationships with the flora, and ethnoastronomy which is chiefly concerned with cultural constructions of the night sky. Similarly, ethnozoology is focused on human interactions with the Animal Kingdom: as a subfield within it, ethno-ornithology is concerned with the same types of relationships with birds. In spite of the academic origins for the use of these terms, the study of the connections between culture and the physical environment has moved rapidly beyond its colonial origins. Today, Indigenous peoples are much more involved in the shaping of research programs that concern their culture and Country. Most studies within the ethnoscience field have tended to focus on Indigenous relationships with the environment, but the same methods of investigation could, with some modification, also be used to study human/environment relationships in Western European cultures.

Across the world, birds are a conspicuous part of the fauna. This means that for environmental researchers an account of the local bird life and its relations with Indigenous people is an obvious partner for the description of local plant use.[8] Relevant to both these fields are ethnoecological studies in Australia that contain both birds and plants as components of the biota.[9] The focus on regional ecology is useful for studies of human interactions with the environment, because there are analogies and close links between cultural and biological diversity, as seen with the synergistic demise of language, culture, biocultural knowledge and biodiversity.[10] In particular, Aboriginal knowledge and experience of birds is prominent in the literature of Indigenous seasons and calendars.[11]

In terms of scholarly interests, there is some complexity exhibited by the various subfields within the ethnosciences. Within ethnozoology, ethno-ornithology focuses on the cultural interactions with what biologists refer to collectively as the avifauna; that is, the birds. Like its better-known sibling fields, it provides the scholarly means to investigate the relationships that people from specific cultures have with an element of their landscape. In the case of ethno-ornithology, one academic view is that it 'explores how peoples of various times and places seek to understand the lives of the birds around them'.[12] Another explanation is that 'ethno-ornithology might be thought of not so much as the study of indigenous or traditional ways of naming and knowing birds and interpreting these as a coherent knowledge (though this aspect is important), but more as the study of the knowledge practices that shape different framings of bird-related knowledges'.[13] The modern scientific concept of 'birds' representing a group of related organisms that can be identified in the hierarchy of a tree is sometimes absent in non-Western systems of ethnobiological classification, which possess different non-evolutionary ways of grouping organisms together.[14] With the aim of gaining an understanding and appreciation of an Indigenous perspective of organisms that modern Europeans define collectively as birds, it is first necessary to investigate how members of an Indigenous culture broadly structure their own view of the world, and more specifically how the knowledge about it is produced, organised, owned and passed on.[15]

For the present-day researcher, the written sources of data on Aboriginal relationships with Australian birds, for any given area, range from the historical to the contemporary, and have been compiled by people who are variously described as explorers, colonists, settlers, folklorists, museum ethnologists, linguists, anthropologists, geographers, historians, biologists and ecologists. Indigenous scholars have also added significantly to the literature. Often, the various professions of the data recorders and researchers, who for the purpose of the current study are all to be considered ethno-ornithologists, are difficult to distinguish and some individuals may be members of two or more of the above-mentioned groups. This wide-ranging background of sources is to be expected, since birds are key elements within a wider system of Aboriginal belief and tradition relating to how cultures are linked to Country. The broad limitations of the Australian historical records relevant to Aboriginal use and perceptions of birds are in general those that are shared with the study of Aboriginal plant use.[16] A particularly severe impediment for the modern scholar drawing upon historical sources is that early European colonists had scant knowledge of the Australian environment and lacked understanding of local Aboriginal cultures.

The extraction of bird knowledge from other aspects of Indigenous understandings of the environment is somewhat artificial, but for the purpose of this book the separation of various aspects will serve to explore the diverse ways that people interact with birds. The information concerning birds that can be gleaned from research with contemporary Aboriginal peoples and their recorded culture is rich and vast, and there are methodological dangers for scholars who selectively extract such data for scientific purposes without consideration of the cultural context within which it was framed.[17] Over the last decade, researchers from a variety of backgrounds have studied what has variously been called Indigenous Biocultural Knowledge, Indigenous and Local Knowledge, and Traditional Ecological Knowledge. They have developed a framework for understanding how knowledge of organisms, such as birds, is holistically incorporated into local cultures.[18] The high growth in this literature has demonstrated the value and capability of Aboriginal environmental knowledge to complement and corroborate more intensive and local scientific studies within the ecology discipline.

Scholars working within the ethnoscience field have been more recently decolonising their practice of recording relationships that cultural groups have with the environment. Indigenous people are now routinely involved at all stages of research, which typically leads to beneficial outcomes for the participants. Western knowledge concerning the physical environment is also another form of ethno-knowledge. In the future, the use of a term as simple as Indigenous knowledge will greatly help in making the study of such things as Aboriginal relationships with birds appear less bound by academic ways of thinking. The cultural context of all recorded data, both in terms of the original Indigenous sources and the non-Indigenous recorders, requires consideration when scholars use it for other purposes. In this book, when the cultural source of the recorded information is actually known, it is acknowledged to region, Country and language.

Aboriginal peoples have gained environmental knowledge, through their own foraging experiences and from teachings by older relatives, that explains within their cultural view of the world the distributions and behaviours of all the local plants and animals within their Country. For the broader Australian community, research into Aboriginal environmental knowledge can inform the understanding and management of pressing environmental issues such as fire control, conservation of threatened species, removal of invasive species, halting the decline in aquatic ecosystems and the mitigation of climate change.[19] When considering Western science and its overlap with Indigenous knowledge and environmental experience, it has been argued that the 'successful integration of the two knowledge systems may improve the conservation of both biological and cultural diversity, while empowering Indigenous peoples'.[20]

The field of ethno-ornithology is interdisciplinary, which as the name suggests means that researchers within it must have good working knowledge of both human cultures and ornithology. In my experience, the ideal scholar in this field is a zoologist or ecologist with anthropological or linguistic training. Donald Thomson from the mid-20th century is a prime example of such a scholar.[21] Anthropologist and ethnobiologist Eugene S. Hunn stated that 'professional ornithologists have discovered that they share with unschooled hunters and farmers a common appreciation of the beauty and fascination of birds and that they also share a common language of equivalent bird names with which to compare their ornithological observations'.[22]

The written sources

In Australian studies of Aboriginal relationships with birds there are two main frameworks for research: studies that relate to human–avian interactions within a specific cultural region,[23] and works that focus on Aboriginal relationships with a particular set of bird species.[24] For introductions to the field, there are compilations of records pertaining to Australia-wide beliefs and practices.[25] Many recent studies are charged with creating greater recognition of the knowledge and rights of Indigenous peoples on the land, and as such they include a conservation aspect.[26] A comparison between studies of cultural relationships with birds from across the world shows that many analogous beliefs and traditions are shared by different Indigenous peoples.[27] In some cases, the present work suggests why this is so.

An apparent problem with many existing sources of data when investigating Aboriginal relationships with birds is the lack of sufficient base knowledge possessed by the recorder. The issue is that 'researchers often do not have the depth of knowledge about birds that local people possess and may not be able to interpret the stories they are told, while local communities are often unfamiliar with printed materials'.[28] European colonists arriving in Australia from the late 18th century were unfamiliar with many Australian bird species. When combined with the rapid decline of local Aboriginal populations, this has led to a shortage of data for regions such as south-eastern Australia, where British settlement first began. The recorder's poor understanding of Indigenous languages is also an impediment.[29] In spite of these major difficulties, the colonist Alfred C. Stone at Lake Boga in central northern Victoria remarked, about learning information from living Aboriginal sources, that 'having gained his confidence it becomes surprising to find the vast knowledge possessed of the flora and fauna of his surroundings, and the tales and sometimes weird traditions of his tribe'.[30]

There is a wealth of recorded ethnobiological information in Australia that is relevant to the study of Indigenous relationships with birds, although there are significant biases. The published ethno-ornithological work across the world, including Australia, has so far largely focused on the cultural roles of birds, with less emphasis on the physical aspects. In Australia at least, this is the reverse situation from its much better-known academic sibling, the study of ethnobotany, which has been biased towards looking at Indigenous relationships with the flora as sources of food, medicine and material culture.[31] In spite of this, practitioners from both fields have had strong antiquarian interests in documenting the past. The fields of ethno-ornithology and ethnobotany shared an imperative to record relationships from the early hunter-gatherer period, to the detriment of studying aspects of the changing relationships that Indigenous peoples have had with the biota since British colonisation, particularly in rural areas.

In this book I have chiefly employed descriptive methods.[32] There is much information provided here about Indigenous relationships with birds, but it is necessary to stress that in relation to Aboriginal traditions I have endeavoured to reproduce so-called 'outside' (public) accounts, in order to fully protect any 'inside' sacred knowledge about birds that contemporary Aboriginal peoples may hold and wish to remain secret. I have therefore relied upon information that can be seen as in the public domain, even if it is hard to find, and have steered away from citing the more recent 'grey literature' generated by native title and cultural heritage practitioners, which often contains information of uncertain status. In relation to Aboriginal accounts of the creation period, I have avoided using 'Dreaming' or 'Dreamtime', unless writing about Central Australia where such terms have been commonly used by both Aboriginal custodians and scholars.[33] Aboriginal peoples in other parts of Australia prefer to use English words, such as 'Story', 'History' and 'the Law', when referring to the ancient cultural traditions concerning their cosmological origins.[34] Of course, in their own languages they have terms for the creation, such as *Wangarr* in Yolngu spoken in north-east Arnhem Land,[35] *Tjukurpa* in Pitjantjatjara/Yankunytjatjara from the Western Desert cultural region,[36] and the *Altyerrenge* in Eastern and Central Arrernte of Central Australia.[37] In this book I have referred to creation traditions as myths in an anthropological sense, but in no way should this be taken to mean that they are lesser accounts of the past than others, such as the perspective offered by Western science. As explained in Chapter 2, I consider that the use of 'myths' as a term is preferable to both 'legends' and 'tales', as described by some folklorists.[38] I have also avoided using 'songlines', as coined by travel writer Bruce Chatwin,[39] due to the associated incorrect inference that all Aboriginal myths are linked into an Australian-wide grid of ancestral tracks.

The bulk of the data in this book comes from either the corpus of published sources, such as books and papers, or from the largely unpublished material in the historical records housed in government-funded libraries and archives. In the case of the latter, I aimed to bring this too often hard-to-find information into the present, where it can be more readily and appropriately accessed by interested lay people and members of Aboriginal communities. In many parts of Australia, particularly where Aboriginal peoples have experienced the full impact of British colonisation commencing in the late-18th century, the knowledge of past relationships with the environment is being actively reclaimed from the past as an element of the local expression of Aboriginal identity. Where appropriate, I have drawn upon my own unpublished field data from the Lower Murray of South Australia, chiefly gathered during my doctoral research during the late 1980s and early 1990s,[40] and included material from other parts of the continent where I also worked, in order to provide an Australia-wide perspective on the topic. In the case of my own fieldwork notes, wherever possible I have acknowledged the living source of the information; however, on a few occasions for cultural and personal reasons their identity must remain hidden.

The historical use of bird names is varied, but whenever possible the recognised standard common names used in this book are those listed in official Australian checklists.[41] For example, when there is sufficient evidence from either its description or through cross-referencing with Aboriginal dictionaries, I have interpreted the somewhat vague term 'eaglehawk' to mean the 'wedge-tailed eagle', and similarly identified the 'seahawk' as the 'white-bellied sea eagle' in some instances.[42] More straightforward has been the replacement of 'plains turkey' with 'Australian bustard', 'native companion' with 'brolga', 'stone plover' with 'bush stonecurlew' and the 'spur-winged plover' with 'masked lapwing'. This is intended to lessen the reader's confusion with colloquial names. All the names for the organisms mentioned in this book are listed in the index.

There is much variation in the naming of Aboriginal peoples, as reflected in the historical records. Whenever possible, the favoured spellings for Aboriginal group and language names, as chosen by contemporary community organisations, are adopted. Examples of the standardisation of names in this book are Arrernte used instead of Aranda, Bunganditj for Booandik, Dhanggati for Thangatti, Diyari for Dieri, Gamilaraay for Kamilaroi, Ganai for Kurnai, Gugu Rarmul for Koko-rarmul, Gupapuyngu for Kopapingo, Karajarri for Karadjeri, Kokatha for Kukata (in South Australia), Ngarrindjeri for Narrinyeri, Nyungar for Noongar/Nyungah, Pitjantjatjara for Pitjandjara, Tiwi for Tewi, Wardaman for Waddaman, Warumungu for Warramunga, and Yuwaalaraay for Euahlayi.[43]

Book structure

This book covers a broad range of relationships that Aboriginal peoples have with birds, on a spectrum from cultural perceptions to the actual physical use of bird materials. In terms of cultural perceptions, the text commences with the role of birds as ancestors (Chapter 2), creators (Chapter 3) and spirit beings (Chapter 4), then describes avian nomenclature (Chapter 5). The physical relationships that Aboriginal peoples have with birds are discussed in relation to the topics of bird foraging (Chapter 6), birds working with people (Chapter 7), birds as food and medicine (Chapter 8) and avian materials used to make material things (Chapter 9). The concluding chapter (Chapter 10) considers the future of Australian ethno-ornithology.

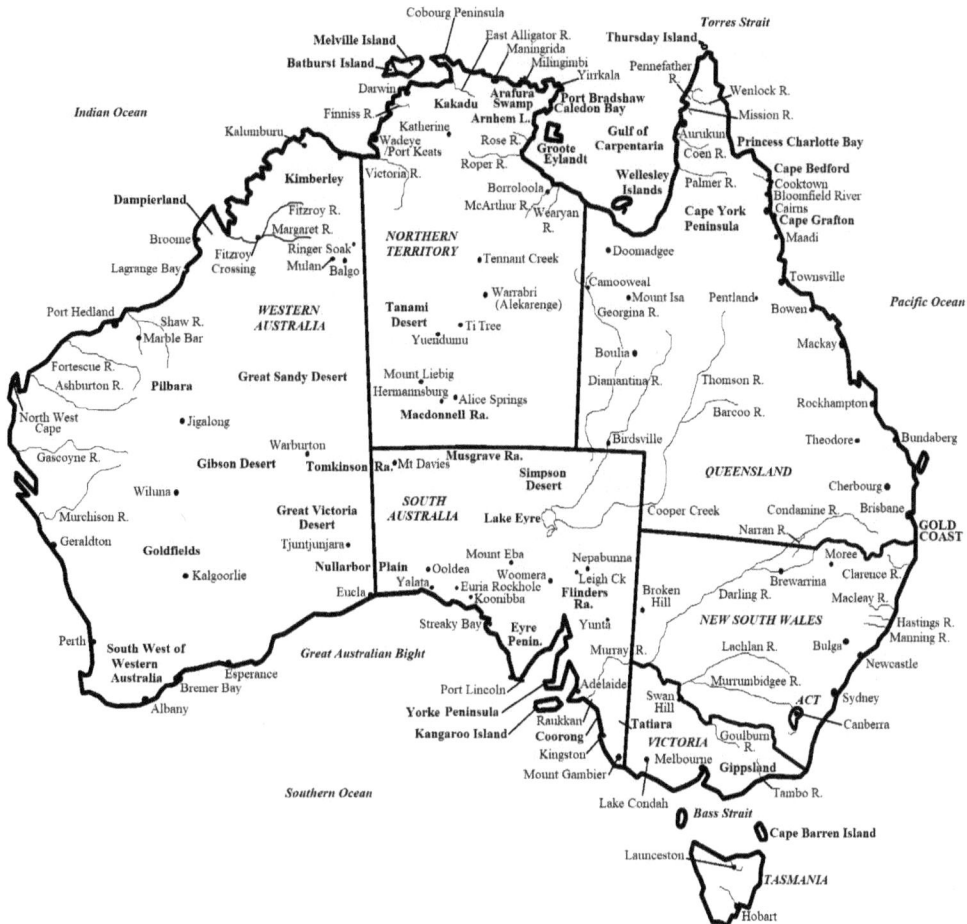

Map of Australia, showing places and regions mentioned in the text. Philip A Clarke, 2021.

Endnotes

[1] For overviews of Aboriginal culture, see Curr (1886–87), Maddock (1982), Jacob (1991), Horton (1994), Berndt & Berndt (1999), Mulvaney & Kamminga (1999) and Clarke (2003a). For overviews of Aboriginal language diversity, refer to Blake (1981), Yallop (1982), Schmidt (1993), Henderson & Nash (2002) and Sutton & Walshe (2021).

[2] Roth (1903), Blythe & Wightman (2003), Gosford (2003, 2009), Tidemann & Whiteside (2010) and Tidemann *et al.* (2010).

[3] Clarke (2008a) and Olsen & Russell (2019).

[4] Thomson (1935, 1939, 1949, 1975, 1983a, 1983b, 1996).

[5] Chewings (1936:9).

[6] Sturtevant (1964) and Amundson (1982).

[7] Barfield (1997:157–160).

[8] Bartholomaeus (2012).

[9] Raymond *et al.* (1999), Puruntatameri *et al.* (2001) and Wynjorroc *et al.* (2001).

[10] Nettle & Romaine (2000), Lepofsky (2009:161) and Grant (2012).

[11] Davis (1989), Reid (1995), Clarke (2009c) and Turpin *et al.* (2013).

[12] Hunn (2010:xi).

[13] Jepson (2010:327).

[14] Bulmer (1967, 1978), Maddock (1975, 1978a), Brown & Naessan (2014) and Clarke (2019a).

[15] Medin & Atran (1999) and Tidemann *et al.* (2010).

[16] For Australian ethnobotany, refer to Webb (1969, 1973) and Clarke (1986a, 2003c, 2014a).

[17] Clarke (2014d).

[18] Ens *et al.* (2012, 2015, 2016) and Hanspach *et al.* (2020). Indigenous Biocultural Knowledge (IBCK) is a modified description for what was widely known as Indigenous Ecological Knowledge or Traditional Ecological Knowledge (see Berkes *et al.* 2000; Huntington 2000), with an increased emphasis on the importance of cultural connections. In Australia, IBCK has been termed Aboriginal Biocultural Knowledge (ABCK) in relation to Aboriginal peoples (Cahir *et al.* 2018a, 2018b, 2018c). In some contexts, it is equivalent to Indigenous and Local Knowledge (ILK), as described by Fernández-Llamazares *et al.* (2021). For a background to ABCK and its associated fields (i.e. Traditional Ecological Knowledge or TEK) in Australia, refer to Shackeroff (2007), Russell-Smith *et al.* (2009), Ens *et al.* (2012), Woodward *et al.* (2012), Leonard *et al.* (2013), Ziembicki *et al.* (2013), Ens *et al.* (2015), Lynch *et al.* (2015), Pert *et al.* (2015) and Cahir *et al.* (2018b, 2018c).

[19] Examples of environmental research incorporating an Aboriginal perspective include Andersen (1999), Bird Rose (1995) and McKemey *et al.* (2020) for fire management; Pursche (2004), Wilson *et al.* (2004), Langton & Rhea (2005), Ens *et al.* (2015) and Davies *et al.* (2018) for biodiversity; Vaarzon-Morel & Edwards (2012) and NM Smith (2013) for exotic species invasion; Barber (2005), Barber *et al.* (2012), Jackson *et al.* (2012) and Berry *et al.* (2018) for wetland and water resource management; and Green *et al.* (2009), Jones *et al.* (2013, 2018) and Low Choy *et al.* (2013) for climate change mitigation.

[20] Sinclair *et al.* (2010:115).

[21] Thomson (1983a, 1996).

[22] Hunn (2010:xi).

[23] Such as Webb (1933), Johnston (1943), Waddy (1988), Blythe & Wightman (2003), Puruntatameri *et al.* (2001), Wynjorroc *et al.* (2001), Turpin *et al.* (2013), Brown & Naessan (2014) and Clarke (2016a, 2016b, 2017, 2018h, 2018i, 2019a).

[24] Berndt (1940c), Hercus (1971), Blows (1975), Lewis (1988) and Gosford (2009).

[25] Gosford (2010) and Tidemann & Whiteside (2010).

[26] Bonta (2010) and Tidemann & Gosler (2010).

[27] Examples with analogous 'bird' beliefs include Ichikawa (1998), Kizungu *et al.* (1998), Gonzalez (2011) and Loss *et al.* (2014).

[28] Ng'weno (2010:103).

[29] P Sutton 1980 (cited in Clarke 2014a:134).

[30] Stone (1911:434).

[31] For overviews of Australian ethnobotany refer to Clarke (2003c, 2007a:146–147, 2008a:150–151, 2012:272–273).

[32] These techniques are utilised by Tidemann & Gosler (2010).

[33] Arthur (1996:27–30,39–42,59–60).

[34] Sutton (1988c:252).

[35] Williams (1986:245).

[36] Goddard (1992:155).

[37] Henderson & Dobson (1994:105–106).

[38] For example, Ramsay Smith (1930) for 'legends' and Waterman (1987) for 'tales'.

[39] Chatwin (1988).

[40] Clarke (1994).

[41] Christidis & Boles (2008) and Gray & Fraser (2013).

[42] According to Ramson (1988:228), the 'eagle hawk' is 'A large bird of prey, usu. The *wedge-tailed eagle*'. The 'Wedge-tailed Eagle, the largest bird-of-prey' is a common totemic ancestor in Aboriginal Australia (von Brandenstein 1982:111,135,142). The identity of 'sea-hawk' is less certain, although when a large bird is involved, in my opinion it is more likely to refer to the white-bellied sea eagle than to a species of osprey.

[43] The tracking of many of these names can be achieved through reference to Tindale (1974).

2
Birds as ancestors

Fundamental to the cultural beliefs of Aboriginal Australia is the concept that there was a creation period in the past, when ancestral beings performed heroic deeds and, in the process, moulded the landscape for the benefit of Aboriginal peoples and laid down for them the laws and customs to be followed.[1] The creation mythology comprises an Aboriginal system of beliefs that offers answers to the great universal religious questions of humankind, such as those concerning the origin, meaning, purpose and destiny of life. It is the driving force behind Aboriginal songs, ceremonial performances and the styles reflected in the material culture. There are also myths about events that take place sometime after the creation, such as those that involve spirit beings (discussed in Chapter 4) and others that relate more specifically to aspects of Aboriginal culture, such as death and historic inter-group conflict.[2]

The Aboriginal creation mythologies explain how a relatively featureless landscape was shaped and given meaning for all those who were to follow. It was believed that the main beings present in the creation period were the spiritual ancestors of both themselves and many of the animals, plants and other objects, such as the moon and sun, as seen today. The early anthropologist and medical practitioner Walter Edmund Roth reported that in northern Queensland 'when a native wishes to speak of the earliest conceivable eras, he usually expresses himself somewhat in the form of "When the animals and birds were all black-fellows"'.[3] These ancestral beings were considered to have possessed all the human traits, virtues, pleasures and vices. They could also die and be transformed into landforms at the conclusion of the creation. Aboriginal peoples perceive their traditions about the creation as collectively including the beliefs in their ancestors, the cultural practices they introduced, and the tangible objects and places that they left behind in the landscape. The ancestors are still considered to influence Aboriginal Australia. Therefore, to many contemporary Aboriginal peoples their creation mythology is of crucial importance for understanding their world, even though the myths are referenced to past events.

In Aboriginal English, terms such as 'dreaming' and 'story' represent a wide range of meanings, but they can loosely be defined as the whole body of mythology and associated belief and custom that provides important insights into Aboriginal religion.[4] Although Aboriginal peoples gained fresh insights into their past and the relevance of the landscape through actual dreams, the dreaming and dreaming ancestors are not the products of those dreams. The use of 'Dreaming' to refer to the creation has most currency in Central Australia;[5] 'Story' and 'History' are used in many other parts of Aboriginal Australia.[6] The creative period itself is sometimes referred to as the 'dreamtime', but since its power is conceived as eternal this term is best replaced by the 'creation', as favoured in this book. In the lay literature, the paths or tracks taken by ancestors are sometimes referred to as 'songlines', following the writings of popular travel writer Bruce Chatwin.[7] However, the use of 'songlines' is not recommended since it implies that all Aboriginal traditions are connected into the one grid – and this is not so.[8] In the study of anthropology, myths are generally defined as the narratives of a culture that chiefly involve the actions and events of supernatural beings, embody popular ideas about the natural world and historical events, and are believed to be true.[9] Across Aboriginal Australia, the creation mythologies are highly diverse, albeit there are some shared characteristics and resonance.

Aboriginal knowledge concerning birds was incorporated into the mythologies that explain the origin of the world and the order of the universe.[10] In 1957, anthropologist Ursula H McConnel described Aboriginal mythology as a 'fund of information' which was required in order 'to maintain the necessary standards of knowledge and behaviour'.[11] More recently, Australian ethno-ornithologists Sonia Tidemann and Tim Whiteside remarked that 'stories are a part of the fabric of Aboriginal culture, often indicating expected cultural behaviour, but also account for plumage characteristics, calls, habitat, food, the relationships between Earth and extraterrestrial objects, and interspecific behaviour of birds'.[12] The myths are therefore repositories of much essential environmental knowledge that Aboriginal foragers required for their subsistence and religious life.[13] Cultural interpretations of birds and their complex roles in the landscape as reflected in myth require close attention to avian characteristics and behaviours.[14] For this reason, research methodologies that involve a focus upon narrative and storytelling will reveal much Indigenous knowledge about birds.[15]

Recording myths

The colonial recording of Aboriginal mythology was severely hampered by the European lack of appreciation for the sophistication of the art and culture of hunter-gatherers. The Aboriginal performances that the British colonists observed became known as 'corroborees', based on the Dharuk word *garabari*, from the Sydney area of New South Wales.[16] Historian Margaret Clunies Ross remarked that while early European descriptions of these events indicated the existence of a type of theatre, 'the songs which accompanied these performances were not generally received with much understanding. They were considered lugubrious, repetitious, discordant, barbarous, and heathen'.[17] Furthermore, the early European reliance upon the use of pidgin English when compiling accounts from Aboriginal sources produced what often appear to be brief and somewhat childish accounts of myth narratives that in reality were both extensive and complex.[18] The proper recording of Aboriginal culture required academically trained anthropologists and linguists, who had the intellectual tools to compile detailed ethnographic accounts.

Myths are not just an Indigenous form of what Europeans see as history, but are expressions of the teller's world view and are shaped by Indigenous notions of time and space.[19] Anthropologist Fred R Myers explained that, for the Pintupi of the Western Desert, their ontology or nature of being as reflected in their Tjukurrpa or 'Dreaming' 'emphasises the relatedness of the cosmos, rather than the opposition of spirit and matter, natural and supernatural, or good or evil'.[20] The early traditional Aboriginal notion of time is cyclical, whereby the divide between past and present is not as great as it is conceived by Westerners, whose understanding involves science and a linear perception of time.[21] The mythology generally emphasises an explanation of the background to the lived present, instead of passing on records of past events that have little or no contemporary relevance to the cycle of life. When explaining Aboriginal cosmology and metaphysics in relation to the northern Rainbow Serpent ancestral spirit, anthropologist Kenneth Maddock explained:

> A first approach would be to lay stress on the cyclicity embedded in the concept and to draw attention to the role of cyclical thinking in Aboriginal thought generally. Thus human spirits are conceived of as engaged in indefinitely repeated rounds of existence; there is a cycle of the seasons marked by the presence or absence of rain and rainbows; and subsection systems, introduced as we have seen by the All-Mother, are descent cycles. The curvilinear imagery of snakes and rainbows might be considered apt to express the abstract notion of cyclicity.[22]

The dynamic nature of Aboriginal myths and the telling of them allows for reconfigurations to take account of the rapidly changing social and cultural environment. For this reason, recorded myths frequently contained new elements, such as the arrival of exotic peoples. Redundant information, such as that concerning an extinct species or a past period of volcanic eruptions, is not as readily passed on. The ease with which exotic influences were incorporated into the Aboriginal mythologies was apparent to the earliest European recorders, who often attempted to edit out the new elements in order to maintain the perceived purity of their account. For instance, colonist Ethel Hassell produced a detailed record of the early beliefs and traditions of the Wheelman people living north-east of Albany in the south-west of Western Australia. She noted that 'the stories I heard began to show the influence of the whites, but the old terms were there and where I could palpably see the whites' influence I have always eliminated it from my legend'.[23]

The expansion of Aboriginal geographical and cultural knowledge, through the movements of people after European settlement, has also altered the telling, and therefore the recording, of many myths. It has also produced some altogether new mythologies.[24] For instance, in response to European colonisation in the Lower Murray during the early 19th century the Aboriginal peoples developed a belief that a cataclysmic event was about to take place, which would transform people into birds. Colonist James Campbell Harwood told of an apocalyptic belief that there was a man with 'an eye like a star' who guarded a large tree which reached into the clouds, where it was being attacked by thunder and lightning.[25] Lower Murray people believed that if this tree was eventually destroyed, then everything 'would be inundated with water and all the population, black and white, would be turned into ducks and swans and other aquatic birds'.[26] This account is consistent with the post-European mythology of the region, whereby Kulda the 'meteor man' came out of Yuki the Southern Cross in the Skyworld to bring smallpox to Aboriginal people and to take the souls of the dead to Kangaroo Island and beyond.[27]

The dynamic quality of Aboriginal myths, and the environmental knowledge within them, renders it dangerous for present-day scholars to treat them simply as largely static records of ancient events. Yet some have done just that; for example, regarding the arrival of people in Australia, extinction of the megafauna, major landscape changes and irregular cosmic phenomena.[28] Scholars who selectively extract 'new' scientific data from myths end up ignoring data that does not fit. An example is the myth involving crows, fires and alleged volcanoes, which is discussed in Chapter 3. Myth narratives, as oral traditions, are subject to constant processes of augmentation which require the excising of inconsistent or redundant information, even when the main themes are retained. Biologist/anthropologist Donald Thomson discussed the role of myth in validating innovations, such as the use of Asian-style watercraft in eastern Arnhem Land. He remarked that:

> The Australian aborigines are extremely conservative, and before absorbing any important element of culture, social or material, they generally give it a place in their myth pattern and bring it within the totemic scheme. This provides a background in tradition and so gives a charter for its use.[29]

The anthropological recording and description of Aboriginal mythology has been influenced by several scholarly debates. An issue in the late 19th century, debated by scholars working predominantly in south-eastern Australia, was whether or not supreme male beings like Baiame, Bunjil, Ngurunderi and other so-called 'All-Father' figures connected to the Sky, were signs of the pre-existence of 'high gods' in Aboriginal mythology, akin to the monotheistic Judeo-Christian God.[30] More recent and nuanced scholarship has considered that, due to early interactions with British colonists, Aboriginal peoples in the newly 'settled' regions embraced aspects of Christianity as a reaction to the formation of new sources of power within their community – to them, the existence of a single god-like being in the heavens was an innovation.[31] Into the early 20th century, it was argued that the Aboriginal moiety system, which divides the community into two halves for such purposes as organising marriages and ceremonies,[32] had evolved from the relatively recent structural merging of two formerly separate 'races', designated the Eaglehawk (wedge-tailed eagle) and the Crow (Australian raven). These, like their totemic emblems, were physically different and had been in conflict.[33] While this proposition appears simplistic today, given what is now known about the great depth of Aboriginal occupancy of Australia,[34] the existence of the wedge-tailed eagle and crow/raven as opposites in many moiety systems has nonetheless been of interest to scholars studying social structure.[35]

Another influence for the recording of Aboriginal myth comes from authors writing for the broader public. They appropriated the idiom of Aboriginal myth for their fictional work, and in doing so have generally added different elements and altered the emphasis of the original narrative.[36] Their popular stories have simplified the myths by reducing the use of Aboriginal language and by stripping out the site-related and kinship-related data. Such stories are of mixed value for researchers who aim to extract useful cultural data. Other writers were less transparent with the treatment of their sources of myth. An example is W Ramsay Smith's book *Myths and Legends of the Australian Aboriginals* (1930). Ramsay Smith provided an account of Aboriginal life and tradition that is now known to have primarily been derived from the wide-ranging writings of the famous Ngarrindjeri scholar, David Unaipon, whose portrait is on the Australian $50 note.[37] Through heavy editing and alteration, Ramsay Smith incorporated a mix of folklore extracted from diverse Aboriginal cultures into a narrative as a 'legend' with an uncertain voice. A reviewer noted, 'It would appear that the author has committed the fatal error of translating aboriginal ideas into European terms'.[38] Because of this, in my opinion, while there is some unique ethnographic information incorporated into Ramsay Smith's work, utmost caution must be observed when referring to it.

The literary blending of materials from Indigenous and non-Indigenous sources renders the new work of little or no value to researchers of the early Aboriginal mythologies. In the process of writing 'tales' – defined as a 'usually imaginative narrative of an event'[39] and 'legend' – as 'a story coming down from the past *especially*: one popularly regarded as historical although not verifiable',[40] the cultural origins of key aspects can remain hidden. An example from 1952 involves an Aboriginal tale which appeared in the *Argus* newspaper of Melbourne. It reads as a local version of Shakespeare's 1597 play *Romeo and Juliet*; however, it lacks any reference to specific Aboriginal groups or places, and the young lovers who resist the chief become parrots after the murder of the hero and the suicide of the heroine.[41] *Aboriginal Peoples and Birds in Australia* generally avoids secondary sources that have assimilated Indigenous bird beliefs through a process of cultural syncretism. Instead, it favours primary sources of Aboriginal knowledge concerning birds.

During the mid 20th century, academics actively recorded Aboriginal explanations of the past as a means of building an Indigenous foundation myth for Australia as a nation. In comparing Aboriginal myths with those of the rest of the world, Ramsay Smith claimed that 'Every race has had a great traditional leader and lawgiver who gave it its first moral training, as well as it social and tribal customs'.[42] According to a gatherer of Aboriginal myths, Charles P Mountford, this literary resource amounted to 'Valuable Knowledge' and:

> All these are treasures of considerable scientific interest as well as excellent entertainment. In this connection it is interesting to recall that Grimm's fairy tales were collected by two grave ethnologists, the brothers Grimm, as scientific material, before they became the property of children of all ages and countries.[43]

The writers of Aboriginal folklore believed that 'All Australians should know how the kangaroo got his long tail, why the magpie is black and white, and why the emu has split feet. The aboriginal knew why'.[44] The aim of building the resource base for a national Indigenous heritage drew upon the work of those who had documented Aboriginal 'dreaming' stories and myths.[45] Their publications, particularly those by Mountford, Katherine Langloh Parker and Alexander W Reed, filled the shelves of many school and public libraries in my own childhood. Back in the early to mid 20th century, it was apparent that the opportunities for recording more directly from the living sources of the early mythology in the temperate parts of Australia, where European colonisation had occurred first and most intensively, were rapidly failing. By the 1940s, it was considered that the process of gathering creation stories direct from knowledgeable Aboriginal people required urgent attention.[46]

The ethnographic recordings of myth produced by anthropologists allows the Aboriginal voice to remain intact, which enables the reader to gain deeper insights into the breadth of

Indigenous relationships with their environment. For instance, anthropologist John Bradley utilised Aboriginal drawings paired with Indigenous myth narratives to illustrate the deep connections that Yanyuwa people had with the 'Brolga Dreaming country' of the southern Gulf of Carpentaria region.[47] Much unique information about birds held by Aboriginal people has been recorded by linguists working with speakers to compile wordlists for dictionaries. When out in the field recording bird mythologies, a partnership between an ornithologist and a linguist will allow a greater depth of Indigenous avian knowledge to be recorded than would otherwise be possible.[48]

Birds in myths

Bird ancestors feature prominently in most recordings of Aboriginal myth. A detailed analysis of myth recordings from across Australia found that the emu appeared most often, followed by the Australian bustard and then the brolga, sulphur-crested cockatoo, willie-wagtail and the Australian pelican.[49] Smaller birds without features like spectacular splashes of colouring and unusual behaviour, such as many of the honeyeaters, thrushes and thornbills, are generally under-represented in the records. It is possible that this reflects the interests and knowledge of the recorders, but in my opinion, it is more likely that these commonly featured bird species are those that Aboriginal people consider to be most important. During the writing of this book, I chose to focus on the species that appeared to be the most significant culturally, and have therefore unavoidably reproduced the above bias.

The mythic birds are ancestors who shaped the world for the benefit of their descendants, the people and animals who remained after the creation period had closed. Because of this, the creation narratives can be described as origin myths. For instance, in the traditions of the Tiwi people living on Melville and Bathurst Islands north of Darwin, Milapukala the sulphur-crested cockatoo woman created a big freshwater lagoon at Milapuru near Cape Fourcroy, and populated the surroundings with possums, wallabies, carpet snakes, various lizards, sugarbag (wild bees) and many other foods.[50] The Tiwi believe that without her actions during the creation, the Country would not have been able to support the people who came later. Similarly, in the Flinders Ranges of northern South Australia, Adnyamathanha people have a creation tradition concerning Murlambada the crested pigeon, who stole large grindstones from Kurukuku the diamond dove and left them in places such Prism Hill and Reaphook Hill, where Aboriginal people later quarried them to make their own grindstones for preparing grass seed.[51] The clattering noise that is heard when a crested pigeon takes flight is perceived by the Adnyamathanha as the rattle of the grindstones it is carrying. When the bird almost topples over during landing, this is attributed to the extra weight being carried.[52]

The actions of the ancestors, who generally had human form during the creation period, were considered to have shaped the physical form of their animal descendants after the creation. For instance, the bald patch under the crest of a galah was considered by Aboriginal people in the Narran River region of northern New South Wales to have been caused by a blow from a boomerang thrown by a lizard ancestor during the creation.[53] In the same region, it was believed that the black swan, which was formerly white, was given its black feathers by the crows and that its prominent red bill was a wound resulting from an attack by a wedge-tailed eagle.[54] It was recorded that Arrernte people in Central Australia believe that men who constantly painted themselves red with ochre during the creation later became crimson chats.[55] Adnyamathanha people in the northern Flinders Ranges have a creation tradition that Awi-irtanha the male mistletoebird was struck on the head by his wife wielding a wooden club, when he was feeding on harlequin mistletoes rather than sheltering his children from an approaching storm. As a result of the blood flow, he has a red chest today.[56] In their myth concerning the theft of large grindstones from Kurukuku the diamond dove, her crying caused the red rings that are now seen round her eyes.[57] The Wik Mungkan people of western Cape York Peninsula in northern Queensland have a myth telling how the black-necked stork ancestor was rubbing red 'clay' or ochre onto his spears when it got on his legs, leaving the bird with red legs after the creation.[58]

Behaviour is also enshrined in the creation myths. For instance, a myth recorded from the Narran River region in northern New South Wales explains why Bibbi the brown treecreeper today is 'always running up trees as if he wanted to be building other ways to the sky than the famous roadway of his Yulu-wirree [rainbow]'.[59] During the creation, the ancestors of many birds of prey were good hunters. In the Wadeye area, south-west of Darwin, Thimanthi the black-breasted buzzard took with him many hooked spears that he had made from the favoured northern ironwood timber as he travelled up Memarl Creek.[60] Myths frequently involve the interaction of several ancestors. For instance, it was recorded from the Princess Charlotte Bay area of north-east Cape York Peninsula in Queensland that the Gugu Rarmul people had a tradition that it was the 'redbill' (possibly the purple swamphen) ancestor who used a log of northern ironwood to make a crocodile, which then ate the 'pygmy goose' (probably the cotton pygmy-goose) that had taken all the lily-seed from the waterhole.[61]

A recorded myth from Salt Creek in the south-east of South Australia explains why night birds, such as owls, are harried by diurnal birds when caught out in the open during the day. Ornithologist Walter J Harvey said that:

> In the very earliest times the [Eastern] Rosella and the Boobook Owl were very good friends, helping each other, where they could, in finding nesting hollows and telling each other of suitable food, but one day they had a severe quarrel and the Owl gave the Rosella a fearful thrashing and pulled its feathers out. The other birds were very annoyed with the Owl when they saw what had been done, so they chased it whenever they saw it, until the Owl decided to feed at night instead of in daylight, when it was worried by the other birds. They also nursed the Rosella, and to make up for the damage done to its plumage the birds all plucked some of their own feathers and stuck them in the Rosella, so now it is one of the gayest birds of the country.[62]

Mythological connections can directly link birds with spirit beings. For instance, during the events of the creation mythology in the Lower Lakes area, the two wives of the male ancestor Nepeli were able to attract Waiyungari the Red Ochre Man's attention by mimicking the sound of the emu.[63] In the Wadeye area, south-west of Darwin, the red-winged parrot is significant as the identity of the two daughters of Kunmanggur the Rainbow Serpent being, and they both carried digging sticks as they hunted.[64] The origin of bird calls is the subject of myths, such as in the northern coastal New South Wales region where linguists recorded a creation myth in the Gadang (Kattang) language concerning two kookaburras that started to laugh when a giant Aboriginal man, a *djagiri*, who was chasing them fell into a river after a rotten log he jumped on had collapsed.[65] In the Gulf Country of north-west Queensland, Roth recorded a myth that a kookaburra had laughed and therefore frightened away emus that hunters were manoeuvring into set nets. 'Of course the blacks were very wild at having lost their intended prey, and therefore killed him. The bird now always laughs at their failure in catching emus'.[66]

For knowledgeable Aboriginal people the evidence for the existence of their ancestors is everywhere, both on the land in the form of sacred myth sites and in the heavens as cosmic bodies they see at night. In Wunambal Gaambera Country of the north central Kimberley, the famous ancient *gwion* (Bradshaw) rock art figures are perceived by the present local Aboriginal custodians to have been drawn by Gwion the sandstone shrikethrush during the creation.[67] Ngurde in Wadeye Country, south-west of Darwin, is the place where black duck ancestors camped during the creation, and today it is a sacred site marked by a stone arrangement.[68] In the northern Flinders Ranges of South Australia, the Adnyamathanha people believe that Yuduyudulya the white-winged fairy-wren shaped the hill they know as Wadna Yaldha Vambata, literally 'boomerang crack hill', with a throw of his boomerang.[69] Yuduyudulya also left seams of exposed quartz across the Flinders Ranges, where his white feathers had dropped.

In the Northern Hemisphere, historian and classical folklorist Adrienne Mayor has convincingly argued that ancient Greeks and Romans uncovered strange and often large fossilised bones that inspired their mythical accounts of giants and monsters, such as griffins and cyclopes, that roamed their country before being killed by the gods.[70] In Australia, exposed fossils likewise provided Aboriginal people with evidence, as defined within their own world view, of the past feats of their ancestors. For instance, colonist and naturalist Thomas Paine Bellchambers wrote that embedded in the limestone cliffs of the Murray River in South Australia was once a large shell-like white fossil, almost 60 cm of it exposed. The local Aboriginal people called it 'Nooree' and 'say it is the remains of a large fish killed and eaten by Nooreela, the river maker, and her piccaninnies [children], and that it marks the line of the former water level'.[71]

In the Cooper Creek area of eastern Central Australia, inland explorer John Walter Gregory argued that if scholars could determine whether the mythic Kadimakara ancestors of the Diyari people were associated with the fossilised bones that Aboriginal people frequently encountered, it would prove that Aboriginal people 'inhabited Central Australia, at the same time as the mighty diprotodon and the extinct, giant kangaroos'.[72] Such a line of reasoning is based on the erroneous assumption that the substance of their mythology would be unchanging and therefore retain ancient records. Contemporary anthropologists prefer a more dynamic model of Aboriginal tradition. Fossils, nevertheless, could be sacred objects for Aboriginal people.[73] Gregory claimed that the Diyari people, as part of their ceremonial cycle, 'make pilgrimage to the bones of the Kadimakara … and persuade them to intercede with those who still dwell in the sky, and control the clouds and rain'.[74] In 1890, Aboriginal people reported to European station owners the existence of giant bones at Lake Callabonna in the north-east of South Australia. This led to staff from the South Australian Museum recovering material from the extinct birds *Genyornis* and *Dromaius*, and several genera of extinct marsupials.[75] Palaeontologist Pat Vickers Rich claimed, in relation to the extinct *Dromornis* and *Ilbandornis* bird fossils recovered from the Alcoota area of Central Australia, that 'Aboriginals had long been aware of deposits in the area'.[76]

Aboriginal responses to finding the remains or imprints of animals no longer present in that form on Earth provided them with a pathway for connecting their creation narratives with extinct animals, in some cases from millions of years ago. Their discoveries of fossilised material, naturally exposed by the elements, were not isolated. As demonstrated with existing collections of australites,[77] it is likely that Aboriginal people have collected many of the fossil specimens now residing in museums. The late 19th century anthropologist Robert Hamilton Mathews recorded from the Gamilaraay people of inland northern New South Wales that 'in former times their forefathers occasionally found huge bones, believed to be those of the *Kurreas* [giant snake-like spirit beings], in the banks of deep, dry watercourses', where they had died of thirst.[78] Langloh Parker gave another example of Aboriginal proof for the creation at the Guddee Spring near Brewarrina in northern New South Wales, where:

> ... every now and then come up huge bones of animals [of the megafauna] now extinct. Legends say that these bones are the remains of the victims of Mullyan, the eagle-hawk [wedge-tailed eagle], whose camp was in the tree at the foot of which was the spring.[79]

The behaviour of the ancestors was also a blueprint for how the Aboriginal people who came later were to act, as the ancestors generally had many human characteristics before being transformed at the termination of the creation period. In *A Tale-type Index of Australian Aboriginal Oral Narratives*, folklorist Patricia Panyity Waterman described the ancestors as either 'people-who-behave-like-animals or animals-who-behave-like-people',[80] but it is often difficult to determine which applies. An example of people acting like animals in Tiwi tradition is the myth of Mudati the black kite man and his wife Kirijuna the dingo woman, who were travelling together across southern Melville Island when they encountered Kirijuna's brother, Jurumu the wedge-tailed eagle man.[81] Mudati told Kirijuna not to look at her brother, but she did. She then took the form of a dingo, who slinked away through the shame. Since then, in the Tiwi culture it is not permitted for siblings of the opposite gender to look at each other once they are past childhood.

Birds retained human-like characteristics after their ancestors were transformed at the end of the creation period, after having performed their mythological feats. During the Tiwi creation, Tukimbini the yellow-faced honeyeater man lived with his many wives at Murupianga on the south-eastern part of Melville Island, and when he was eventually transformed into a bird he kept the yellow marks on the sides of his face, which were the ornaments he wore as a man.[82] Apart from physical features, some birds also retained human-like behaviours from the creation, such as *gindaja* the southern cassowary in the rainforest region inland from Cairns in northern Queensland. He is said by Yidiny-speakers to have a human-like behaviour of biting the red wattles hanging down the front of his neck, as if they were an ornament, before going into battle[83] – in some parts of Australia, Aboriginal peoples would put charms, sometimes placed inside 'biting bags', in their mouth when going into battle.

Aboriginal peoples referred to their traditions of the creation period to explain close ecological alliances between particular plants and animals. In the Ngurunderi creation mythology, which gave an account of the formation of the Murray River and the making of its fishes, a group of people who were hiding in the reeds were turned into *waitji* birds (*wetjungali*, superb fairy-wrens).[84] In a version of the *waitji* mythology recorded from Ngarrindjeri man Henry Rankine in the 1980s, the superb fairy-wren ancestor was found to be cheating during a flying competition and was punished by other bird ancestors through never again being allowed to fly higher than the lignum (also called *waitji*), which are tangled bushes that grow around swamp verges.[85] After the creation, wrens lived among the 'whispering bushes' in constant fear of being caught by the descendants of other ancestors, particularly eagles, hawks and owls. This account is structurally similar to a myth recorded in the south-west of Western Australia, whereby Chiriger the splendid fairy-wren had managed to capture blue from the sky by hiding in the feathers of Warlitje the wedge-tailed eagle when it flew high.[86] From the other side of the continent, recorded from the Dhanggati people of the Macleay Valley in north-eastern New South Wales, there is a related myth concerning a competition among the birds to see who could fly the longest. It was won by the wren 'or sparrow' who hid on the back of an 'eaglehawk', which due to its prominence as a large high flier was probably the wedge-tailed eagle.[87] In that account, due to the anger of the other bird ancestors at being tricked, the wren hid and now nests in a hole in the ground.

Myths reflect Aboriginal observations of specific bird behaviours, and the sites associated with the bird ancestors are typically those favoured by their animal descendants. Anthropologist Paul Memmott noted that in the North Wellesley Islands in the Gulf of Carpentaria, 'Usually there is a distinguishing physical feature at a story place [sacred myth site for an ancestor]; for example … a clump of rocks containing birds nests ([topknot] pigeon or *banbaji*, sea hawk [white-bellied sea eagle] or *yarakara*)'.[88] In my experience of mapping Aboriginal Country in northern Australia, brolga myth sites are often lagoons where the birds are likely to be found today. This has also been the experience of other anthropologists. For instance, John Bradley found that for the Yanyuwa people living at Borroloola in the Gulf of Carpentaria, Burangkul the brolga is a major ancestor for the Mambaliya-Wawukarriya semimoiety (a division within the moiety in some Aboriginal societies) and is associated with two lagoons, known as Wubunjawa and Lurriyarri, that cross the plains west of the Wearyan River.[89]

The dancing behaviours of birds are major themes in many recorded myths and, due to the importance of bird ancestors in the creation, many of the ceremonies that Aboriginal peoples hold feature dancers performing as birds and re-enacting their feats. Beyond Aboriginal ceremonial life, the myths are rich sources of environmental knowledge. In an ethno-ornithological study, Tidemann and Whiteside collected over 400 myth narratives from across Aboriginal Australia that primarily concerned birds. The authors determined that:

> They tell about social proprieties, morals, relationships, the landscape and law ... some of the stories summarized ... demonstrate that knowledge about certain birds preceded that 'discovered' by Western scientists. Ornithologists may have benefited (and may still benefit) from making themselves familiar with the indigenous body of knowledge in preparation for their own studies.[90]

Totemic beings

In Aboriginal Australia, each individual possessed links through descent to one or more totemic ancestors. Totemism is defined as a belief system in which people are treated as having kinship with ancestral beings, which are usually animals or plants, that as totems are thought to interact with their human relatives and serve as an emblem or symbol to the group.[91] The word 'totem' is a mid 18th century borrowing, originally by American English, from the Ojibwa language spoken by the Algonquin people of Canada.[92] In this language, the expression *ototeman* translates to 'He is a relative of mine'. It refers to a connection to a clan whose members are all considered to be descended from a remote common ancestor with an animal form.[93]

Although the term 'totem' was appropriated from a non-Australian source, it has been extensively used to describe the social structures of people in Aboriginal Australia.[94] The view of anthropologist Adolphus Peter Elkin was that totemism is the 'key to the understanding of aboriginal philosophy of life and the universe – a philosophy which regards man and nature as one corporate whole for social, ceremonial and religious purposes'.[95] The totemic system was ordained by the ancestors, and as Waterman observed, 'Children enter this system at birth from the spirit world'.[96] Aboriginal people will often express a close totemic connection with a particular species and with other people who are connected to it by saying that they were all of the same 'flesh', 'flavour' or 'meat'.[97] The totemic connection was the basis for forms of ritual imitation and representation, as illustrated by anthropologist/linguist Peter Sutton's observations among the Wik people in the north-west of Cape York Peninsula in the 1970s:

> Middle-aged and older people, when rising and stretching in the morning, may make the noises of their major totems, refreshing themselves with their power. Thus members of clan 4 have been heard to get up saying /kang kang kang kang!/ ([Magpie] Goose), members of clan 6 /pubbbbbbb!/ (Shark's tail hitting water), and members of clan 12 /korr korr korr korr!/ (Brolga).[98]

The cosmological association of totemic ancestral beings and people provided each clan, as a descent group, with a world view that it shared with others within the broader culture.[99] The species associated with each clan as a totem also had cultural relevance to people across a larger region than just that of the clan's estate. For instance, in a study of the Yarralin people in the Victoria River region of the Northern Territory, anthropologist Deborah Bird Rose observed that:

> Emu people share their flesh with emus. If an emu person dies, other people are reluctant to shoot emus because this group has suffered a loss. In this way a specifically human or country viewpoint is enlarged; the enlargement is never uniform for every identity is cross-cut by others.[100]

Australian totems can be seen in terms of what Bird Rose described as a 'three-way relationship between the people, the species, and the country',[101] giving them all a shared identity. This was illustrated at Borroloola in the Gulf of Carpentaria during an Aboriginal land claim hearing in 1992, when senior Yanyuwa woman Annie Karrakkayn said that *a-alanthaburra* the chestnut rail 'lifts up' the mangroves and *a-kuthayikuthayi* the red-capped plover guards the sea.[102] She explained the significance of the red-capped plover to her culture:

> When we hear that bird we know she is holding all the sea country; we know the country is good. In the old days, old people would hear her call and they would call out to her. They would say, 'Yes, you are there, dweller of the coastal country. Allow us to harpoon a dugong, a turtle, a fish. Make the country good for us. She is our countryman.'.[103]

In some Aboriginal communities, each clan was grouped into one of two moieties, which among other things served as a basis for organising marriages.[104] In such systems, membership was worked out through descent, with each person permitted to marry a person of the opposing moiety. In many places, the moieties were represented by bird ancestors, under which other totemic beings were allocated to one or the other moiety – as was virtually the whole world. For instance, among the Nyungar people in the south-west of Western Australia, the moieties were Maniychmat (western corella) and Wordungmat (Australian raven).[105] Across south-eastern Australia the main moieties were the Mukwara (wedge-tailed eagle) and Kilpara (Australian raven).[106] In the north central Kimberley, Wunambal Gaambera society has two moieties: Bunarr, represented by *woday* the spotted nightjar; and Boomalarr, being *jirrin.girn* the Australian owlet-nightjar.[107] It was recorded that in south-east Cape York Peninsula, the Muluridji people possessed moiety names that reflected the dichotomy of darkness/night and light/day. They were respectively named after the 'night-owl' (Kokku-rokku, Wallar) and 'day-bird' (Mirriki, Dabu), although it is difficult to be certain about the exact species involved.[108] Dichotomies were apparent in other totemic systems. In 1934, anthropologist Caroline Tennant Kelly conducted fieldwork at the Cherbourg Aboriginal Settlement in south-east Queensland with Bidjera people, who were originally from the upper reaches of the Warrego River in southern Queensland. She was told for the moieties 'that "nearly always cold skin went Wuturu and feathers Yangaru"; so that Wuturu had water, lizard, frog etc., and Yangaru has emu, duck, native companion [brolga] etc.'.[109]

During Aboriginal ceremonies, these totemic connections were expressed in the body designs of performers, who mimed their ancestors in dance.[110] Anthropologist Alfred William Howitt described in detail the dances performed during an extended ceremony of the Yuin people, from the southern coast of New South Wales. He recorded that of the 'totem dances some were merely the magic dance to the name of the totem. Others were prefaced by pantomimic representations of the totem animal, bird, or reptile'.[111] After describing a dance mimicking the behaviour of the 'wild dogs', he remarked that 'Another was the crow dance, in which men, with leaves round their heads, croaked like those birds, and then danced the owl dance, in which they imitated the hooting of the Takula, owl; the lyre-bird dance, and that of the stone-plover [bush stonecurlew]'.[112] At Coranderrk Aboriginal Mission, east of Melbourne, the famous late 19th century Aboriginal artist William Barak, from the Woiworung group with the Kulin people, often featured animals in his watercolours; for example, with a superb lyrebird as an ancestor in the centre of a ceremony.[113]

The totemic relationship between people and their animal kin gave structure to Aboriginal life, to the extent that it was routinely acknowledged outside of formal occasions. For example, linguist Julie Waddy who worked among the Anindilyakawa of Groote Eylandt in the northern Gulf of Carpentaria remarked that 'The relationship with biological taxa in particular is personified so that a man who sees, for example, *wurruweba* a red-winged parrot flying overhead, might say, "There goes my brother-in-law!"'[114] Such close associations between artist and subject as a theme is the basis for much Aboriginal art.[115] Other birds have more universal importance across the wider group. For instance, among the Yanyuwa people of the southern Gulf of Carpentaria region, Bradley explained that the brolga 'is an important Spirit Ancestor and its trumpeting call and dancing displays make it an emotional favourite with all age groups. The sacred but public funeral rites associated with this bird are also amongst the most spectacular in the Yanyuwa repertoire of ceremonies and dances'.[116]

The following is an example of how totemism worked within a particularly regional community, and illustrates the roles that key bird species had within it. The mechanics of how totemic systems worked varied among Aboriginal groups across Australia, but the fundamental relationships remained the same. I have chosen the Ngarrindjeri of the Lower Murray due to the prominence of birds in their cultural traditions and because of my own familiarity with the Lower Murray region and its complex Aboriginal culture through my doctoral studies in the late 1980s and early 1990s.

Lower Murray clans

The Ngarrindjeri people in the Lower Murray of South Australia lived in a region dominated by water, and consequently their Country was rich in birdlife. Several related dialects were spoken among the group, and people lived in foraging bands composed of members with mixed clan membership. The Ngarrindjeri possessed what anthropologists have termed as patrilineal clan totemism, with membership to each clan determined by the father's identity.[117] Each Lower Murray clan, known as a *lakalinyeri*, possessed one or sometimes several totemic emblems – *ngaitji*, which were generally birds, other animals or plants, but sometimes entities such as the sun or atmospheric phenomena like the wind.[118] It was claimed by a researcher at the South Australian Museum, Norman B Tindale, that the *ngaitji* 'most of whom are considered to be birds (e.g. crow, eagle, silver gull, pelican), although when the events of the stories are taking place they usually are manifesting more of their human attributes than of their bird-like ones'.[119] The Lower Murray system comprised many *ngaitji* ancestors, and this has been termed a 'Spirit Protectors Complex' by anthropologists Ronald M Berndt and Catherine H Berndt.[120] The Ngarrindjeri saw the *ngaitji* as a source of personal strength. For instance, songs relating to the *ngaitji* gave power to healing rituals. In the *nuri* (pelican) healing song it was said that the actions of the bird in bending its neck into the water to drive fish forward were symbolic of badness being driven out of the patient.[121]

The names of some Ngarrindjeri clans related directly to bird *ngaitji*, such the Wutaltinyerar clan from *waltauteringgi* (banded lapwing) and the Turion clan from *turi* (Eurasian coot).[122] Places were also associated with the *ngaitji*. For instance, on Narrung Peninsula, Lawareangar (Lawari-angar, Lowanyeri) was a swamp place that literally meant 'place of grey geese [*lawari*, Cape Barren geese]'.[123] I have had personal experience in observing large numbers of the bird at that place. Many of the Lower Murray sites of significance mapped by Tindale and his colleague Henry K Fry during the mid 1930s were linked to various bird *ngaitji* and associated with origin myths.[124] The *ngaitji* were present during the creation and were perceived as being among the ancestors of living people. Their role, according to Ronald M Berndt and colleagues, 'was to humanize the natural environment and establish a workable relationship of a spiritual kind between themselves and human beings to come'.[125] The *ngaitji* as the 'totem' in such systems 'guards the totemite' or clan member.[126] The prominence of avifauna among clan identities has been noted in other parts of southern Australia.[127]

In Ngarrindjeri society of the Lower Murray, the *ngaitji* were important elements in each individual's identity. Jacob Karruck Harris from the Wutaltinyerar clan described them as a 'signature or a coat-of arms'.[128] On occasion, and particularly in song, people would refer to a certain person by the name of their primary *ngaitji*.[129] Anthropologists consider that many of the clans bearing the same *ngaitji* were once part of a previous super clan, which segmented due to weight of numbers.[130] When two clans with adjoining territory had a close relationship, they were said to be '*tauwali* (plural *tauwalar*) to each other' and therefore acted in some situations as a single entity.[131] Anthropologist Alfred Radcliffe-Brown stated that:

> When two clans are connected in this way the members are regarded as brothers and sisters to each other and may not intermarry; nor would the two clans fight against each other ... When two clans are *tauwali* to one another there seems to be a tendency for each to claim the totems of the other.[132]

The totemic identity of each person limited their selection of potential marriage partners, with people of the same *ngaitji* not permitted to marry.[133] Individuals also had recognised close relationships with the *ngaitji* of relatives from outside their father's clan. This meant that, in addition to not marrying within their own clan, a person was also not allowed to marry someone from the clans of their mother's mother, mother's father or father's mother. The *ngaitji* were ritually consulted by the community when determining whether or not there had been a recent sorcery death. For instance, in 1861 at the Point McLeay Mission, the missionary George Taplin remarked that 'They say that Louisa was millined [sorcery involving a club] and ngadhungied [sorcery involving fat mounted on a bone] too, and that her ngaityengk [*ngaitji*], viz. a little duck and a lizard, were seen scratching inside her stomach after her death, as a sign that this was the case'.[134]

The connection between people and their *ngaitji* was personal and intimate. In some cases, particularly when an animal species was involved, people avoided killing and eating their *ngaitji* although they would allow others within their community to do so.[135] This prohibition custom, which has been recorded elsewhere in Australia,[136] would have contributed to Taplin's remark in 1859 that 'I may as well say that I find the different classes among them have singular rules about eating … It is also thought that if any of the natives eat mountain ducks they will become grey'.[137] While greyness in hair may have been attributed by Aboriginal people to several causes, on one occasion it was believed that two men had turned prematurely grey and then bald soon after having eaten their *ngaitji*.[138]

Clans held ceremonies in honour of their *ngaitji*. In the Lower Murray, Manangki clansman Albert Karloan participated at a Tenetjeri (Caspian tern, 'red-beaked gull') ceremony with the Kandukara (Kanma-indjeri) clan at the Murray Mouth in the 1880s.[139] The Berndts were told that it:

> Consisted of singing and dancing which reproduced the behaviour pattern of this bird … With arms outstretched and then at their sides the dancers mimicked the gull diving into the water. The owner of the *ngatji* [*ngaitji*] cried out like the bird first and was then followed by all the other participants. Then he sang as the actors danced the flight of the gull.[140]

The Australian bustard (turkey) dance that was performed in the Lower Murray in the late 19th century, having come downstream from the upper reaches of the Murray River, involved an object that symbolised the bird's tail-feathers but was actually made of red-ochred human hair cord and sticks.[141] Another ceremony that came downstream involved a pole, decorated with emu feathers and a boomerang, which was placed in the middle of a canoe.[142] The *ngaitji* were treated as a source of immense power in ritual life. Across south-eastern Australia, traditional healers, sometimes referred to in English as 'clever men', 'doctors' or 'medicine men', typically summoned their own totemic 'friend' or 'protector' when ritually treating their patients.[143]

As has been recorded in other parts of Aboriginal Australia,[144] the totemic clan structure in the Lower Murray was malleable, with modifications due to changing population numbers. Ngarrindjeri people interpreted their first direct contacts with Europeans in accordance with their world view, which gave prominence to the *ngaitji*, spirits and creator beings.[145] This is apparent in the reminiscences of Jenny Pongi (Pundji), who recalled when the first European explorer arrived by boat at the Murray Mouth in 1830. According to a former colonist who knew her, Jenny had said 'how she ran away with her folk when frightened by the approach of [Charles] Sturt down the Murray River. They thought he was a spirit and associated him with the Seagull [silver gull] totem'.[146]

*****

Through the totemic system, Aboriginal people see themselves as having a shared origin and identity with many of the animals and plants of their Country. This is particularly so with birds. This belief added to the Aboriginal perception that both they and the totemic ancestors were holistically part of the landscape, in contrast to the belief systems of the European colonists who regarded much of Australia as a 'wilderness' which had not previously been shaped by humans.[147] The myths are repositories of environmental knowledge, providing a framework within which the distribution and behaviour of organisms, such as birds, could be understood and remembered. This is apparent with the narratives for the Yawulyu songs of Central Australian women, that record in detail the knowledge concerning edible seeds.[148]

Aboriginal peoples believe that during the creation the bird ancestors possessed many human-like characteristics, and afterwards their descendants were divided into people and bird species. Birds are powerful totemic entities. They have the ability to fly up to Skyworld, and many of them exhibit behaviours, such as brolga dancing and eagle hunting, that resonate with the actions of people. The physical characteristics of birds also record some of their human-like qualities. The next chapter investigates further the creative power of birds.

Endnotes

1. Strehlow (1970), Hiatt (1975a, 1975b), Myers (1986: Ch.2), Sutton (1988a, 1988b, 1988c), Berndt & Berndt (1989), Morphy (1998:149) and Smith (2021).
2. Tindale (1937a:107–108) provided a description of the different types of song, as expressions of the myth narratives, within the south-eastern South Australia Aboriginal community.
3. Roth (1903:15).
4. Sutton (1988b:14–16) and Clarke (2003d:382–383).
5. Spencer (1896:51), Stanner (1953), Maddock (1982:105–120), Sutton (1988b:14–19), Bird Rose (1992:42–57) and Arthur (1996:27–28).
6. Sutton (1988c:252).
7. Chatwin (1988).
8. Jones (2017) and Nicholls (2019).
9. Barfield (1997:334).
10. For examples, refer to published collections of mythology such as Isaacs (1980), Waterman (1987) and Berndt & Berndt (1989).
11. McConnel (1957:162).
12. Tidemann & Whiteside (2010:153).
13. Clarke (2018a, 2018b, 2018c).
14. Sault (2010).
15. Tidemann *et al.* (2010:4).
16. Ramson (1988:172) and Dixon *et al.* (1992:152–153). Other written versions of corroboree include coroborey, corrobbaree, corrobara, corroboree, corrobori and corrobory.

[17] Ross (1986:232). The example Ross used was Eyre (1845:Vol. 2:233–234).

[18] Strehlow (1947:xviii–xx).

[19] Clarke (1995, 2018f, 2018g) and Sutton (1988c).

[20] Myers (1986:52).

[21] TenHouten (1999), Janca & Bullen (2003), Iwaniszewski (2014:3–4) and Clarke (2018g).

[22] Maddock (1978b:115).

[23] Hassell (1975:210–211).

[24] Bird Rose & Swain (1988) and Clarke (1996b:84–89). For examples of new traditions, Mackinolty & Wainburranga (1988), Bird Rose (1988) and Beckett (1994:106–111) described Aboriginal accounts of non-Aboriginal ancestors such as Jesus, Captain Cook and Ned Kelly. Saethre (2007) provided an account of UFO beliefs being incorporated into the Aboriginal belief system of a remote Central Australian community.

[25] JC Harwood (cited in Tindale 1930–52:193–194).

[26] JC Harwood (cited in Tindale 1930–52:193).

[27] Tindale (1931–34:232, 1937a:111–112) and Clarke (1997:137).

[28] Sutton (1988:251) and Clarke (2014a:76, 2018c:51, 2019b:171). Examples of extracting scientific data from myths include Tindale (1974:119) and Flood (1983:29–30,146–147) for the arrival of Aboriginal ancestors and megafaunal extinctions. Dixon (1972:29) considered a myth as a record of Ice Age land bridges and crater lakes. Nunn & Reid (2015) and Nunn (2020) found evidence in myth of sea level changes. Similarly, Isaacs (1980:87–89) and Cane (2002:91–92) saw evidence in myth for the formation of the cliffs along the Great Australian Bight. Tindale (1959:46–47, 1974:50,119), Isaacs (1980:29–31) and Wilkie *et al.* (2020) equated fire myths with volcanoes. Tindale (1937b, 1974:135) and Hamacher & Norris (2011:104,109) interpreted a myth as reporting the 1793 eclipse.

[29] Thomson (1949:7).

[30] Kolig (1992). See also Berndt (1974:27–31), Swain (1990, 1993: Ch.3) and Thomas (2011:273–274,279–285).

[31] Kolig (1992:29).

[32] Elkin (1931), Fry (1950), Sutton (2003a: Ch.7) and Koch *et al.* (2018).

[33] Mathew (1898, 1910).

[34] Clarkson *et al.* (2017), O'Connell *et al.* (2018) and Bradshaw *et al.* (2021).

[35] Radcliffe-Brown (1958), Roheim (1971) and Blows (1976).

[36] Clarke (1999b:60–63).

[37] Gale (2000), Jones (1990) and Shoemaker (2000).

[38] Telamon (1930).

[39] Merriam-Webster Dictionary (https://www.merriam-webster.com/dictionary/tale).

[40] Merriam-Webster Dictionary (https://www.merriam-webster.com/dictionary/legend).

[41] Anonymous (1952:23).

[42] Ramsay Smith (1930:17).

[43] Mountford (1941a).

[44] Mountford (1943).

[45] Such as Langloh Parker (1905, 1953, 1991), Ramsay Smith (1930), Mountford (1948, 1965, 1976b), Wilson (1950), Mountford & Roberts (1965, 1969, 1971), Reed (1978, 1980), Roberts & Roberts (1975) and Berndt (1988). Note that Catherine Eliza Somerville Stow (1 May 1856–27 March 1940) wrote as 'Katherine Langloh Parker' (Muir 1990).

[46] Mountford (1941c).

[47] Bradley (1988:32–38).

[48] Diamond (1991), Tidemann & Gosford (2006) and Fleck (2007).

[49] Tidemann & Whiteside (2010:158, Table 12.2).

[50] Mountford (1958:53).

[51] Tunbridge (1988:22–23).

[52] Tunbridge (1988:22–23,44).

[53] Langloh Parker (1953:94).

[54] Langloh Parker (1953:124).

[55] Spencer & Gillen (1899:652).

[56] Tunbridge (1988:17).

[57] Tunbridge (1988:22).

[58] McConnel (1957:84–86).

[59] Langloh Parker (1953:100). Identification of Bibbi (Bibi) provided by Ash *et al.* (2003:36).

[60] Hardwick (2019a:64).

[61] Roth (1902:7). There are several other possible 'redbill' species. In the Lardil language spoken on Mornington Island of the Wellesley Islands in the Gulf of Carpentaria, *kethuku* the 'redbill duck' or 'black redbill' (unknown species, Memmott 2010:123) is 'associated with low water' (Ngakulmungan Kangka Leman 1997:151), while *wiriwir* the 'brown redbill' is a 'curlew' (Ngakulmungan Kangka Leman 1997:281).

[62] Harvey (1932:226).

[63] Tindale (1935:268,271), Berndt *et al.* (1993:229) and Clarke (1999b:53).

[64] Stanner (1963:98) and Hardwick (2019b:61, 2019c:19).

[65] Holmer & Holmer (1969:37–38).

[66] Roth (1897:126).

[67] Karadada *et al.* (2011:87).

[68] Hardwick (2019a:38).

[69] Tunbridge (1988:120–121).

[70] Mayor (2011).

[71] Bellchambers (1931:112).

[72] Gregory (1906:7, see also 8,74,80–85,108,145,224,231,353).

[73] Nobbs (1989) described a fossilised tree as the 'pillar of the sky' for the Diyari people. Long (2002: Ch.1) reported the theft of rare dinosaur footprints on a beach near Broome, a violation of an Aboriginal sacred site.

[74] Gregory (1906:4).

[75] Rich (1979:2–3). See Stirling & Zeitz (1896a, 1896b).

[76] Rich (1979:4).

[77] Clarke (2018j:123–127).

[78] Mathews (1898c:118).

[79] Langloh Parker (1905:134). See also Mathews (1898b:92) and Isaacs (1980:18). Ash *et al.* (2003:106) identified *maliyan* (i.e. Mullyan) as the wedge-tailed eagle.

[80] Waterman (1987:11).

[81] Mountford (1958:55–56). Mountford refers to the black kite as the 'fork-tailed kite'.

[82] Mountford (1958:38–39). Mountford described Tukimbini as a 'brown honeyeater', which based on the bird's description I have identified as the yellow-faced honeyeater.

[83] Dixon (1991:157).

[84] Berndt (1940a:173) and Berndt *et al.* (1993:124,237,452–453).

[85] Clarke (2007a:15–16).

[86] Hancock (2014).

[87] N Holmer 1963 (cited in Waterman 1987:124).

[88] Memmott (2010:63). The birds were identified in the Lardil dictionary, with *banbaji* being a 'pigeon with topknot' and *yarakara* a 'large sea hawk, fishing hawk' (Ngakulmungan Kangka Leman 1997:63,292).

[89] J Bradley 1987 (cited in Sutton 1988a:222) and Bradley with Yanyuwa Families (2010:97,100,102).

[90] Tidemann & Whiteside (2010:154).

[91] Durkheim (1915: Ch.4) and Barfield (1997:468).

[92] Chamberlain (1902:263) and Lang (1902).

[93] Morris (1987:270).

[94] Mathews (1897b), Radcliffe-Brown (1929), McConnel (1930), Elkin (1933, 1934, 1964: Ch.7), Thomson (1933), Sharp (1939), Stanner (1965) and Berndt & Berndt (1999: Ch.7).

[95] Elkin (1933:131). See also Piddington (1932).

[96] Waterman (1987:15).

[97] Howitt (1904:145–146,202,241,243,249,282–283,305), Berndt (1974:27), Elkin (1977:67[endnote 18]), Berndt *et al.* (1993:81,83,122,198) and McConvell & Ponsonnet (2018:301,307).

[98] Sutton (1978a:215).

[99] Berndt & Berndt (1999:41–43) and Sutton (2003a:39–42,46,49,51,99,152,155–157).

[100] Bird Rose (1992:221).

[101] Bird Rose (1996:28).

[102] A Karrakkayn (cited in Bradley 2010:248).

[103] A Karrakkayn (cited in Bradley 2010:166).

[104] Warner (1937:31–33), Elkin (1964:121–125,138–139,173–174), Berndt & Berndt (1999: Ch.) and Sutton (2003a:202).

[105] Bates (1901–14:74–77) and Abbott (2009:213).

[106] Fison & Howitt (1880:322–324), Blows (1975) and Clarke (2018a:6–10).

[107] Karadada *et al.* (2011:79).

[108] McConnel (1939:70).

[109] Tennant Kelly (1935:464–465). The author's name was sometimes given as Tennant-Kelly. Note that Cherbourg was earlier called Barambah, and then for a while was known as Cherburg.

[110] Memmott (2010:36).

[111] Howitt (1904:546–547).

[112] Howitt (1904:547).

[113] Sayers (1994:16). See W Barak, Coranderrk Mission, Victoria, 1890s (South Australian Museum Archives, reproduced in Clarke 2000:22).

[114] Waddy (1982:75).

[115] Sutton (1988b) and Morphy (1998: Ch.5).

[116] Bradley *et al.* (2006:89).

[117] Elkin (1933:122–123).

[118] For descriptions of the *ngaitji* refer to Meyer (1846:186–187), Fison & Howitt (1880:307–308), Taplin (1874:1–2,63–64) and Berndt *et al.* (1993:122–123,197–199,249–51,470–473,417–419). Alternative spellings of *ngaitji* include *ngatji*, and *lakalinyeri* has also been written as *lakatindjeri*.

[119] Tindale (1938:18).

[120] Berndt & Berndt (1974:302).

[121] Berndt *et al.* (1993:215).

[122] Tindale (1987b) and Berndt *et al.* (1993:207–208). The Wutaltinyerar are sometimes recorded as the Wuthawuthindjeri, and the Turion are sometimes described as the Turaorn.

[123] Taplin (1874:130). See also FE Mann (cited in Padman 1987:7), Tindale (1987b) and Berndt *et al.* (1993:326–327).

[124] Tindale (1934b).

[125] Berndt *et al.* (1993:243).

[126] Elkin (1933:115).

[127] The prominence of bird totems in south-eastern Australia is demonstrated in accounts by Smyth (1878:Vol. 1:xxvi,431), Worsnop (1879:9) and Stone (1911:457). A description of moieties and totems from the south-west of Western Australia by von Brandenstein (1977) also shows a predominance of birds.

[128] Harris (1894–95:21 September 1894).

[129] Berndt *et al.* (1993:218–220).

[130] Berndt *et al.* (1993:26).

[131] Brown (1918:232).

[132] Brown (1918:232).

[133] Berndt *et al.* (1993:25,31–33).

[134] Taplin (1859–79:17 December 1861).

[135] Taplin (1874:63–64) and Berndt *et al.* (1993:122–123,198).

[136] Spencer & Gillen (1904: Ch.10) and Elkin (1933:115).

[137] Taplin (1859–79:28 May 1859).

[138] Tindale (1934–37:39).

[139] Berndt *et al.* (1993:431).

[140] Berndt *et al.* (1993:210).

[141] Berndt *et al.* (1993:211–213, Fig. 31).

[142] Berndt *et al.* (1993:213).

[143] Radcliffe-Brown (1923:442) and Elkin (1977:74–79).

[144] Elkin (1933), Rose (1947), Worsley (1955) and Maddock (1978a).

[145] Clarke (2007b).

[146] HF Dodd (cited in Tindale 1938–56:260).

[147] For discussion on the cultural concepts of 'wilderness' in Australia, refer to Mulvaney (1991) and Griffiths (1996:273–274).

[148] Curran *et al.* (2019).

3

Birds as creators

The ancestors of those animals that Europeans class together as birds are considered by Aboriginal peoples to be major actors in the original creation. Among the pantheon of ancestors, ancestral birds created both landforms and other species, and after the creation they remained important as totemic species. The following is a small selection of examples from a vast record of birds in the creation mythologies of Aboriginal Australia.[1] Many of these accounts incorporate environmental knowledge, as seen with mythological accounts of the form and behaviour of birds resulting from the actions of ancestors during the creation. Chapter 2 explained how, in comparison to present-day bird species, the creation ancestors were an amalgam of various physical and cultural attributes – sometimes described as being more human, at other times more animal-like. As described in Chapters 2, 4 and 7, the powers that the bird ancestors possessed in the creation could later be accessed in a variety of ways by their human descendants.

The following account of the bird creators gives the likely species that each ancestor became, according to Aboriginal sources. This prescriptive interpretation of the original set of ethnographic sources is not always straightforward, as on occasion errors with translation may occur. For instance, in the Tjilbruki (Chirr-bookie, Tjelbruke, Tjilbruke) creation mythology of the Lower Murray in South Australia, the ancestor was first described by Norman B Tindale in his field notebook from the 1930s as 'like a blue crane, same size, feathers golden colour on chest',[2] which was consistent with how other ethnographers had recorded the identity of this ancestor as a 'blue crane' or just 'crane'.[3] The brolga, which is sometimes called a 'crane', would have been a possible identity for Tjilbruki, but Tindale had noted that the bird was 'found anywhere along the coast on stones'. This is more consistent with an identity as the white-faced heron, also commonly called the 'blue crane'.

In 1987, when Tindale published a full account of the Tjilbruki mythology, the species was strangely identified by him as the 'glossy ibis'.[4] No explanation for the transition from 'crane' to 'ibis' was given, although it may have been due to a hunch – he acknowledged that 'the notes are widely scattered in his journals and in part therefore [they] have been linked together from personal recollections'.[5] The glossy ibis, however, appears to be a poor fit in terms of his original recording. Tindale's observation of the 'golden colour on chest', which is not present on the glossy ibis, is perhaps a reference to the chestnut nuptial plumes on the breast of the white-faced heron. Since Tindale's 1987 publication, the identity of Tjilbruki as the glossy ibis has been largely accepted by both researchers and the contemporary Kaurna community of the Adelaide Plains.[6] I provide this example as a demonstration how the identity of ancestors as depicted in the scattered records of a myth can, in certain situations, be open to reinterpretation.

Wedge-tailed eagle

The wedge-tailed eagle, which in early publications was generally referred to as the 'eaglehawk', is present as a major ancestor in a considerable number of myths recorded from across Aboriginal Australia.[7] As with many myths, the wedge-tailed eagle is portrayed as being male although, as with some other raptors, today the female is noticeably larger than the male when they are observed as a pair.[8] The birds are brown when immature but become progressively darker as they age.[9] In my experience, Aboriginal people often refer to the species as the 'black eagle', as distinct from the 'white eagle' that is generally the white-bellied sea eagle.

The wedge-tailed eagle is described as existing in a state of perpetual conflict with other birds, particularly ravens and crows, and it was said that as an ancestor the bird could self-heal and only rarely suffered defeat.[10] The ancestral wedge-tailed eagle is often described as a kinsman to the ravens and crows but nevertheless their enemy, leading to the behaviour of these species today. Among the Wiilman people of the wheatbelt in the south-west of Western Australia, settler Ethel Hassell recorded that:

> Waalich [wedge-tailed eagle] and Wording [Australian raven] are kin but are not
> fond of each other in spite of their relationship. They live near each other and try
> not to quarrel, but they have never got over an ancient feud.[11]

In many regions, the wedge-tailed eagle was seen as a major creator. For instance, artist Butcher Joe Nangan in the south-west Kimberley gave an account of Warakarna the wedge-tailed eagle man who was married to two sisters: Ngakalalya the Major Mitchell cockatoo, who was a good forager; and Nyunbara the crow, who was lazy and a poor food collector.[12] Due to jealousy, Nyunbara killed Ngakalalya one day when they were alone, and hid her body behind some termite mounds. Eventually, Warakana discovered what had been done. He speared Nyunbara to death, and threw her body into the fire. Due to these events, it is believed that Major Mitchell cockatoos have pink feathers that are blood-stained, while all crows are black due to being burnt.

For the Nyungar people in the south-west of Western Australia, the wedge-tailed eagle ancestor devised the moiety social structure through which all marriages were organised. Here, anthropologist Daisy M Bates reported that:

> The eaglehawk was sometimes called 'Mamangurra,' and was supposed by
> the southern coastal natives to have made all living things and divided them
> into Noyyung or Ngunning [moieties]. He was himself both Noyyung and
> Ngunning. He had a wife in the squeaker crow [grey currawong], appropriately
> named Bella. Many of their legends have the eaglehawk as the central figure, but
> animals, birds and reptiles figure in all native legendary here.[13]

In the Tjukurrpa (creation) of arid western South Australia, Walawuru the wedge-tailed eagle hunts the children of Nganamala the malleefowl and Karrpitji the bettong.[14] Further south in the same broad desert region, Bates wrote in 1918 that Waldya the wedge-tailed eagle 'is always the bird who brought the first fresh water to the natives in th:gur [Dhugur] or "dream times."'[15] According to Bates' account published in 1947, during the creation the Waldya ancestor was camped at Euria (Yuria) Rockhole on the west coast of South Australia with his wife, Wirlu.[16] Waldya was nearby at the Moonaba waterhole, intending to spear an emu, when Gunggara the owlet nightjar ancestor sneaked into Waldya's camp and stole Wirlu away to become his wife.[17] The crows, Garn.ga, were uncles to Waldya but they mocked him as he sharpened his spear. Gunggara already had a wife, who was Yagana the white-bellied sea-eagle.[18] Yagana always sheltered among the foliage, while Wirlu preferred to build her hut in hollows of stony places. Waltja chased Gunggara and his wives across Country, using the big rain clouds from the west to drive them. After many events, Waldya speared Gunggara and took Wirlu back to Euria Rockhole, where she was beaten and cried 'weeloo'. Waldya was transformed into a wedge-tailed eagle and Wirlu became a bush stonecurlew, which today still cries.

The Waldya mythology is associated with a string of sites. The name of Waltabie Well, located between Yalata and Coorabie in western South Australia, was derived from Waltha'bi, meaning 'Eaglehawk Water'.[19] Anthropologists Ronald M Berndt and Catherine H Berndt recorded Waldya as having a 'Cult Totem' with a 'Horde country' on the 'South coast',[20] and its own ceremony and a 'profuse mythology'.[21] According to the mythology, at a place called Gabi Rana a pair of wedge-tailed eagle ancestors made a nest on the top of a granite boulder, where after the creation two large stones represent their eggs and other stones the leaves and branches of the nest.[22] It was recorded in 1942 that 'Each year members of the *Waltja* [Waldya] Cult lodge visit this site and anoint the stones with red-ochre or blood'.[23] In 1952, Tindale recorded the 'Story of origin of fire among Wirangu Tribe' from Rev Carl Hoff of the Koonibba Mission in western South Australia.[24] In this account, the wedge-tailed eagle was involved in a fight over 'firestones' with the carpet snake ancestor at Euria Rockhole. The bird grabbed the stones and flew off south-east towards Port Lincoln. The bird's plumage was blackened by the event, and during its flight some of the heated stones were dropped along the way, thereby giving Wirangu people access to fire. Tindale suggested that this narrative might have been a creation myth based on the memory of an australite shower, since early 20th century scientists considered (erroneously, as discussed later in this chapter) that these Australian tektites had existed only a few thousand years.[25]

During my own western South Australian fieldwork in 2017, Wirangu people informed me that the Waldya sites were generally hilltops, as this is where the wedge-tailed eagles as birds still spend much of their time. An account of Waldya the wedge-tailed eagle, published in 2010, was recorded from Wirangu man Alan Wilson. It links the track of this ancestor to waterholes associated with prominent granite outcrops across the west coast of South Australia. According to him, 'the eagle came down, the Waldya eagle you know, come down from the north and it sat on those rocks down at Murphy's Haystack [near Streaky Bay], and you can still see the blood on there … then it flew down towards Marble Range way area', near Port Lincoln.[26]

Across Aboriginal Australia, the wedge-tailed eagle ancestor is prominent in the recorded mythologies concerning the Skyworld.[27] The wedge-tailed eagle's abilities of high flying and far sight, as well as its large form, led to this bird being admired to such an extent that along the Murray River of Victoria it was claimed that the supreme male ancestor Nureli (Nooralie) was like an eagle.[28] A colonist in Victoria stated that an Aboriginal tradition held that 'before the blacks took possession of the earth, birds were the monarchs of the universe, and the eaglehawk [wedge-tailed eagle] seems to be regarded by the aborigines as the chief ruler'.[29] This was recorded in many parts of south-eastern Australia, although overstated by late 19th century scholars who supported the existence of an All-Father figure in the 'high god' debate, as discussed in Chapter 2. In central Victoria, the Woiwurrung people considered Bunjil the wedge-tailed eagle to be the creator of all people,[30] and there is a painting of him in human form, along with his two dogs, on the wall of a rock shelter in Black Range, near Stawell in western Victoria.[31] A cave painting on the wall of Baiame Cave near Bulga in central eastern New South Wales features a large possibly male figure with large penetrating eyes, extended wing-like arms with fingers that look like primary feathers, and feather-like projections under the upper arms.[32] He is believed to be Baiame (Baiamai) the Great Spirit, who is sometimes associated with the form of a large eagle.[33]

As Australia's largest raptor species, which ranges across much of the continent, the association of the wedge-tailed eagle with ancestral beings is widespread. The associated mythology frequently involves other bird ancestors, apart from the crow/raven. For instance, in the south-west Kimberley there is a tradition recorded from Butcher Joe Nangan that the wedge-tailed eagle was treacherously wounded in the foot by the willie-wagtail using a hidden stake. All the other bird ancestors gathered round to help, and as the stake was pulled free:

> ... all the filth, blood, red, green and yellow pus and rotten flesh sprayed everywhere, covering everybody. Until that time, all birds were white. The matter from the wound gave them the colours we see today.[34]

Crow/raven

Ornithologists classify Australian species of crows and ravens in the *Corvus* genus, although to Aboriginal people and the lay observer they were generally indistinguishable and in the Australian historical literature were often just called 'crows'.[35] Most southern ethno-ornithological sources simply refer to a 'crow', rather than to the Australian raven which would in many cases have been more biologically correct.[36]

The ravens and crows collectively form another group of bird ancestors frequently mentioned in the recordings of Aboriginal mythologies, and are typically portrayed as difficult characters who readily take offence.[37] In a version of the myth told in the Lower Murray region of South Australia, Marangani the Australian raven was described as being generally disliked due to his habit of calling every man he met '*ronggi*' (brother in-law), as a tactic for gaining access to their women.[38] Threatening behaviour was acknowledged by other Aboriginal groups. On the Yorke Peninsula of southern South Australia, the Australian raven ('crow') in the Narangga language is called *kua* or *kuwa*.[39] *Kua-milatu* meant 'one who takes by force, like a crow', derived from the words *kua*, meaning 'crow' and *milla*, meaning 'taken by force'.[40] In the mythologies of the Yanyuwa people of the southern Gulf of Carpentaria, the 'crow' (Australian raven/Australian crow) clashes, sometimes violently, with the wedge-tailed eagle, collared sparrowhawk and spotted nightjar ancestors.[41] The distinctive behaviours of the ravens and crows in relation to other bird species has undoubtedly contributed to the widespread distribution of this theme in Aboriginal myth.

In southern temperate Australia, the epic struggle between the crow and the wedge-tailed eagle was a dominant mythological theme in societies that possessed the moiety or dual social organisation.[42] For instance, in the south-west of Western Australia, the crow was placed in a moiety opposite the wedge-tailed eagle.[43] The perceived need for strict marriage laws is a theme in many raven/crow and eagle myths. In the Marangani mythology recorded from the Lower Murray, the crow was camped with bird ancestors at Rufus River, a tributary of the Murray River near the South Australian/New South Wales border.[44] In this account, Marangani was annoyed that he had been denied access to women, so he took the two young sons of his two nieces and hid them in a large tree that he had sung into existence. The children were eventually rescued by Tjuit (Tuit), described as the ancestor of 'a black-and-white whispering reed bird'. In the Murray River version of essentially the same myth, the two nieces are young girls in the care of Kanau the wedge-tailed eagle, who is married to the sister of Waku the crow.[45] In this myth, Walpu was an initiate anointed with red ochre. He was able to traverse the barrier of galls jutting out on the trunk of the large tree containing the boys, in order to rescue them. In what appears to be another related account of the eagle and crow mythology from the Madimadi people along the lower Murrumbidgee River, it is the brown treecreeper ancestor who rescues the child who the crow had abandoned in the tree.[46]

Across the inland regions of South Australia, detailed accounts of the crow and wedge-tailed eagle mythology were recorded from the Ngadjuri people of the mid north[47] and the Adnyamathanha of the northern Flinders Ranges.[48] There is also a recording of the crow and wedge-tailed eagle epic from the arid Arkaba area of the Central Lakes region in northern South Australia.[49] Here, during the creation Wildu (Wildoo) the wedge-tailed eagle ancestor was growing tired of being mobbed by crows (Wocalla), who were back then as 'white in colour as the younganna (Lake Hope cockatoo [probably little corella])'.[50] So, it was believed that:

> On the (Stewart's Creek) Howieandina Run there is a great black ironstone range containing a roomy cave, and it was into this cave that Wildoo, after carrying thither a fine heap of dry sticks, inveigled his persecutors when buccola (frost) came earlier than usual and caught the crows before they had got their new (yarldoo) white feather pelthas (clothes) fairly finished.[51]

Continuing with the Arkaba myth, Wildu suggested to the Wocalla, who were cold, that they all go along to 'that nice warm undenya wurley (stone house)' and the eagle said that he would be along too once his 'carlbee (feathers) are all right'.[52] The Wocalla fell asleep on the sticks. Later at night, once Peera (Moon) had provided some light, Wildu took a firestick and set the bed on fire. All the feathers on the crows were burnt off, and when they eventually grew back 'they were blowarn (black), and so they have continued to be ever since'. The cave became known as 'Woceala an Pindina', meaning 'the place of the burning of the crows'.[53]

In New South Wales, related raven/crow and eagle narratives have been recorded from the Narran River region in the northern parts of that state.[54] There are similar accounts recorded from Maraura people at the confluence of Murray and Darling Rivers,[55] and from the Bagandji along the Darling River.[56] According to the Bagandji, in the creation the wedge-tailed eagle stole a large Murray cod from the 'Ancestral Crow', who was 'a great old man, an ancestor that walked on this earth and created the hill country; [while] the mudlark [magpie-lark] created the Darling River'.[57] In Aboriginal beliefs recorded in the southern parts of Victoria, Kilparra the crow and Mokwarra the wedge-tailed eagle were constantly fighting with each other. When they eventually stopped, they each became the totem for a moiety marriage section.[58] In northern Victoria along the Murray River there was a similar myth, and here the marriage sections were Kil-parra the crow and Mak-quarra the wedge-tailed eagle.[59]

In the Lower Murray, the raven was generally despised, among other things possibly for its habit of staying in the vicinity of burial grounds and stripping the dried flesh off the bones of the dead placed on funeral platforms.[60] According to Tangani man Milerum (Clarence Long), *marangani* the 'crow' (Australian raven) was the 'bird that eats anything'.[61] These negative beliefs for ravens/crows are analogous to those connected with these birds and other dark medium-sized scavenger birds in different parts of Australia. For instance, in Wik Mungkan language spoken on western Cape York Peninsula, the name for the black butcherbird is *panch thukkay*, which translates to a 'bird not eaten' (*panch*) and 'lice' (*thukkay*), which the bird can give to people.[62] The crow/raven is often and widely said to be a spirit familiar for a sorcerer.[63] One of the recorded names for the crow in a southern Western Desert language was *goona ngalgu*, said to mean 'excrement eater'.[64] According to Charles P Mountford, the poor standing of crows was apparent in the way Pitjantjatjara people read animal tracks on the sands. He noted that the 'bombastic gait of the crow … that black robber of the Outback, was reflected in his swaggering tracks'.[65] Bobby Brown stated that in his Antikirrinya language, spoken in the central Western Desert:

> *Wati kaarnka* 'crow man' and *kungka kaarna* 'crow woman', are used for 'light-fingered' people or thieves. *Kaarnka* (a term for Crows and Ravens, *Corvus* spp.) is an inquisitive scavenger that might pick up just about anything it sees and eat it or take it away, even if it is not food.[66]

Since crows/ravens have a wide range of food, they are naturally attracted to places where people live. In the Wadeye area of the tropics, south-west of Darwin, it is recorded that 'Crows are despised and not eaten even though it is said their flesh is free from the taste of garbage out bush. Crows eat the garbage around the [Port Keats] Mission'.[67] On Cape York Peninsula, biologist and anthropologist Donald Thomson observed that the crow is 'A scavenger always in the vicinity of native camps during the south-east [wind] season'.[68] Crows are also seen as potential sorcerers by many groups. When a crow alighted near a camp in the south-west of Western Australia and started to cry out, it would be seen as a sorcerer.[69] Here, if someone then got sick after such a visit, it would be blamed on sorcery from the Aboriginal group whose Country lay in the direction to which the crow flew. Crows have some human-like traits. The Kuungkari people in the Thomson River area of south-west Queensland believed that during the creation Wakarn the crow ancestor chased away the spotted bowerbird, who was teaching people how to walk like a bird. Instead, Wakarn taught them to walk like him so as not to frighten the game away.[70]

In Aboriginal Australia, there are many myths that explain the black colouring of different types of bird. A myth recorded from Palmer River in northern Queensland described the 'crow' gaining its dark plumage when she 'covered herself in excreta to make herself look fearsome'.[71] The burning of ancestors was another means by which birds became black. The Adnyamathanha people of the northern Flinders Ranges have a tradition concerning Wildu the wedge-tailed eagle and how he tricked other bird ancestors into going into a cave, outside which he later made a large fire. Here, linguist Dorothy Tunbridge recorded that:

> The first birds to get out of the cave were the cockatoos [little corellas]. They were able to keep away from the smoke, so they are still white. The [Australian] magpies and the willy wagtails [willie-wagtails] were closer to the fire, and were rather badly burnt. This is why they have a lot of black on them. A long time after the big fire and smoke, the crows [Australian ravens] got out. They were black all over. Before all this happened, they were white.[72]

Crow/raven species are also disliked elsewhere in the world. In the Congo in Africa they 'are disliked because they are "polluted," feeding on human wastes and other dirty things'.[73] Europeans in Australia have also had poor opinions about these Corvids, as shown by a record of hard-pressed South Australian settlers being followed by crows through the bush as if waiting for them to perish.[74] It is perhaps the closeness of these species to humans that makes them so despised.

Emu and brolga

Recordings of the joint emu and brolga myths are widespread across Aboriginal Australia,[75] with anthropologist Ursula H McConnel remarking that 'The story of the jealousy between the emu and the native companion [brolga] occurs wherever these birds are found. The emu, who lost her children through neglect, steals those of the native companion, who has now only two'.[76] These myths are structured round the dichotomies: flightless/flight ability, large clutch/small clutch, and inland/coast. In the case of the myth as told by the Wik Mungkan people on western Cape York Peninsula in Queensland, the emu and brolga are in different moieties.[77] It has been noted that when both of these protagonist bird ancestors are of the same gender, they tend to be female in the south and male in the north, possibly in respective opposition to the All-Father ancestors in the south and the All-Mother ancestral being in the north.[78] The brolga is sometimes replaced by other large bird ancestors in myths which are otherwise the same, such as those accounting for the emu ancestor losing her wings and the Australian bustard reducing her clutch size to two.[79] In the Gulf Country of north-west Queensland, a structurally related recorded myth concerns the emu and water-hen (probably the purple swamphen), who had her eggs stolen.[80]

The Lower Murray mythology of the emu and brolga ancestors is based in the southern parts of the region, particularly in the Country surrounding present-day Kingston.[81] In the south-east of South Australia, this myth was sometimes used to identify local Aboriginal groups, with the Moandik (Meintangk) people referring to coastal clans as Porolgi (brolga), while inland clans were Pindjali (Peindjali, emu).[82] The brolga was the *ngaitji* (totem) of some clans based at the southern end of the Coorong, as well as at Kingston.[83] According to Liwurindjeri clan man Thralrum (Mark Wilson), it was Lower Murray tradition that if any man from brolga country was 'knocked down and bruised by one from another tribe, and left on the ground unconscious, the Brolgas come down, lift him up, and show him the road home'.[84]

The main unifying theme in the various recorded Lower Murray accounts of this mythology is that a pair of emus tricked a pair of brolgas into killing all but two of their own chicks. In the Lower Murray, Thralrum said that after the brolgas lost most of their chicks, 'In their deep grief, they put their heads in the fire, and rubbed hot ashes on their heads and necks, and that is why brolgas now do not have any feathers on their heads or neck'.[85] The brolgas sought revenge. By pushing their long sharp beaks into the ground in the Kingston area, they caused a flood that drowned the emus, which as flightless land birds were unable to fly away. The account of this mythology recorded in the Yaraldi dialect of the Ngarrindjeri language in the Lower Murray emphasised the female gender of the main birds involved.[86] It has been suggested that the 'Emu wives' of the Waiyungari (Mars) mythology of the Lower Murray were connected with the emu and brolga myth.[87]

For Aboriginal peoples, evidence of the creation events was all round them. On the edge of the sea at the southern end of the Coorong, north of Kingston, the large boulders known as the Papajara (the Granites) are associated with the drowned emu ancestors, and some of the chicks are associated with other granite outcrops at Katera, Kingston and Bosan Point on the eastern side of Cape Jaffa.[88] It was recorded from Thralrum that:

> You can see to this day the young emus in the water, and the father and mother emu on the beach in the form of granite rocks a few miles north of Kingston … The Brolgas, knowing that the emus, if they escaped, would hunt them up to kill them, flew up into the air, and, like the pelican flew up in circles getting higher and higher, until they reached the sky, and found it a good country to live in. So they stopped there. You can now see them at night in the form of two pieces of cloud at the end of the Milky Way.[89]

Other sites in the south-east of South Australia were connected to the emu and brolga mythology too. Cadara is a hill south-west of Kingston that was called Teriteritjngola, which reportedly meant 'willie-wagtail lookout'. This was a reference to how the willie-wagtail who 'by his chatterings warned the emu people when they were about to be cut off by the rising tide during a visit to Baudin Island; the flooding of the land by the sea was due to the evil genius of the brolga'.[90] It was asserted that this creation event had transformed the emus into scrub birds – no longer closely associated with the coastal wetlands.[91]

In the upper south-east of South Australia, it was a tradition of the Moandik people that a myth track of Ngurunderi's two 'nephews' came from the Lower Murray to their north and briefly connected with that of the emu and brolga in their Country.[92] Tangani-speaking people living on the Coorong knew of several songs associated with the emu and brolga mythology from the south-east of South Australia. One described how an enraged male emu was trapped on Baudin Rocks and was displaying his anger at having been tricked by his long-time foes, the brolgas.[93] Another two recorded songs concerned a pair of emu ancestors who made their nest in the bracken near trees where the wedge-tailed eagle had nested.[94]

For Aboriginal peoples, the physical appearance of emus and brolgas provides evidence for the deeds of the ancestors during the creation. In a myth of the Nunggubuyu people of the Rose River in eastern Arnhem Land, the emu and brolga fought and it was said that the 'marks that brolgas nowadays have on the heads and emus on their backs reflect the violence of their conflict'.[95] Anthropologist Frederick D McCarthy recorded a myth from the Minwula people of Cape York Peninsula that described how:

> ... after the emu had laid her eggs in a big nest the brolga stole all but two eggs. The emu looked everywhere for them, and then realised that the brolga had taken them. She told everyone in the camp what had happened. In a fight with the brolga, the emu hit her over the head with a stick which caused her hair to turn white and then rubbed red berries on the brolga's head in anger.[96]

Emus and brolgas also appear separately as ancestors in some other myths. In north-west Victoria, the two ancestral Brambambult brothers created the first cock and hen emus by killing the giant emu-like Ngindyal monstrous being then dividing its feathers into two heaps, which were transformed into the birds.[97] Through sorcery, these brothers enabled emus to have a clutch of many eggs, rather than the one only possessed by Ngindyal. A variety of creation ancestors are variously involved with creating the emu's plumage. Adnyamathanha people in the Flinders Ranges have a tradition concerning Pupu the rat who poured burning coals onto the chest of Warratyi the emu, giving the bird its dark plumage.[98] When the Pupu was chased into a cave, Warratyi tried to dig her out. In the attempt, the sharp stones sliced the emu's feet into three sections with toes.

Emus are pre-eminent in Central Australian Aboriginal ceremonial life.[99] Across Australia, emu symbolism is frequently seen in a wide variety of paintings and engravings on rocks, artefacts and earlier even on tree trunks.[100] Concerning the emu, late 19th century European anthropologist/medical officer Herbert Basedow remarked that 'Speaking generally, there is perhaps no other creature living which figures so frequently in aboriginal art, both on the cave wall and in the dance, as the great struthious bird of Australia'.[101] Emus are also prominent in the mythic landscape of northern Australia. In a Nunggubuyu myth, the emu receives its burnt complexion as punishment meted out by the 'Thundering Gecko' ancestor.[102] In what appears to be a Top End myth, the emu and crow ancestors fell out when camping under a bark shelter, leaving the emu with burnt-looking feathers and the crow with a hoarse voice caused by a cold.[103] According to a myth of the Dalabon people in southern Arnhem Land, during the creation the emu ancestor swallowed a *ngalwad* or sacred stone, which is how it came to lay eggs[104] and presumably became more bird-like. In northern Queensland, the emu ancestor is paired with an Australian pelican in a myth from Pennefather River, and with a waterhen in a myth from Boulia.[105]

The emu's symbolic significance in some areas was greater than its overall importance to the subsistence economy. Ceremonies centred on the emu ancestor were widely recorded from across Aboriginal Australia. In Central Australia, Arrernte people held Arleye (Erlia, emu) 'increase ceremonies' for the benefit of maintaining the population for this bird in their Country.[106] At Jessie Gap in the Macdonnell Ranges, it is an Arrernte tradition that an emaciated emu turned into stone here after death.[107] Among the Wiilman people in the south-west of Western Australia, Hassell stated that:

> It is interesting to note that the emu, which has much less value to the native than the kangaroo, is found in a great many legends and the kangaroo in comparatively few. The emu furnishes food fat; with which the natives rub their bodies; and feathers for ornaments.[108]

Emu ancestors have been linked to ochre deposits, which were the source of ceremonial paint. Nyungar people in the south-west of Western Australia believed that red ochre deposits were formed from dripping blood, as Jitti-jitti the willie-wagtail chased Wej the emu across Country to the place where he eventually died, which was at Koolbing.[109] In the Adnyamathanha tradition of the northern Flinders Ranges in South Australia, it is believed that the ceremonial ochre extracted from the Pukardu mine near Parachilna is the 'blood' of the emu ancestor.[110] Here, according to a man who Europeans knew as 'King Harry':

> It is said that in remote times two wild dogs chased an emu from the direction of the Parachilna Gorge to the Elder Range, passing two prominent peaks which have ever since been known as the Two Dogs. They ultimately captured the bird on the slope of another hill now known as Emu Hill. There the bird was killed by the dogs, and its blood is supposed to have penetrated the clay and given it a red colour.[111]

There is an association between emus and the Australian form of tektites, known as australites, which are used in healing rituals in Central Australia and the Western Desert.[112] The flanged button-shaped australites are known as 'emu stones',[113] and geologists believe that they were formed when a meteorite struck somewhere in south-east Asia ~0.793 million years ago and produced a strewn field of tektites across much of Australia.[114] The Diyari people of Cooper Creek had traditions concerning australites being associated with the emu ancestor, and as such the stones were described as *warukati-undru*, 'pertaining to emus'.[115]

Here, according to McCarthy, the act of the emu losing their eyes was said to have occurred during the creation period:

> After lighting a ring of fires around a water-hole, the hunters threw emu's eyes (australites or tektites of meteoric origin) at the birds to confuse them and cause them to run into the water, where they were easily killed. The australites were believed to be the eyes of ancestral emus which lost them while searching for food, but imbued them with a magical control over the living birds.[116]

Scholars once believed that Aboriginal people and emus were both enlarging the spatial range of australites. Basedow noted that:

> Their universal distribution has, no doubt, been assisted by the agency of the native and the emu (in the form of gizzard stones). The natives call obsidian bombs *Pandolla* and *Kaleya korru*, the latter meaning emu eye. They are collected by the medicine men of the tribes, and applied in the healing of sickness.[117]

Brolgas and Aboriginal people have an acknowledged close relationship, which is expressed by the bird's historic Australian English name, 'native companion'.[118] An early 20th century newspaper correspondent remarked that 'It is said that some of the movements in corroborees among certain aboriginal tribes are based upon the dancing antics of the native companion'.[119] Concerning brolgas and their 'corrobborrees', or Aboriginal-style dances, colonist AJ Campbell noted in 1893 that 'During these performances, we are told, the birds go through grotesque movements, as well as graceful kinds of minuets, courtesying to each other and gambolling in a remarkable manner'.[120] Another colonist, writing with the *nom de plume* of 'Bill Bowyang', observed in 1929 that 'The native companion is a very sociable bird, and is not averse to the company of man'.[121] In 1951, travel writer Julitha Walsh observed that in the west Kimberley, 'Often, on an open plain one saw the timid "brolgas," or native companions – tall grey birds with long thin legs – which often hold "corroborees" of their own dancing majestically in longlines'.[122] In the interior of Australia, it was the settlers' tradition not to shoot brolgas, as it was said that 'You'll have a run of stiff luck. The blacks say those birds are in league with Mooras (devils)'.[123]

Bird fishers

The recorded mythologies have many accounts of birds who, like their human descendants, were fishers. The cluster of different waterfowl species featured in the same myth is a reflection of how these species are observed to closely interact with each other ecologically as a community, in a human-like manner. Published versions of a fishing legend from the Lower Murray have described how in the creation period all birds were men.[124] In brief, this mythology recorded that fishing bird ancestors, such as the silver gull, Caspian tern, shag, hoary-headed grebe (diver) and Australian pelican, lived at Tenetjanual (Murrayville) in western Victoria, where there are two hills alongside a lagoon. The path from here to the

Lower Murray can still be readily identified, as it follows the zone of calcareous soils which has open scrub growing on it.[125] According to anthropologist Alison Harvey, the bird group, having fished out the surrounding lagoons, travelled west towards Lake Alexandrina in South Australia:

> The pelicans and all the other birds, as they travelled through the mallee scrub, made a great 'pad' by treading down the vegetation. This is now a line of clear spaces in the mallee, patches of claypan and samphire swamp running south-west from the Wimmera to Lake Alexandrina.[126]

At Lake Alexandrina, the fisher bird ancestors made their fishing nets and avoided the areas of saltwater, which pushed in from the sea. Along the lakeshore they travelled in company with land birds, among them the white-backed magpie who was in possession of firesticks. The fishing birds were annoyed with the magpie, who repeatedly refused to make fire, so they gave him only the *thukeri* (*tukkeri*, silver bream) from their nets – a fish with excessively bony flesh. Harvey recorded that as a result:

> The magpie, becoming angry, began to belabour the other birds with the fish he had been given. And so arose the magpie's characteristic action of chasing the smaller birds. He hit the crow's eyes with his firestick, and gave him his 'smoky' eyes. The pelicans, who previously were coal black, became splashed with white where the scales of the *tukkeri* [*thukeri*], wielded by the irate magpie, stuck to their bodies.[127]

A myth recorded from the Murrumbidgee region in New South Wales concerns the ability of birds to fish. In this narrative, Goolay-yali the pelican had a large fishing net which he would secretly regurgitate from his body when needed.[128] Here, the ability to make the nets from strips of kurrajong bark was eventually shared with other ancestors, but the pelican retained the practice of drawing in the nets. Katherine Langloh Parker recorded that:

> ... the Daens [Aboriginal people] tell you that if you watch the Goolay-yali or pelicans fishing, you will see that they do not dip their beaks straight down, as do other fish-catching birds. The pelicans put their heads sideways, and then dip their long pouched bills, as if they were going to draw a net. Into these pouches go the fish they catch, and then down into their nets, which they still carry inside them.[129]

For Aboriginal peoples, the present physical form of birds is tangible 'proof' of events taking place in the creation. The Mawng people of coastal western Arnhem Land say that the forked tail of the frigate bird was originally a pair of firesticks when he had more human-like traits during the creation.[130] In the Murray River area of South Australia, the settler

John P Bellchambers pointed out the similarity between the pelican's upper bill and a spearthrower to an Aboriginal person, who was reported to have replied:

> 'Oh, that phellar [fellow], he bin keep his wommera [spearthrower], that's all; he bring him along.' This meant to imply that from the time when the pelican was a blackfellow he had managed to hang on to his wommera during his transformation.[131]

Creation myths combine explanations concerning the behaviour, form and preferred habitat of bird fishers. For instance, in the Flinders Ranges the Adnyamathanha people have a myth about the dispute over water that led to the black duck man attacking the Australasian grebe man with boomerangs, causing the latter to dive. When the boomerangs had all been thrown, the duck man started swimming and the grebe man became a diving bird.[132] At Princess Charlotte Bay on north-east Cape York Peninsula, early anthropologist and doctor Walter Edmund Roth recorded a myth about the eastern osprey and the pheasant coucal ancestors being in conflict over poisoned fish from an inland waterhole.[133] In this narrative, the result was that the pheasant coucal made floods that drove the eastern osprey down to the sea, where he was thereafter to remain. The pheasant coucal had his fishing spears hidden in the tree canopy, where he was later to be seen always searching for them. At Pennefather River on north-west Cape York Peninsula, Roth recorded another account about the 'first fish hawks', which were eastern ospreys. Roth said:

> Two men were hunting turtle, but one always caught males, and other one females. Now this was just what each did not want, for the one who caught bucks was anxious to get eggs, and he who caught females wasn't. Naturally, they got very angry, and began swearing, words soon giving way to blows. They knocked each other over the faces, and broke their noses (the present hooked beaks), while the cuts over their bodies gave rise to feathers – like wooden chips when a log is cut. They are fish-hawks now, and cannot hunt turtle any more, but have to content themselves with fish.[134]

Bird ancestors were implicated in the present-day distribution of fish species in some areas. There is a tradition held by the Walmajarri people of the Great Sandy Desert of Western Australia that during the creation Jalka the great egret threw various fish, such as barramundi, away from the lakes as far as Turkey Creek: the white rocks on the nearby hills are where some of the fish landed.[135] It was said that only the spangled perch, Kimberley catfish and rainbow fish were left in the northern lakes.

Capturing fire

In the mythologies of Aboriginal Australia, bird ancestors are frequently associated with fire and credited with releasing it to prepare the world for the people who were to follow.[136] For instance, in the Narran River region of northern New South Wales, a firestick was stolen from

Bootoolga the crane and Goonur the bettong by Biaga the hawk, who set fire to the grass as he fled.[137] In western South Australia, it is a creation tradition that Kipara the bustard kept the ability to make fire to himself, by hiding the flint stones beneath the sea.[138] Anthropologist Scott Cane recorded that:

> The Wati Kipara Tjukurrpa [bustard man creation myth] involves the travels of an old bustard who attempts to steal the world's fire and drown it in the ocean at Madura, on the shores of the Nullarbor Plain. The bustard is watched by his two sons, who hover in the sky as stars. They snatch the fire before it is drowned in the sea and drag the Old Man out of the water, north across the Nullarbor Plain and back to the Spinifex homelands [such as Tjuntjuntjara], where he slowly dies … His sons then take the fire and distribute it, with great celebration, to all the inhabitants of the earth.[139]

Various birds were players in the corpus of fire creation mythology from the Lower Murray. An account of a Ramindjeri myth, published by missionary Heinrich AE Meyer in 1843, is localised to the upper Hindmarsh River area. It concerned Kondoli the whale ancestor and the origin of fire, and said that at the termination of a ceremony the decorated young male performers, 'who were ornamented with tufts of feathers, became cockatoos, and the tuft of feathers being the crest'.[140] There was a fight over the power to make fire, and Rilballe the brown skylark speared Kondoli in the neck, giving the southern right whale its distinctive blowhole. The property of fire was placed into a grasstree, from where it could be released by using the dried flower stalks in a fire-drill.[141] Other bird species are involved with other recordings of what is structurally the same myth. A Ramindjeri version recorded from a Yaraldi-speaker chiefly involved Krilbali the brown skylark and Kuroldambi the owl (barn owl), along with the whale.[142] According to a Yaraldi version in the Lakes area of the Lower Murray, it was the skylark and willie-wagtail who conspired to spear Kondoli and steal his fire.[143] In the latter account, the fire released by the spearing was spread throughout the country by the skylark flying off and dropping the sparks onto the ground, where they were transformed into flints, from which fire can be made by percussion. In another recorded Yaraldi account, which is essentially the same as above, it was Krilbali who speared Kondoli.[144] As with many myths, the Kondoli narrative was told as a song.[145]

Birds with red splashes of plumage are particularly prominent as participants in fire creation myths. For example, in a myth from the Bunganditj people of the Mount Gambier area in the south-east of South Australia, Tatkanna the 'robin redbreast' ancestor stole the power to make fire by secretly thrusting a dry flower stick from a grasstree into a fire made by Mar the sulphur-crested cockatoo. A fight then broke out between the ancestors as fire escaped into the surrounding scrub.[146] A recorded myth from the Swan Hill area of central Victoria involved Pupperrimbul, a 'little bird with the red patch above the tail [possibly the beautiful firetail, diamond firetail or even red-browed finch]', who as a human during the creation made the sun by casting a prepared emu egg into the dark void.[147] In western Victoria, Tirrtu the 'crows' (Australian ravens) from the Grampians had sole possession of fire until it was stolen by Yuuloin keear the 'fire-tail wren' (possibly a firetail or finch) and by Tarrakukk (named from 'its cry') a large 'kestrel hawk' (nankeen kestrel).[148] The Gunai of eastern Victoria have a creation tradition that the 'fire-tail finch' stole a fire-brand from two women, and when he was transformed into a bird the red spot remained on his tail where he had carried it.[149]

In the Lake Eyre district of eastern Central Australia, a recorded Diyari myth concerned the Mura-mura ancestral woman named Kuti, who fought and killed a Nardu woman in order to steal a firestick.[150] Kuti became a black swan, who had a red edging on the inside of her beak where the fire had burnt her. The injured condition of the black swan is also a feature in its name, *kool-jak*, in the Nyungar languages, in which it was translated as 'pus on beak'.[151] The Mak Mak Maranunggu people of the floodplains south of the Finniss River in the Northern Territory believe that in the creation fire was stolen by A-titit the brown goshawk from Puley Puley the Rainbow Serpent, who had been trying to extinguish it in the sea and thus prevent its use by other ancestors and their descendants.[152] At Cape Grafton on the east side of Cape York Peninsula, it was Bin-jir Bin-jir the red-backed wren who flew to the Skyworld and brought back the first fire hidden under his tail.[153] A Wik Mungkan myth from western Cape York explains that the bush stonecurlew's mournful call at night means 'I can't see! I can't see! I'm going blind!'[154] In this narrative, the bird made unsuccessful attempts at starting a fire with firesticks, but eventually had to walk about in darkness with his eyes protruding from the strain of his exertion.

As mentioned above in relation to species of crows and ravens, exposure to extreme heat was a major Aboriginal theme describing how birds and other animals gained their black or white colouring. In western Victoria, Goruk the Australian magpie cooked a dog he had speared, and while it was still hot he pulled it out of the earth oven and slung it across his back. However, 'the cooked dog was hotter than Goruk realised, and here and there his skin was burnt white, as can be seen on every magpie today'.[155] In this region, Bunjil the wedge-tailed eagle ordered his sons, who were Djurt-djurt the nankeen kestrel and Thara the collared sparrowhawk, to set fire to the Country in order to burn Balayang the bat, who was left black.[156] Bare patches of skin on birds were also attributed to the ancestors being burnt. A recorded western Victorian myth concerned a bird ancestor, who probably became the Major Mitchell cockatoo, having his fire stolen by the collared sparrowhawk. The 'cockatoo still has a red crest to show where he kept the fire, and a bare patch underneath it, where the fire once burnt him'.[157]

The theme of burning in relation to how birds obtain their plumage colouring is common in northern Australian mythologies. In the Endeavour River area of eastern Cape York Peninsula is a record of a myth concerning how a white cockatoo, who was asleep in the canopy of a tree, became a black cockatoo when burnt by a fire set by the frill-necked lizard.[158] In Wik Mungkan creation mythology from western Cape York, the magpie-lark and willie-wagtail are white birds who fought each other with burning ashes from the fire, giving them black patches that they retain today.[159]

Given the continent-wide theme of birds being associated with fire, it seems remarkable that Tindale proposed that the ancestors' campfires recorded within the eaglehawk and crow myth narrative of the Mount Gambier area in South Australia were symbolic of volcanic vents, and that they were 'racial memories' passed down through Aboriginal tradition of an actual volcanic eruption, commencing at 4710 BP [Before Present] and with the last minor recurrence being 1410 years BP.[160] If Tindale was correct, which I doubt, this would mean that the myth had been orally passed, largely intact, down through 188 to 56 generations to the present – a truly remarkable achievement. The estimated number of generations is based on an average 25 years of age for each parent. Other problems with the proposal is that the track of the eaglehawk and crow mythology extends much further than the Mount Gambier area into non-volcanic regions, and that knowledge of it appears to have been subjected to movement in the historical period.[161] Superficially, Tindale's theory appears to be well-meaning through his acknowledgment that elements of Aboriginal myth are equivalent to historical fact. Unfortunately, by asserting that themes remain largely unchanged within the myth for millennia he is locking Aboriginal tradition into what Europeans have in the past considered to be its 'primitive' form, which modern anthropology does not accept.

Birds, minerals and gemstones

The corpus of recorded mythologies in Aboriginal Australia is replete with narratives concerning the production of valuable minerals and gemstones. Perhaps, given the spectacular colouring of many birds, an association with precious minerals and gemstones should not be unexpected. When Southern Pitjantjatjara people speak of the myth track for the Nyiinyii the zebra finch men they refer to the scatters of flaked white chalcedony, an opal-like stone, on the ground at certain sites across western South Australia, such as at Wilson Bluff and at the base of the Nullarbor cliffs.[162] The zebra men had carried the chalcedony from the desert, and for Aboriginal people the scatters were an important source of stone for making tools with sharp blades.

As mentioned in Chapter 2, there are documented Aboriginal creation narratives that clearly have incorporated European-derived elements. These myths are still culturally important because they are consistent with the Aboriginal view that their own ancestors created the world and left things of value that would be of crucial benefit to the people who were to come. An example of such a myth is from the northern Flinders Ranges of South Australia, where Adnyamathanha people have a tradition concerning Yurlu the red-backed kingfisher, who created the seams of coal at Leigh Creek by burning dry mallee sticks as a sign that he was on the way to a ceremony.[163] The Leigh Creek coal, known by local Aboriginal people as 'Yurlu's Coal', was mined during the 20th century for use in the production of South Australia's electricity.

The Adnyamathanha people also have a myth concerning Marnbi the common bronzewing, which connects the northern Flinders Ranges with places as far away as Broken Hill to the east and Mount Isa to the far north.[164] The myth track for this ancestor may have also been connected to the southern Flinders Ranges region, where a related name, *manbi*, was recorded for this species in the Ngadjuri language,[165] and *marnpi* in Barngarla and Nukunu.[166] In the adjacent Yorke Peninsula region, the common bronzewing was termed *manpi* in the Narangga language.[167] In the Adnyamathanha account, a hunter had netted many bronzewings at Baratta, north-west of Yunta, but a wounded Marnbi escaped. The locations where Marnbi stopped became deposits of gold from the blood, and seams of quartz from the feathers. European miners later dug Marnbi's 'blood' as gold from mines at these places. The Marnbi is also credited with making gemstones. Tunbridge recorded that:

> A long time ago Marnbi the pigeon threw an *ardla wirdni* (firestick). It went a long, long way. It landed over Coober Pedy way [north-west of the Flinders Ranges]. When the firestick came down and hit the ground, *ardla virldi* (sparks) flew in all directions. These sparks are the opal at Mintabie and Coober Pedy.[168]

In the Western Desert of Central Australia, ringneck parrots are associated through the creation mythology with two gemstones – australites and chrysoprase. During a South Australian Museum fieldtrip to the north-west of South Australia in 1963, Tindale was at Mount Davies in the Tomkinson Ranges and noted in his journal:

> According to [Aboriginal patrol officer, Walter (Wally)] MacDougall's story the Njinggar or Ice men made a black hail of australites fall from the sky to kill ring necked parrot [*patilpa*] men as punishment for killing all the *waro* or wallabies [*waru*, black-footed rock wallaby].[169]

On a later fieldtrip, to Woomera in western South Australia in 1965, Tindale again interviewed patrol officer MacDougall about australites. He recorded a truncated version of the australite myth involving ringneck parrots and rock wallabies:

> MacDougall found that the Mt. Davies aborigines called these australites *ko:di* and not *japu*. In their myth the *waro* or rock wallaby quarrelled with the ring-necked parrots who were killing many *waro* without cause. A *waro* being caused black hail to fall and this killed all the ring-neck parrots while the *waro* hid safely in their rock shelters.[170]

There is another ringnecked parrot myth narrative from this part of the Western Desert, recorded in the 1960s by Bob Verburgt, who was then working as superintendent at Amata for the Department of Aboriginal Affairs, when out on a trip with Pitjantjatjara men near Mount Davies in the north-west of South Australia. It was on this trip that the valuable green gemstone known as chrysoprase was discovered, and plans were made for Aboriginal people to mine it. The mythological account of the creation of the chrysoprase is similar to that of the ringneck parrot given above, but on this occasion without any mention of the Ice men, rock wallabies or australites. Tindale was willing to link the early ringneck parrot myth to the 'racial memories' of an australite shower, as he done with the crows' fires and volcanoes, but in this version the mention of chrysoprase, with a geologically ancient origin, would not have fitted. Verburgt said:

> The old men who were with me told an interesting story as to how the green stone came to be. In the Dreamtime they said the green parrots [i.e. ringnecks]

used to come to the area to drink from the rock-holes in the early morning and later afternoon. They would come in their thousands, swooping and squawking before settling down to drink. One day a fearful storm came up with strong winds, thunder and lightning. This frightened the parrots and they took to the air in panic. They were circling around the hills as a burst of hail thudded into the ground. The hail was so large that it knocked most of the green parrots onto the ground, transforming them into stone.[171]

Celestial beings

The perceived existence of the heavens as an image of the terrestrial landscape is common across Australia.[172] This Skyworld is perceived as a region which, to some extent, obeys the same laws as those on the terrestrial landscape. The means by which ancestors reached the heavens towards the end of the creation has been described as like an 'Indian rope trick'.[173] The Skyworld was the place where unborn human souls resided and was the destination of souls from the deceased.[174] Anthropologist Ralph Piddington, who worked with the Karajarri people of Lagrange Bay in northern Western Australia, explained that:

> Among the Karadjeri the mythology of the heavens plays an important part in the beliefs connected with immortality. It is generally believed that the sky consists of a dome of a very hard substance (rock or shell), the stars representing the *bilyur* (spirits) of dead men and women.[175]

The Skyworld was perceived as a utopian paradise, which the Aboriginal informants of colonist James Dawson in western Victoria described as 'a beautiful country above the clouds, abounding with kangaroo and other game, where life will be enjoyed for ever'.[176] Similarly, the Diyari of Cooper Creek in eastern Central Australia believed that their *kunki* or 'medicine-men' were able, 'by means of a hair-cord, go up to the beautiful sky-country, which is full of trees and birds, and drink its water, "from which they obtain the power to take the life of those they doom."'[177] It is likely that 'high', in relation to the heavens, may have only been considered to be the height of a tree or at most a hill, because in some Aboriginal myths the Skyworld was reached by the throw of a spear.[178] There is other evidence for the close proximity of the Skyworld and its inhabitants. Howitt recorded that in western Victoria there was a Wotjobaluk legend which 'runs that at first the sky rested on the earth and prevented the sun from moving, until the [Australian] magpie (*goruk*) propped it up with a long stick, so that the sun could move, and since then "she" moves round the earth'.[179]

For Aboriginal peoples, the cosmic bodies were rich with meaning, in most cases associated with ancestors who had originated on Earth during the creation. The myths for each constellation frequently intersect with others, as they do with sites on Earth. For example, in the Narran River region of inland New South Wales, the Pointers were seen as white cockatoos, who are always trying to reach their roost in the white gumtree, the Southern Cross.[180] This interweaving of accounts about the stars is illustrated in the reminiscences from settler T Giles, which concerned a Manggurupa clansman named Billy Poole, from the Lake Albert area of the Lower Murray. On one occasion:

> When around the camp fire at night he [Billy Poole] told me the names of stars, and, more over, of constellations. He pointed out one group as an old man kangaroo with his arm broken; another group was a turkey [bustard] sitting on her eggs, the eggs being our constellation Pleiades, another a Toolicher, a small and very prettily marked kangaroo [toolache wallaby] peculiar to the district; another an emu and so on.[181]

The linking of seasons with the movements of celestial bodies, many of them seen as birds, is common across Australia.[182] For instance, among the Tangani-speaking clans of the Coorong, the 'doctor men' interpreted the appearance of Lawarikark, seen as Vega in the constellation of Lyra, as the sign that it was the nesting time for the *lowan* (malleefowl).[183] These birds were known to make a harsh scolding call when racking plant debris into mounds in which to incubate their eggs. Aboriginal peoples therefore considered that the malleefowl ancestors in the Skyworld would also be quarrelsome.[184]

There is a malleefowl myth, similar to the Coorong version, recorded from the Swan Hill area of central Victoria, in which Lyra represents a flying *lowan*.[185] The acronychal (at dusk) rising of Vega in mid August signals the time when the malleefowl begins building its nest mound.[186] Another example of egg-laying being linked to celestial movements comes from south-west Victoria, where it was recorded that 'when Canopus is a very little above the horizon in the east at daybreak, the season for emu eggs has come'.[187] Canopus will rise heliacally (at dawn) in early May, as seen from south-west Victoria, which coincides with the emu egg-laying season and the acronychal rising of the celestial emu in the Milky Way.[188]

At the close of the creation period, Lower Murray people believed that the crow (Australian raven) climbed into the Skyworld to become a far-distant star seen in the south-east during autumn. Autumn was known as Marangalkadi, which is a time when stars of this name appear,[189] and the name reportedly meant 'pertaining to the crow'.[190] According to Yaraldi tradition, the autumn stars are low in the south-eastern sky precisely because it was to the south-east of the Lower Murray that Marangani entered the Skyworld. Tindale recorded a Yaraldi 'rutting season':

> When Marangali [Marangani] the 'Crow' star is in the zenith (not yet identified) women and all animals seek their mates. When a woman was reported to be 'going wild' and taking two different men on successive nights old men shook their heads and took younger men out of the huts and pointing to Marangali said 'When Marangali is up there women are always like that'.[191]

Marangani the 'crow' (Australian raven) had complex relationships with living people, and he was considered to be important for fishing. It was stated by anthropologist Ronald M Berndt and his co-workers that:

> Crow, metamorphosed as a star … is propitious for fishing and is responsible to the ebb and flow of the tides … Crow appears in June–July (winter) when fishing is sporadic, since he spends little time in the sky and most in the huts of various women: he disappears in October, coinciding more or less with Waiyungari [planet Mars].[192]

The celestial identity of the 'crow' is suggested by the reminiscences of a Goolwa settler, who claimed that the Lower Murray people 'had great reverence and awe for the heavenly bodies, and had names for certain stars, calling the morning star "Murrunungu" – the crow'.[193] In the adjacent Adelaide Plains region, the 'autumn star' (Parna) has been identified as Fomalhaut, based upon its heliacal rising in mid March during a time of increased rainfall.[194] Birds could also be associated with particular weather events; for example, Aboriginal people on the Adelaide Plains claimed that Karndo was 'the name of a large bird, said to be the author [i.e. creator] of the thunder'.[195]

Across Aboriginal Australia, ancestors and spirit beings linked to birds of prey and crow/raven species were commonly believed to reside in the Skyworld. For instance, in the Diamantina River area of south-west Queensland, it was tradition that Mars was Kooridala the wedge-tailed eagle, Sirius was a hawk and another star was Wakerdi the crow.[196] In Victoria, it was recorded that the 'Eagle (*Quarnamero*) is now the planet Mars, and justly so, because he was warlike, and much given to fighting. The Crow (*Wagara*) is a star, and smaller stars are set about him, and those represent his wives'.[197] In the Adnyamathanha language spoken in the northern Flinders Ranges, Wildu is the name for both the wedge-tailed eagle and the planet Mars.[198] The Yuwaalaraay-speaking people of central New South Wales considered that Mullyan the wedge-tailed eagle ancestor was represented in the night sky by the Corona Australis (Southern Crown) constellation.[199] It is recorded that in the south-west Kimberley, Jarinkalang the giant wedge-tailed eagle is seen as the Southern Cross, and the nearby False Cross represents its tracks across the Skyworld.[200]

Gaining the ability to fly during the creation is linked, in some myths, to the deeds of ancestors, who were then more human-like. A Guugu-Yimithirr myth, localised in the Endeavour River area of eastern Cape York Peninsula, described how an ancestor known as 'old Gujal' the eagle first grew his wings by accidentally eating a winged beetle that had been one of the grubs gathered by his magpie wives from the base of a 'grasstree palm' (Johnson grasstree).[201] In this case, the beetle was probably a longicorn, based upon its documented identity as an edible wood grub in northern Queensland.[202] A favoured food that was perceived as having transformative powers, wood grubs, as larvae of moths and beetles that fly, appear in other myth narratives concerning birds and flight. In the Ararat area of south-western Victoria, the Pirt Kopan Noot dialect speakers of the Tjapwurrung language had a creation tradition whereby Gneeanggar the eagle, who was 'queen' of the

Seven Sisters, was being pursued across Country by Waa the 'crow' (Australian raven).[203] Gneeanggar was tricked into being caught, after Waa turned himself into a white wood grub that she extracted with a small wooden hook. Waa then turned back into a 'crow' and in this form carried Gneeanggar into the Skyworld, where she can be seen as Sirius and he as Canopus. The remaining sisters became the Pleiades. A neighbouring Aboriginal group, the Kurnkupanut (Kuurn Kopan Noot) dialect speakers of the language spoken in the Warrnambool area, knew the Pleiades as Kuurokeheear, and described them as a 'flock of cockatoos' that went up into the sky.[204]

The presence of emu and brolga ancestors in the Skyworld appears to be a common cosmological theme across south-eastern Australia. The emu, in particular, is perhaps the commonest bird motif in Aboriginal art due to its large size and ceremonial importance.[205] In the south-west of Western Australia, the emu ancestor was seen as a dark spot in the Milky Way.[206] Basedow published a sketch of '"Dangorra", the great emu in the southern sky', as seen in the dark spaces of the Milky Way by Aboriginal groups living to the south-west of Darwin.[207] The Aboriginal association of the emu ancestor with the Coalsack is widespread across much of the continent. Astronomers Ray and Cilla Norris stated:

> The Coalsack is the head of the emu. Stretching away to its left you should be able to see its long dark neck, round body (near Scorpius), and finally (towards the horizon) the legs. This Emu in the Sky features in the songs and stories of Aboriginal groups right across Australia, from Western Australia to New South Wales, although it's not universal, and detailed interpretations differ.[208]

The Gunditjmara people in south-west Victoria believed that the larger Magellanic Cloud was the 'male native companion' or 'gigantic crane', in other words a male brolga, while the smaller cloud was the female equivalent.[209] Similarly, in the Narran River area of northern New South Wales, the brolga and her daughter were seen as the Magellanic Clouds.[210] In the south-east of South Australia, Moandik man Alf Watson claimed that the emu ancestor involved in the fight with the brolga was seen in the night sky as a black patch in the Southern Cross.[211] It was reported that in western central Victoria it was a belief that Tchingal the emu resided in the dark section (the Coalsack nebula) under the constellation Crux (the Southern Cross).[212] A similar version has been recorded in the Gamilaraay and Yuwaalaraay languages of northern central New South Wales.[213] In southern central New South Wales, there is a record of a tradition that during the epic fight between Dinewan the emu and Bralgah the brolga on a large plain near the Murrumbidgee River, the brolga seized one of the emu's eggs and tossed it up into the Skyworld. Here, 'it broke on a heap of firewood, which burst into flame as the yellow yolk spilt all over it, which flame lit up the world below [as the sun], to the astonishment of every creature on it'.[214]

The Lower Murray people believed that there were two Prolggi the brolga ancestors in the sky, having got there after fighting with Pindjali the emu ancestor, who became the Magellanic Clouds.[215] In the Lower Murray perspective, there were probably other emu ancestors seen in the night sky, as the black patch to the west of Ngurunderi's canoe

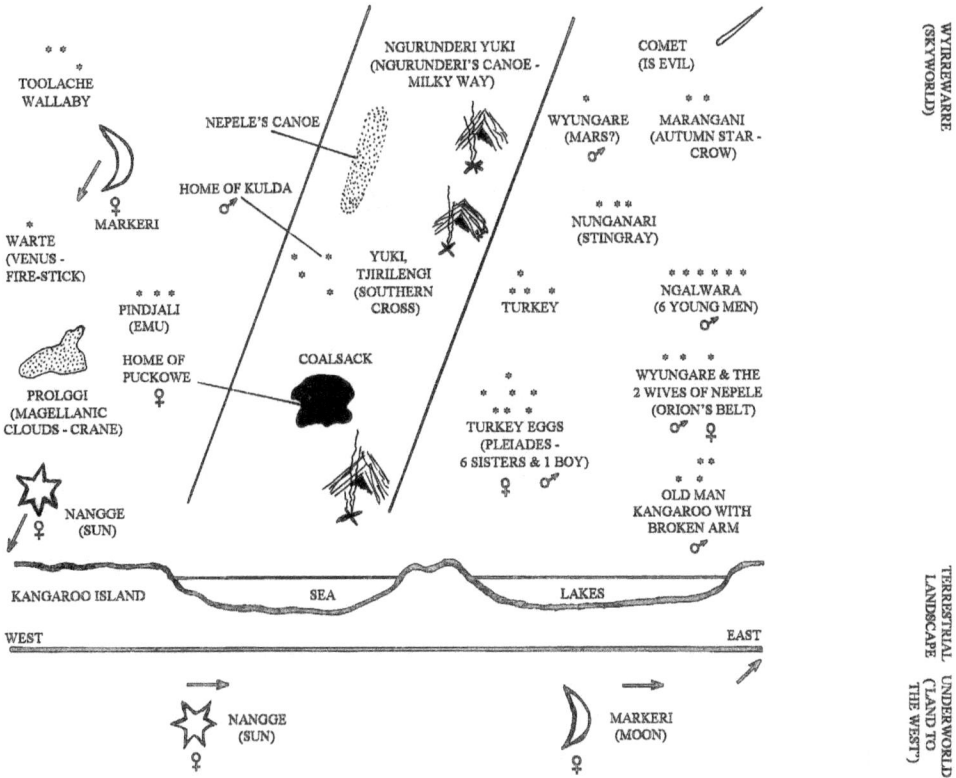

TOOLACHE
WALLABY

NGURUNDERI YUKI
(NGURUNDERI'S CANOE -
MILKY WAY)

COMET
(IS EVIL)

NEPELE'S CANOE

WYUNGARE
(MARS?)

MARANGANI
(AUTUMN STAR -
CROW)

HOME OF KULDA

MARKERI

NUNGANARI
(STINGRAY)

WARTE
(VENUS -
FIRE-STICK)

YUKI,
TJIRILENGI
(SOUTHERN
CROSS)

TURKEY

NGALWARA
(6 YOUNG MEN)

PINDJALI
(EMU)

HOME OF
PUCKOWE

COALSACK

WYUNGARE & THE
2 WIVES OF NEPELE
(ORION'S BELT)

PROLGGI
(MAGELLANIC
CLOUDS - CRANE)

TURKEY EGGS
(PLEIADES -
6 SISTERS & 1 BOY)

OLD MAN
KANGAROO WITH
BROKEN ARM

NANGGE
(SUN)

KANGAROO ISLAND

SEA

LAKES

WEST

EAST

NANGGE
(SUN)

MARKERI
(MOON)

WYIRREWARRE
(SKYWORLD)

TERRESTRIAL
LANDSCAPE

UNDERWORLD
('LAND TO
THE WEST')

Map of the Skyworld from a Lower Murray perspective, drawn by the author based on the literature. Note the presence of bird ancestors as astronomical features. Philip A Clarke, 1994.

(the Milky Way) was also said to be an emu.[216] As with their terrestrial counterparts, these celestial beings migrated according to the season. It was recorded from Yaraldi-speakers living round the Lower Lakes that 'The Native Companions [brolgas] (husband and wife) may be seen rising in early spring in the south-eastern sky, disappearing to the south-west. They were said not to rise too far into the sky "because they belonged to the South-East"'.[217] In another account, the brolgas are seen lying to the south-east in the winter sky, in spring they are south of the Milky Way and then shift towards the western sky in summer.[218] From a Lower Murray Aboriginal perception, the weather is largely produced by ancestral forces in the heavens, with the south-west wind being known as *parolgamai*, in reference to the brolga.[219]

*****

Aboriginal recognition that birds and people share many behavioural attributes is apparent with the prominence of birds as major actors in the foundation mythologies. The mournful cries of some species, such as the bush stonecurlew, reflect the sadness experienced by their ancestors. The colouring of birds, such as the crow/raven, black swan and firetail, is seen as evidence for certain events that took place during the creation, such as being wounded through fighting or coming into contact with fire. From their heavenly homes in the Skyworld after the creation, bird ancestors are considered to still influence all life on Earth. The performance of ceremonies in honour of them is one way to access their power. Aboriginal peoples who have a deep understanding of the corpus of creation mythology hold much environmental knowledge about key bird species, in addition to what they discover through their own observations of birds when foraging. In the next chapter, the role of birds as spirit beings is investigated.

Endnotes

[1] Waterman (1987) provided a detailed catalogue of myths, or 'tales', from Aboriginal Australia, organised by theme.

[2] Tindale (1934–37:52).

[3] Clarke (1996b:86, 2016a:282, 2019a:141) has provided a summary of records relating to Tjilbruki.

[4] Tindale (1987a:9,11).

[5] Tindale (1987a:5).

[6] Amery (2016:115,121–122,225).

[7] Clarke (2018a:6–10).

[8] Slater (1970:247).

[9] Slater (1970:247).

[10] Berndt & Berndt (1989:187,192).

[11] Hassell (1934b:331).

[12] BJ Nangan (cited in Akerman 2020:128–129).

[13] Bates (1906).

[14] Cane (2002:105–106).

[15] Bates (1918c:160). Waldja is a modern rendering of Bates' Walja. In the literature, other variations of Waldya include Waldya, Waldja and Waltja. My own fieldwork in western South Australia has confirmed that Waldya, often described as an 'eaglehawk', is the wedge-tailed eagle.

[16] Bates (1947:245–246). Bates' Weeloo is a variation of Wirlu.

[17] Gunggara is a modern spelling of Koongara, as written by Bates. In the original text Koongara was described as a 'little hawk', but in local languages 'Koongra' has been recorded as 'Nightjar' (Sullivan 1928:169) and *gunggara* as the 'spotted nightjar' (Miller *et al.* 2010:43).

[18] Garn.ga is a modern rendering of Bates' Kaan'ga, as is Yagana from Yanguna. In the original text Yanguna was a 'white cockatoo', but elsewhere 'Yakana' was recorded as a 'White-breasted Sea-eagle [white-bellied sea-eagle]' (Sullivan 1928:166) and *yagana* as a 'sea-eagle' (Miller *et al.* 2010:90), while the 'Cockatoo-Parrot' was 'Kooyilgurra' (Sullivan 1928:169) and the 'cockatoo (general)' as *gagalya* (Miller *et al.* 2010:32).

[19] Hercus & Potezny (1999:171).

[20] Berndt & Berndt (1942:144).

[21] Berndt & Berndt (1943:44).

[22] Berndt & Berndt (1943:49).

[23] Berndt & Berndt (1943:49).

[24] Tindale (1940–56:203).

[25] Clarke (2019b:168).

[26] A Wilson (cited in Letch 2010:11–13).

[27] Johnston (1943:279), Clarke (1997:130,137, 2018f:258, 2018g:276) and Johnson (1998:20,50,56,71–72,84,114–115,118,124–125). For outlines of the mythology refer to Mathew (1899), Tindale (1939) and Clarke (2016a:280,287,289, 2018a:4,6–8).

[28] Bulmer (1855–1908:40).

[29] Anonymous (1888).

[30] Smyth (1878:1:423–428) and Howitt (1884:192, 1904:492).

[31] Massola (1968:198).

[32] Mathews (1893:355) and Martin (2011:237–241,273).

[33] Forbes *et al.* (2019:219) and Karskens (2020:488,506,509,511–512). For an account of Baiame (variously Baiami, Baiamai, Byamee etc.), refer to Swain (1993:127,130–131,137–140,144–146,277). Karskens (2020:505–506) considered Koen to be another name for Baiame.

[34] BJ Nangan (cited in Akerman 2020:86).

[35] Ramson (1988:182).

[36] For Corvid distributions, refer to Blakers *et al.* (1984:644,648).

[37] Berndt & Berndt (1989:187). Waterman (1987:107–110) has provided a detailed listing of crow/raven myths.

[38] Berndt *et al.* (1993:163,223,240–243,458).

[39] Tindale (1936c:62).

[40] Tindale (1936c:62).

[41] Bradley (2010:35–37,42–43,47–53,233,235,249). The sparrowhawk is referred to as the 'chicken hawk'.

[42] Bulmer (1855–1908:41–44, 1887:39–40), Smyth (1878:1:430,450–452), Mathew (1899), Hassell (1934b:331–334), Massola (1968:92), Hercus (1971), Blows (1975) and Ash *et al.* (2003:135).

[43] Von Brandenstein (1977:172). For the mythology, refer to Bates (1992:162–163), which was originally published in 1928.

[44] Berndt *et al.* (1993:241–242,459–462).

[45] Tindale (1939).

[46] Hercus (1971:139).

[47] Tindale (1937b:151–152).

[48] Mountford (1937, 1941b).

[49] Bruce (1902:176–178) and Cooper (1952).

[50] Bruce (1902:176). Possible identification of Younganna provided by Cooper (1952:80).

[51] Bruce (1902:177).

[52] Bruce (1902:177).

[53] Bruce (1902:178).

[54] Langloh Parker (1953:60–66,124).

[55] Tindale (1939).

[56] Hercus (1982:246–257).

[57] Hercus (1982:249).

[58] Mathew (1899:15).

[59] Smyth (1878:1:423–424).

[60] For accounts of 'crows' in burial grounds, refer to Angas (1847b:124,fig. opp.71, 1877), Stirling (1911:11) and Hemming *et al.* (2000:31).

[61] Milerum (cited in Tindale c.1931—91b). For a biography of Milerum, refer to Tindale & Pretty (1980:51) and Tindale (1986).

[62] Kilham *et al.* (1986:169,221).

[63] Refer to Chapter 4 of this book.

[64] Bates (1928).

[65] Mountford (1948:117–118).

[66] Brown & Naessan (2014:26).

[67] Hardwick (2019b:66).

[68] Thomson (1935:76).

[69] Bates (1901–14:231).

[70] Tindale (1938–60:25). Tindale referred to 'bower bird', which is most likely to be the spotted bowerbird.

[71] Mjöberg (1918:276).

[72] Tunbridge (1988:28, see also 24 for species identifications).

[73] Ichikawa (1998:106).

[74] Bellchambers (1931:91).

[75] Bulmer (1855–1908:45), Anonymous (1925:8), McConnel (1957:91–94), Massola (1968:36–37,49,66), Maddock (1975), Austin & Tindale (1985), Berndt & Berndt (1989:182–183), Clarke (2018a:11–12) and Hardwick (2019b:62). Waterman (1987:55–56,58) has provided a detailed listing of emu and brolga/bustard myths.

[76] McConnel (1957:91).

[77] McConnel (1957:91).

[78] Maddock (1975:118).

[79] Anonymous (1931), Hassell (1934b:328–330), Mountford (1948:154–155, 1976b:48–51), Langloh Parker (1953:1–5), Mountford & Roberts (1971:32–33), Tunbridge (1988:33–34) and Karadada *et al.* (2011:72).

[80] Roth (1897:125).

[81] Tindale (1931–34:207–209, 1936a, 1938–56:33–61, 2005:363) and Berndt *et al.* (1993:15,164,239–240,456–458).

[82] Tindale (1976).

[83] Berndt *et al.* (1993:240).

[84] Wilson (1937).

[85] Wilson (1937). See also Tindale (1931–34:207–209) and Harvey (1932:226).

[86] Berndt *et al* (1993:239).

[87] Roheim (1971:292). See also Tindale (2005:368).

[88] Tindale (1931–34:192, c.1931–91a, 1934–37:55, 1938–56:59) and Berndt *et al.* (1993:15,240).

[89] Wilson (1937).

[90] Tindale (c.1931–91a).

[91] Berndt *et al.* (1993:239).

[92] Tindale (1934–37:58). Tindale referred to the Moandik as Meintangk.

[93] Tindale (1941b:236).

[94] Tindale (1941b:238).

[95] Van der Leeden (1975:80).

[96] McCarthy (1965:19).

[97] Mathews (1904:367).

[98] Tunbridge (1988:30–32).

[99] Basedow (1925:218,274–280,372,374,377–378, Plates XXXIV,XXXV) and Spencer & Gillen (1927:1:56,275).

[100] Basedow (1925:300,309,311,317,330,334,339–340,344,348–349, Figs18,38,39,44, Plates XLII,XLIX) and Attenbrow (2010:146–148).

[101] Basedow (1925:334).

[102] Van der Leeden (1975:61,64).

[103] B Harney (cited in O'Bader 1945).

[104] Maddock (1975:105).

[105] Roth (1903:12–13) and Mjöberg (1918:275–276).

[106] Spencer & Gillen (1927:1:154–157) and McCarthy (1965:17).

[107] McCarthy (1965:18).

[108] Hassell (1934a:233).

[109] Bates (1912).

[110] Jones (2007:347–353). See also Bruce (1902:83–84), Basedow (1925:316–317), Elkin (1934:174,187–188) and McCarthy (1965:18).

[111] Anonymous (1904).

[112] Clarke (2019b:160–162).

[113] Scott & Scott (1934) and Clarke (2018j:116–117,120).

[114] Lee & Wei (2000) and Clarke (2018j:121).

[115] NB Tindale (pers. comm. Baker 1957:3,21).

[116] McCarthy (1965:17).

[117] Basedow (1905:89).

[118] Ramson (1988:427).

[119] Anonymous (1911a).

[120] Campbell (1893).

[121] Bowyang (1929).

[122] Walsh (1951).

[123] HJL (1927).

[124] Harvey (1939, 1943). See also Taplin (1879:39).

[125] Griffin & McCaskill (1986:52–55).

[126] Harvey (1943:109).

[127] Harvey (1943:111).

[128] Langloh Parker (1953:90–92).

[129] Langloh Parker (1953:92).

[130] Singer *et al.* (2021:63).

[131] Bellchambers (1931:139).

[132] Tunbridge (1988:21).

[133] Roth (1903:12). Roth refers to the eastern osprey as a 'fish-hawk'.

[134] Roth (1903:12).

[135] Doonday *et al.* (2013:133).

[136] Smyth (1878:1:458–462), Maddock (1970), Hallam (1975: Ch.14), Isaacs (1980:102–108) and Clarke (2018a:12–14).

[137] Langloh Parker (1953:39–42).

[138] Mountford (1948:49–53) and Tindale (2005:362). A variation of Kipara is 'Keibara'.

[139] Cane (2002:87).

[140] Meyer (1846:204). See also Johnston (1943:256). A spelling variation of Kondoli is 'Kondole'.

[141] Clarke (2001a:23).

[142] Tindale (1934–37:181–184).

[143] Berndt *et al.* (1993:235–236,450–451).

[144] Tindale (1930–52:272).

[145] Tindale (1931–34:252–253).

[146] Smith (1880:19–22). See also Clarke (2001a:26, 2015d:275).

[147] Stanbridge (1857:138) and Smyth (1878:1:432).

[148] Massola (1968:37–38) and Dawson (1881:54,Appendix lii).

[149] Massola (1968:81). See also Mountford (1948:50).

[150] Horne & Aiston (1924:141–142).

[151] Von Brandenstein (1988:85,154). Other spelling variations of *kool-jak* include *gol-jak*, *kool-jark*, *kuljak* and *quulytyaaaqq*.

[152] Bird Rose *et al.* (2002:18–19). This source used the common name, 'chicken hawk'.

[153] Roth (1903:11). See also Mjöberg (1918:284).

[154] McConnel (1957:61).

[155] Massola (1968:24).

[156] Massola (1968:44,50–53). Note that Thara is described as a 'quail hawk', which is probably the collared sparrowhawk.

[157] Massola (1968:10).

[158] Gordon (1979:16–18).

[159] McConnel (1957:74).

[160] Tindale (1974:50). See also Tindale (1959:46–47, 1974:119).

[161] Clarke (2016a:289, 2018b:8).

[162] Cane (2002:90–91).

[163] Tunbridge (1988:141). For the Ngadjuri language spoken in the southern Flinders Ranges, Berndt & Vogelsang (1941:7) recorded the word for 'king-fisher' as *julu* (i.e. *yulu*).

[164] Tunbridge (1988:64).

[165] Berndt & Vogelsang (1941:10). This source described *manbi* as a 'wood-pigeon', which, due to the other cognatic terms, seems more likely to be a common bronzewing than another pigeon species.

[166] Schürmann (1844:pt.2:28) for Barngarla and Hercus (1992:23) for Nukunu. Simply described by Schürmann as 'pigeon' and by Hercus as 'bronzewing pigeon'.

[167] Tindale (1936c:63) and McEntee *et al.* (1986:21).

[168] Tunbridge (1988:81).

[169] Tindale (1963:177,179). In another Western Desert myth, the *waru* are killed by hailstones created by Gumba the night owl (southern boobook) (Berndt & Berndt, 1989:189–190).

[170] Tindale (1964–65:683).

[171] Verburgt (1999:53).

[172] Clarke (1997, 2009b, 2014b, 2014c, 2015a, 2015b).

[173] Fred S (1893).

[174] Clarke (1999b:160–161, 2007b:148–149, 2014b:313).

[175] Piddington (1932:394).

[176] Dawson (1881:51).

[177] Elkin (1977:107–108), who quoted from O Siebert (cited in Howitt 1904:359).

[178] Clarke (1999b:53–56,58–59).

[179] Howitt (1904:427). The bird term *goruk* is related to *kurruk*, *kuruk* and *kururk* as recorded for 'magpie' in other Aboriginal languages in western Victoria (Blake 1998:45).

[180] Langloh Parker (1953:9–10).

[181] Giles (1887).

[182] Clarke (1990:6, 1996b:75, 2015b).

[183] Tindale (2005:26).

[184] NB Tindale (cited in Clarke 1997:137).

[185] Stanbridge (1857:139).

[186] Hamacher (2012:84–86).

[187] Dawson (1881:75).

[188] DW Hamacher (pers. comm.).

[189] Berndt *et al.* (1993:21,76,240).

[190] Taplin (1879:126).

[191] Tindale (1930–52:266).

[192] Berndt *et al.* (1993:367).

[193] Atkins (1911).

[194] Hamacher (2012:80–82, 2015).

[195] Teichelmann (1857).

[196] Duncan-Kemp (1933:122).

[197] Smyth (1878:1:431).

[198] McEntee & McKenzie (1992:125).

[199] Langloh Parker (1905:108), who refers to the Yuwaalaraay as the Euahlayi. Ash *et al.* (2003:106) identified *maliyan* (i.e. Mullyan) as the wedge-tailed eagle.

[200] BJ Nangan (cited in Akerman 2020:131). The Southern Cross is referred to as the 'Crux'.

[201] Gordon (1979:7).

[202] Lumholtz (1889:154), Campbell (1926:410), Roth (1897:93), Tindale (1966:182) and Bassani *et al.* (2006:28).

[203] Dawson (1881:100) and Massola (1968:37).

[204] Dawson (1881:100).

[205] McCarthy (1965:20–21).

[206] Bates (1992:166). Originally published in 1927.

[207] Basedow (1925:315,Fig.17, see also 332–333,349).

[208] Norris & Norris (2009:5–6).

[209] Dawson (1881:99).

[210] Langloh Parker (1905:108, 1953:132,217).

[211] Tindale (1934–37:58,60). Tindale referred to the Moandik as Meintangk.

[212] Massola (1968:21–24) and Stanbridge (1857:139).

[213] Austin & Tindale (1985) and Fuller *et al.* (2014). Note that some authors have referred to the Gamilaraay as Kamilaroi, and Yuwaalaraay as Euahlayi.

[214] Langloh Parker (1953:7). The brolga ancestor in this myth is identified elsewhere by the author as Bralgah (Langloh Parker 1953:228).

[215] Berndt *et al.* (1993:15,164,456–458,Fig.25) and Taplin (1879:133).

[216] Berndt *et al.* (1993:229).

[217] Berndt *et al.* (1993:240).

[218] Tindale (1938–56:57).

[219] Tindale (1938–56:59–61).

4

Birds and the spirit world

Australian Aboriginal peoples are part of a world where it is believed that birds can communicate with humans in cryptic ways. Bird spirits are seen as being in possession of supernatural powers, which include being prescient. This enables them to warn people of impending danger and foretell future events.[1] They are also believed to carry deceased souls, and in some cases bring misfortune. In comparison to other more cryptic animal groups in Australia, most avian species are highly mobile, easily observed or heard, and exhibit strong seasonal behaviour. Some birds, such as magpies and crows, possess complex social behaviours akin to those of people.[2] Others have human-like features, like the forward-facing eyes of owls, or uninhibited behaviours, such as those of the willie-wagtail, that bring them into close proximity to people. These are all attributes supporting the Indigenous perception that certain bird species, as spirit beings, are capable of practising magic on people. It is an Aboriginal belief that people could project their own soul spirit into the form of animals, particularly birds.[3] There is no clear distinction between birds and spirits. As discussed in Chapters 2 and 5, somewhere in between the spirit birds and extant birds there are also extinct birds, that may or may not relate to one or both of these categories.

Many of the world's human cultures share beliefs about the supernatural roles of birds.[4] For instance, in New Britain off the east coast of Papua New Guinea the islanders throw stones at owls, as they regard those birds as a death omen.[5] Similarly, the San of the Central Kalahari Desert in southern Africa recognise most birds as being intelligent and 'credited with thought processes and values comparable with those of man'.[6] The San believe that birds are capable of practising sorcery against animal enemies and rivals, and have behaviours that can be used to predict future events. The Mbuti of the Congo in Africa believe that owls must be avoided due to their role of 'watchman' for a witch or sorcerer.[7] The Mbuti traditions about birds are analogous to those in Aboriginal Australia, as according to anthropologist Mitsuo Ichikawa they recognise:

> … the role of birds as the mediators between human society and the invisible world. The birds are believed to convey information on otherwise unaccountable causes of illness, unpredictable distribution of animals and their behavior in the forest, unexpected failure of hunting, sudden visit of a guest, and other events which the Mbuti feel require some kind of explanation.[8]

In Western European cultures, birds have featured prominently in traditions concerning death and the spirit world.[9] In western Hungary, a stone was thrown at an owl because its hooting was thought not only to forecast death, but also to cause it.[10] The late 19th/early 20th century folklorist James G Frazer remarked, in relation to the deceased, that 'Often the soul is conceived as a bird ready to take flight. This conception has probably left traces in most languages, and it lingers as a metaphor in poetry'.[11] Certain avian characteristics of behaviour and appearance add significantly to the strength of these beliefs. For instance, it has been observed by religion scholar Christopher M Moreman that:

> Blackbirds in particular are often the victims of profiling, as their color is associated with evil and death. In Wales, there is a saying that a crow flying over a house portends a death within … Ravens may predict a death by croaking … A Danish story holds that a raven's appearance in town portends the death of the local priest.[12]

Spirit beings

In Aboriginal Australia, there is a wide range of spirit beings, many of them seen as separate entities from their ancestors although they were present during the creation and are considered to be still present on Earth today. They are generally seen as living independent lives from people, and yet because of their behaviours and the places within the landscape where they dwell, are treated as having the potential of being monstrous. In a study of monsters around the world, anthropologist Yasmine Musharbash noted that the descriptions are highly dependent upon the humans they haunt.[13] Typically, they are seen as a physical amalgam of people and animals, but often have human-like behaviours. The manner in which spirit beings are feared is indicative of how each culture relates to the landscape and it defines the safe and dangerous places within their culture.[14]

Europeans have been fascinated with Indigenous beliefs in spirit beings, so there is much Australian literature that features them.[15] The record of spirit beings suffers from the same problems as that of myths (see Chapter 2), with non-Aboriginal writers deliberately or inadvertently engaging in syncretism when describing early Aboriginal beliefs. The desire to integrate Indigenous beliefs with the findings from science continues today, as seen with the correlations made between fossil species and spirit beings. Writers, who describe themselves as cryptozoologists, have equated the fossils of early hominins, such as *Homo floresiensis* and *H. erectus* of South-east Asia, with the oral tradition of 'yowies' or 'hairy men' in Australia.[16] It is my experience that contemporary Aboriginal peoples, who peripherally engage with science via the media, will sometimes draw links between recently discovered palaeontological species and their own mythical spirit beings.[17] This is to be expected with members of society who are actively embracing 'new' knowledge within their own realm and who are deeply involved in the rebuilding of an Indigenous identity.

It is widely believed by Aboriginal peoples that birds are among the spirits who were able to connect directly with the supernatural forces that governed the universe.[18] Many of them have been regarded as possessing powers that had influenced ancestors during the creation. Spirit beings can take several forms. Some are believed to be shapeshifters and others to wear cloaks, made from materials such as spiderweb, which give them the power of invisibility. The spirits, or at least the cloaks they wear, are often birds, which provides them with the power of flight.[19] Bird species behaviour is crucial in determining the likely spirit associations. For instance, members of the bowerbird group make bowers within which to perform and house special objects: because this behaviour related to the actions of a sorcerer they were treated as spirit beings (see Chapter 6). In some cases, the avian identity of the spirit bird is obscure, due to its perceived clandestine habits. Such an example is a spirit bird that the Yanyuwa people of Borroloola in the southern Gulf of Carpentaria region of the Northern Territory call *a-kuwaykuwayk*, although reportedly no one has even seen it, as it is involved with activities such as sorcery.[20] Anthropologist John Bradley explained that 'if it could be seen, it would resemble a bush [Australian] bustard with shining white feathers. People say that it is only during the wet season that the bird can be heard calling out, and then only at night'.[21]

In the Lower Murray of South Australia, it was recorded that 'A mularpi [*muldapi* spirit] can travel as a whirlwind; sometimes one sees a mularpi as a bird'.[22] Here, Ngarrindjeri people regard with suspicion birds such as *muldhari* the white-backed magpie, *ritjaruki* the willie-wagtail and *menmenengkuri* the welcome swallow, on the basis that they might be spirit beings or sorcerers.[23] As their preferred habitats are open areas, these species are frequently seen in fields and near homes. The magpie may exhibit aggressive behaviours towards people during its breeding season, as many of us have experienced as children walking home from school.[24] Since swallows spend much of the day on the wing catching flying insects, they are naturally drawn to the activities of humans.[25] In my own ramblings through grasslands, I have frequently experienced swallows fly low past my feet, presumably attracted by the insects I have disturbed. According to anthropologists Ronald M Berndt and Catherine H Berndt, in the case of magpies and swallows the Lower Murray people believed 'that there was always the danger of temporary illness being caused by these birds, through one's own carelessness'.[26] Here, care is taken to burn cut hair, due to the perceived risk that if hair was taken by one of these birds it would be used to make a nest and be 'warmed' during the nesting period.[27] It was reported that it was 'this warming of something belonging to or associated with a victim that caused his or her illness and that was an integral part of the technique of sorcery'.[28]

In Aboriginal Australia there are beliefs in spirit beings that share characteristics with birds, although they would not normally be classified as such by scientists. The presence of emu-like features is a major theme in early accounts of the bunyip, the infamous water spirit from south-eastern Australia.[29] In 1845, a newspaper based in the Port Phillip district, which is now part of Victoria, reported that there had been an investigation of the bunyip, which included interviews with Aboriginal people. It asserted:

> The Bunyip, then, is represented as uniting the characteristics of a bird and an alligator. It has a head resembling an emu, with a long bill, at the extremity of which is a transverse projection on each side, with serrated edges, like the nose of the stingray. Its body and legs partake of the nature of an alligator. The hind legs are remarkably thick and strong, and the fore legs are much longer, but still of great strength. The extremities are furnished with long claws, but the blacks say its usual method of killing its prey is by hugging it to death. When in the water it swims like a frog, and when on shore it walks on its hind legs with its head erect, in which position it measures twelve or thirteen feet [3.7–4 m] in

height. Its breast is said to be covered with different coloured feathers; but the probability is that the blacks have not had a sufficiently near view to ascertain whether this appearance might not arise from hair or scales. They described it as laying eggs of double the size of the emu's egg, of a pale blue; these eggs they frequently meet with, but they are 'so good for eating,' the black boys set them up for a mark, and throw stones at them.[30]

In 1847, Lieutenant Governor Charles Joseph La Trobe reported from the Port Phillip district that the bunyip was reputed to be 'a fearsome beast, as big as a bullock, with an emu's head and neck, a horse's mane and tail, and seal's flippers, which laid turtle's eggs in a platypus's nest, and ate blackfellows when it tired of a crayfish diet'.[31] A colonist visiting Challicum Station in western Victoria in 1850 was taken to a cluster of three large waterholes that, according to local Aboriginal tradition, was the site where a bunyip had once been speared and then dragged dead on to the grass, where its emu-like outline was marked out and the turf within it cleared.[32] In 1857, an item in a Melbourne newspaper claimed that a traveller along the Murray and Goulburn Rivers had seen 'no less than six of these curious animals'. It was reported that:

> Mr Stocqueler informs us that the bunyip is a large freshwater seal, having two small paddles or fins attached to the shoulders, a long swan-like neck, a head like a dog, and a curious bag hanging under the jaw, resembling the pouch of the pelican. The animal is covered with hair, like the platypus, and the colour is a glossy black.[33]

The recognition of bird-like features in spirit beings is a common theme across Aboriginal Australia. In the creation beliefs of the Wiilman people from Bremer Bay in the south-west of Western Australia, the Jannock were 'evil spirits' with 'heads like Carrack

'Bunyip', Toor-roo-dun. Drawn by Kurruk, an Aboriginal person of Western Port, Victoria, in the 19th century. Smyth RB (1878) *The Aborigines of Australia*, Vol. 2, Fig. 244.

(black cockatoo), feet like Watch (emu), legs like Oma (brush kangaroo [western brush wallaby])'.[34] Emu features are particularly common among spirits. From south-western Victoria, colonist James Dawson described the *meeheeruung parrinmall* (mihirung) as a large flightless bird that was no longer present, but was said to have resembled a 'big emu' and to be able to kill a man with a single kick.[35] It was a tradition of the Yuwaalaraay that certain waterholes in northern central New South Wales were occupied by a dangerous 'debbil-debbil' (devil) spirit known as *gouage*, or the 'water emu spirit', which had the form of a featherless emu that travelled each night to crouch in his 'sky-camp' located within the Coalpit near the Southern Cross.[36] The late 19th century anthropologist Robert Hamilton Mathews described a Gamilaraay myth from inland northern New South Wales of an Aboriginal 'tribe' called the Dhinnabarrada, meaning 'foot' (*dhina*) 'split' (*barra*), who had 'the body of a man, and the legs and feet of an emu'.[37] Mathews went on to explain that they had 'forked feet, and never went about singly, but in little mobs, and subsisted upon grubs', and 'if they succeeded in touching a man's feet they will be transformed into emu's feet, like their own'. In the Boggabilla area of north-eastern New South Wales, Aboriginal people believed that Kurrea, an enormous snake-like creature, had children, called the *gowarke*, and each of them 'resembles a gigantic emu with black feathers and red legs'.[38]

In Aboriginal Australia, any bird acting in a peculiar fashion is treated as if it might be a dangerous spirit. For instance, it was recorded in the south-west of Western Australia that 'If any animal or bird is observed to act strangely and not in accordance with its usual habits, there is magic connected with it and it is left untouched'.[39] It is believed that the spirit identity maybe disguised. In the south-west Kimberley, Aboriginal artist Butcher Joe Nangan described a spirit having the 'coat' of an Australian pelican, that is removed to reveal the Rainbow Serpent as a snake.[40] In a western Arnhem Land creation myth, two men tried to kill Gurugadji the emu man, who escaped by jumping into a waterhole, where he transformed into a Rainbow Serpent and later killed his attackers, before going to live in the Skyworld.[41] Certain birds are more likely to attract the attention of Aboriginal observers. In the northern Flinders Ranges, the Adnyamathanha people classified the brown goshawk and collared sparrowhawk together as *mura*.[42] They are apparently 'not likeable birds'. The *mura* is:

> Often seen sitting atop a post or dead tree nodding its head. Believed to be casting spells and can make babies sick. Babies are often covered over with a blanket if possible while this bird is about.[43]

In my field experience in the Lower Murray, a bird of any species acting strangely could be taken as a portent. For instance, in 1987 a Ngarrindjeri woman living at Raukkan said that she had been warned about her sister's tragic death in Adelaide by a common starling, which is an introduced species and therefore not one of the *ngaitji* as recorded by early ethnographers.[44] The starling had made a bang by flying into her kitchen window, and this incident was followed a little later by a stone or other object thrown against her bedroom window. The woman expressed a belief, often voiced when engaged in telling *gupa* yarns (ghost stories), that there were powerful forces in a spirit world that would sometimes affect those in the living world.

Aboriginal peoples listen to strange bird calls with considerable interest. The Wardaman people living south-west of Katherine in the Northern Territory consider that the bird *gerdog*, which an ornithologist would classify as several species of friar bird, sometimes 'sings' out various names in the local Aboriginal language.[45] The Wunambal Gaambera people in the north central Kimberley have a similar belief concerning the silver-crowned friar bird or 'swearing bird', which they say utters uncouth words in their local language.[46] The nocturnal call of birds is particularly poignant. For instance, on the Adelaide Plains in the late 1830s, Aboriginal people told missionaries Christian G Teichelmann and Clamour W Schürmann that *karkanya*, 'a species of hawk', was a bad omen. Although they did not identify the species, the same name, *garkanja* (*karkanja*) in the neighbouring Narangga language of York Peninsula, has been identified as the nankeen kestrel.[47] The Adelaide missionaries said that:

> The voice of this bird in the night the Aborigines take as a prognostication that one or more of their number will soon die, particularly children, the souls of whom he is believed to take away, after which they grow ill. The name of this bird is derived from the ominous sound of its voice.[48]

Birds, with the power of flight, were ideal beings to carry human souls up into the Skyworld, which was sometimes described by early European recorders as the 'land of the dead' or the 'land to the far west'.[49] In the 1840s, Schürmann was at Port Lincoln on Eyre Peninsula when he recorded from the local people that:

> ... when a man dies, his soul goes to an island, where it lives in a state so ethereal that it requires no food. Some say that this island is situated towards the east, others towards the west; so that they either do not agree about the locality or believe in the existence of more than one receptacle for departed souls. On its passage to its new habitation a species of Red-bill [Caspian tern], a bird frequenting the sea-beach, and noted for its shrill shrieks during the night, accompanies it.[50]

Ancestral birds were also recorded as interacting with the souls of deceased people on their journey to the land of the dead. Mathews recorded that in the Mount Coolangatta area of south-east New South Wales, the soul of a dead man went on a long track to the other side of the sea, where he then:

> ... passed through a narrow, rocky gorge, with scrub growing on either side, in which were some [Australian] king parrots of gigantic size, who tried to bite him with their strong beaks, but he defended himself with his shield, and succeeded

in getting through the pass. Upon this the parrots set up a great chattering, similar to that made by these birds in their haunts.[51]

Across Australia, Aboriginal death spirits generally have the form of birds with one or more of the following attributes: human-like faces, particularly large front-facing eyes such as with owls; mournful calls, such as with the bush stonecurlew and the tawny frogmouth; activity associated with night-time, like nocturnal hunting and calling of the southern boobook; big high-flying species, for instance the wedge-tailed eagle; and birds with specific unusual movements, for example the willie-wagtail and welcome swallow. A similar situation exists in other parts of the world, with Moreman stating that apart from the universal associations of owls with human death, 'Other birds – like the curlew, nightjar, and bittern – are also related to death not only because they are active at night, but also because they have uncanny voices'.[52] In the early Aboriginal world view, many bird species were perceived as travellers between the psychic and terrestrial parts of the landscape due to their ability to fly, which has enhanced the perception that they are spirit beings.[53] As discussed further in Chapter 7, some birds were seen as helpful to people by providing a warning of an impending calamity, such as from those species that possessed totemic links to the clans. Aboriginal peoples as Traditional Owners state that their knowledge of local spirit beings, and the associated protocols when dealing with them, is a demonstration of their rights and interests in Country.[54] Birds are potentially powerful. Spirit beings, who are not normally directly associated with birds, wore feather ornaments, as shown in the Kimberley of northern Western Australia by the rock paintings of Wanjina wearing on their heads red-tailed black cockatoo plumes, that may have represented lightning.[55]

Owls, frogmouths and nightjars

Aboriginal peoples greatly feared particular bird species due to their alarming habits. The Berndts were told that 'Because owls haunt caves and dark places along the banks of the River Murray (South Australia), the Jaralde [Yaraldi-speakers] regarded these creatures with fear'.[56] The screeching of the 'night-owl' (barn owl) was considered by Lower Murray people to be a sign that something was wrong.[57] This is consistent with Aboriginal tradition in the Tatiara district east of the Lower Murray, where the name of the barn owl '" Ti-ya-tinity" means cry or call of pain or woe or anguish'.[58] At Encounter Bay in South Australia, colonist Richard Penney recorded the term *muldaubie* (*muldapi*), as a generic term for bad spirit, 'corresponding in some degree to our Satan', and stated that:

> They believe that he appears at night when the moon is up, in the evening or just before the dawn of day, in the form of the screech-owl [barn owl], although he assumes occasionally other appearances. Those to whom he appears in dreams or who see his form almost infallibly die.[59]

A recorded Aboriginal belief in western Victoria was that owls were used by a malevolent spirit as familiars to watch over people who strayed from the camp during the night.[60] Here, colonist James Dawson observed that 'When one of these birds is heard screeching or hooting, the children immediately crawl under their grass mats'.[61] In spite of owls being an evil omen, Dawson also claimed that:

> The fern owl, or large goatsucker [spotted nightjar], belongs to the women, and, although a bird of evil omen, creating terror at night by its cry, it is jealously protected by them. If a man kills one, they are as much enraged as if it was one of their children, and will strike him with their long poles.[62]

The barn owl (night owl) and the southern boobook (mopoke) are prominent in the recorded Aboriginal mythologies, often in conflict with other ancestors. In the Tatiara district, which crosses the border of South Australia and Victoria, Dyuni-dyunity the night owl and his sons were said to have been cannibals during the creation. The ancestral Brambambult brothers eventually killed them, and their burnt camp was left behind in the form of a rocky outcrop.[63] From south-west Victoria is the recorded mythology of Gartuk the southern boobook, who attacked the Brambambult brothers by releasing storms from his kangaroo skin bags.[64] In the Narran River region of northern New South Wales, Mooregoo the southern boobook refused a request from Bahloo the moon man for one of his possum pelts, and as a result Bahloo caused the whole country to be flooded and the bird to drown.[65] The Wik Mungkan people of western Cape York Peninsula in northern Queensland have a creation myth concerning Ngorka the southern boobook, who was attacked by his wife Wata the green tree-ant, and because of this was forced to make his home in a hollow at the top of a tree.[66] According to anthropologist Ursula H McConnel:

> The blood ran down his forehead; it made the red gum [kino]. His forehead had become quite flat. That is why the mopoke [southern boobook] now has a flat forehead and why he flaps his wings up and down and fluffs up his feathers 'all flustered' when he alights on a branch.[67]

In my experience, the Aboriginal spirit in the form of a bird that is most commonly spoken about in the Lower Murray is the *mingka*-bird, which is often described as being owl-like.[68] Ngarrindjeri people believe that the *mingka* visits their houses at night, making a noise like a baby crying. Children doing forbidden things at night, such as crying aloud, whistling, or even putting hats or toys on their head, can bring on a visit by a *mingka*. It is said that the spirit 'will steal a baby's breath if it hears one crying'. The 'breath' here is understood by Ngarrindjeri people to be the infant's soul. The association of this bird with death is strong in all accounts of it. The *mingka* reportedly flew about with the head of a recently dead person on it, which is similar to the account from eastern Arnhem Land of a *marrngit* or sorcerer taking the form of a kingfisher flying about bearing a man's head.[69] The *mingka* is also an omen creature. Ngarrindjeri oral history contains accounts of community members hearing the *mingka* during the night, then being told of the death of a close friend or relative the next day.

The identity of the *mingka* spirit in the Lower Murray is uncertain, as it has variously been recorded as a generalised owl, Australian owlet-nightjar, southern boobook and tawny frogmouth.[70] Perhaps, through the Ngarrindjeri folk classification system, all such identifications are correct. There are related beliefs from other parts of Australia. In the Top End, Aboriginal children regard the tawny frogmouth with much fear.[71] Apart from the large human-like eyes of the tawny frogmouth, its behaviour is also Aboriginal-like, with the Wardaman people living in the Northern Territory claiming that the species has the ability to 'dance at night time'.[72] The tawny frogmouth has a monotonous call of 'oo-oo-oo-oo ...' that is soft but penetrating, and repeated many times.[73]

Nightjars are generally treated by Aboriginal peoples in a manner similar to owls and frogmouths. Among the Walmajarri people of the Great Sandy Desert in Western Australia, it is said that *pangarna* the spotted nightjar 'calls out at night 'oot ... oot ... oot ... tuk tuk tuk'. When this bird calls out it tells that somebody might be sick or if they have died'.[74] On the west coast of South Australian, anthropologist Daisy M Bates recorded that:

> The wailing nightjar, called 'moor'dinya,' utters the most weirdly melancholy cry
> of any bird I have ever heard, and might easily give rise to an Australian banshee
> legend, for the sound is just such a long-drawn wail as the supposed sound of the
> Irish banshee when 'keening for a death.' Moor'dinya was a marrailya (sorcerer)
> in the dream times, and so his cry is never mimicked by the natives, but he is not
> a bird of ill-omen to them, and they do not fear his cry.[75]

In the south-west Kimberley, it is believed that a variety of spirits can take the form of nocturnal birds, and during night-time call out news to the *jalnganguru* or a 'clever man' doctor.[76] In the Cairns rainforest region of northern Queensland, there is a sinister 'black night bird' known as *ngubal*, which may well be a species of nightjar, owl or frogmouth – or all of them. According to the Yidiny-speaking people, 'This bird is said to be a "blind person"; if you follow it the bird may make you go blind'.[77]

Wedge-tailed eagle and white-bellied sea-eagle

Large eagles were universally feared. In the Potaruwutj language of the Tatiara in the south-east of South Australia, *minkar* is the recorded name for both the wedge-tailed eagle and the white-bellied sea-eagle.[78] The *minkar* is described as a 'being, sinister, who may assume form of totem animal'.[79] The association of large birds of prey with carrying the soul spirits of the dead was also recorded on the Adelaide Plains in South Australia.[80] Among the Pitjantjatjara people of the Western Desert is the belief that 'medicine men' in the creation were able to send out their spirits in the form of a wedge-tailed eagle when bringing back the *wanambi* or Rainbow Serpent to their Country.[81] As discussed in Chapter 2, the battles between the wedge-tailed eagle and other species, such as crows, are a major part of the creation mythology across Australia.

As with many other Aboriginal spirit beings, it is children who are at greatest risk from them. There is a myth from the Lake Boga district of northern central Victoria about a wedge-tailed eagle ancestor taking a baby boy away from his mother, then flying up to his eyrie.[82] As the biologist and anthropologist Donald Thomson recorded from northern Queensland, 'The belief, widespread in the Australian bush, that this bird [wedge-tailed eagle] may take babies, also finds expression in the mythology of the Kandju [Kaantyu] tribe of the interior of Cape York Peninsula whose company centres about the Coen and Upper Batavia [Wenlock] Rivers'.[83] In relation to the Tjukurrpa or creation mythology of the southern Western Desert, Walawuru the wedge-tailed eagle chased the children of other ancestors. Anthropologist Scott Cane observed that this 'Tjukurrpa conveys several obvious warnings about child care and parental responsibility'.[84] In the mythology of Mullyan (Mullion) the wedge-tailed eagle in the Barwon River area of central New South Wales, Aboriginal people were caught and fed to eaglets and the hen bird.[85] In 2020, when I was conducting fieldwork inland from Townsville in northern Queensland, I was told by Jangga people of an occasion many decades ago when a wedge-tailed eagle there had tried to fly off with a small male child playing outside, but was stopped by the victim's siblings. In south-west Victoria, colonist James Dawson was told that a wedge-tailed eagle had 'carried off' an Aboriginal baby outside the family hut'.[86] Given that the bird can reportedly carry as much as 5 kg in weight,[87] such accounts must be considered plausible.

The danger posed by large eagles is reflected in Aboriginal creation mythology. For instance, in the south-west Kimberley there is an account from Butcher Joe Nangan of Jaringkaluny the wedge-tailed eagle, who would hunt people and their babies for food. It was killed by two boys, who were the Australian owlet-nightjar and the spotted nightjar.[88] Here, it is said:

> When people hear the noise '*jirrpi-jirrpi-jirrpi*' in the night, they know it is the spirit of Jaringkaluny the giant eagle hunting, and they must cover up any small children. If a child dreams of Jaringkaluny and a willy-willy starts, the child will get sick.[89]

As accomplished flyers, eagles were seen as being capable of entering the Skyworld as well as the Underworld.[90] In 1960, Norman B Tindale was doing fieldwork on Mornington Island within the Wellesley Islands of the Gulf of Carpentaria in Queensland, where he recorded that the 'sea hawk' (white-bellied sea-eagle) was the 'leader' of the sea and 'lives under the water of the ocean and comes out to fly over it'.[91] It is easy to see how the renowned hunting prowess of such large raptors would have engendered such beliefs.

Willie-wagtail

This insect-eating bird is common across the whole of Australia. It is active during the day and appears to be attracted to people and their livestock, possibly because of the insects that are stirred up as a person or large animal walks through the grass. For this reason, Europeans in parts of Australia have called this bird the 'shepherd's companion' or 'shepherd's friend'.[92] The highly erratic flight movements of the willie-wagtail on low and open areas, such as around houses and cars, has sharpened its joint status as a creation ancestor and an omen species. In the Great Sandy Desert of Western Australia, the Walmajarri people have a tradition concerning this common bird that involves strong winds and the formation of the lakes.[93]

There is a creation myth in the Upper Condamine of south-east Queensland that involves the willie-wagtail ancestor, with is quick movements, being able to successfully dodge the weapons thrown at him by the wallaroo ancestor.[94]

In the Princess Charlotte Bay area of northern Queensland, there is a belief that the willie-wagtail ancestor made a crocodile, and for this reason it is said that he is often seen today in the reptile's company.[95] North of Mission River on the western side of Cape York Peninsula, it was said by the Thaynakwith people that:

> The willy wagtail is known as the little messenger bird who tells us about a loved one passing or giving a message from the spirit world. Our story of the legendary *pwi' dharridha* [willie-wagtail] tells of him planting date palms and always looking back to see they were in a straight line.[96]

In many parts of Aboriginal Australia, the willie-wagtail is considered to be a spirit bird, which I have sometimes heard described as a 'devil bird'. During my own fieldwork in the Lower Murray, the most commonly held belief in daytime omens concerned *ritjaruki* the willie-wagtail.[97] When this bird was observed persistently making strange erratic movements near a person's house, it was seen as a message that someone had died. Tapping on a window was taken as a particularly bad sign. The messages delivered by the willie-wagtail were completely unwanted – Ngarrindjeri poet, Margaret Brusnahan, described the bird as like an 'angel of death'.[98] This belief has a connection to earlier times. Explorer/artist George French Angas, during his 1844 trip to the Lower Murray, noted the tradition associated with the observed behaviour of a bird that appears to have been a willie-wagtail. He said:

> An elegant species of flycatcher, of a black colour, which continually hovers about in search of insects, performing all manner of graceful manoeuvres in the air, is regarded by them as an evil spirit, and is called *mooldtharp* [*muldapi*], or devil. Whenever they see it, they pelt it with sticks and stones, though they are afraid to touch or destroy it.[99]

The belief in the willie-wagtail omen was widespread across Australia. For instance, in the south-west of Western Australia the observance of a willie-wagtail flying into camp was a portent of visitors arriving.[100] The behaviour of this bird is explained in the creation mythology of the Wiilman people of this region, in which Chitter-chitter the willie-wagtail ancestor was attacked by Nornt the black snake. Because of this he 'never got over his fright and cannot sit still. He is perpetually on the move and is never still long enough for a snake to charm him'.[101] According to Tindale, among the Alyawarr people of Central Australia it was believed that a pair of willie-wagtail ancestors could be seen in the night sky in what Europeans call Sagittarius, and that:

> On Earth the busy chattering of these birds often warns game of the presence of hunters, and they are also disliked from their supposed complicity in spreading

gossip and disturbing people as they noisily flutter about the camping places of the aborigines.[102]

In the southern Western Desert, the willie-wagtail 'is usually characterised as being a gossiper, or rather devious in his/her social relationships'.[103] In a myth recorded by the Berndts from Ngalia and Antikirrinya people camped at Ooldea in the 1930s, Djinta-djinta the willie-wagtail 'was out gossiping some distance away from his own camp when he heard thunder', but he was struck by a hailstone and killed before reaching the safety of his home.[104] The Antikirrinya still believe that when they see a willie-wagtail it is 'telling you that someone's coming. Messenger bird'.[105] Similarly, Pitjantjatjara people of the central Western Desert consider the willie-wagtail to be a guardian who would call at night if enemies were approaching.[106] They also believed that the bird should never be killed or even have a stone thrown at it, because if its spirit was angered it could create devastating wind storms.

The willie-wagtail is widespread across Australia, and its fearless and frenetic behaviour contributes to it being perceived as a harbinger of news in many regions. On Yorke Peninsula, the willie-wagtail was described as a 'message-carrier'.[107] During my fieldwork on the west coast of South Australia, from the 1980s until the present, Aboriginal people from Koonibba have spoken of willie-wagtails bringing them bad news, and therefore they slap their hands and throw stones to rid themselves of these birds round their homes. Adnyamathanha people of the northern Flinders Ranges in South Australia would not kill *indarindari* the willie-wagtail, and it was recorded that in earlier times a 'Person who killed *indarindari* had a mark (*walkari malka*) cut into the shoulders'.[108] Museum researcher Charles P Mountford wrote that the Adnyamathanha people treated the willie-wagtail as a 'harbinger of death' and:

> … consider that he is a liar and mischievous tell-tale. Should the women learn any of the secrets of the men, even in small matters, the men believe that it is the willy-wagtail who told them. Because of his bad character in these matters, the men always hunt the willy-wagtail away before they discuss the secret matters of the tribe.[109]

In the north central Kimberley, Wunambal Gaambera people believe that the willie-wagtail 'can do good and bad things, sometimes they can lead people astray'.[110] Wardaman people south-west of Katherine in the Northern Territory noted that the willie-wagtail is a 'busy and noisy little bird' that often scares away other game when hunters are approaching.[111] The Jawoyn people of the Katherine area appear to have less stressful encounters with this bird, although its connection with the spirit world is still apparent. Wiynjorrotj and her fellow community members stated that 'He calls when people are approaching and warns you when danger is approaching. If you listen closely to its call it sounds like it is talking language'.[112]

Stonecurlews and American golden plover
The range of the bush stonecurlew across the continent is much reduced, although it is frequently mentioned in the historical records since its range formerly included much of

the continent.[113] The beach stonecurlew is mainly restricted to the coastal areas of northern Western Australia, Northern Territory, Queensland and New South Wales,[114] so when earlier accounts from across inland regions and southern Australia refer to just a 'curlew', they most likely concern the bush stonecurlew. It remains likely, however, for reasons discussed in Chapter 5, that in areas where both bird species occur the local Aboriginal people treat them as the same entity. Ornithologists have described the call of the bush stonecurlew as an 'Eerie, mournful " ker-loo"',[115] with the beach stonecurlew having a similar call but 'higher-pitched and harsher in tone'.[116] These calls caused much alarm for European settlers. Ornithologist Herbert T Condon remarked that the bush stonecurlew (his 'stone plover') is one of those species that can:

> ... induce fear and foreboding in the lay mind. Some years ago, at Stanley Flat, near Clare, South Australia, there were persistent reports of a 'ghost' which sat on a road bridge and prevented travellers from crossing at night – with 'huge outspread wings' and 'uttering a wailing cry.' The tale was investigated and the 'ghost' proved to be a Stone Plover.[117]

The bush stonecurlew was an omen of death across much of Aboriginal Australia.[118] Death is a theme in the bush stonecurlew myth recorded in the 1840s from the Nauo people of Eyre Peninsula in South Australia.[119] In this creation account, Welu was a warrior who killed all of the Nauo people, except the young men Karatantya and Yangkunu, who fled to the top of a tree. Welu attempt to reach them, but fell to the ground where he was killed by a wild dog. After the creation, Welu became a bush stonecurlew, while the men were transformed into two species of hawk. In what appears to be a structurally similar myth recorded from the Narangga people of Yorke Peninsula in South Australia, Winda the barn owl man lived in the cliffs on the eastern side of Yorke Peninsula with his two hunting dogs, and nearby lived Wudlaru the bush stonecurlew and his wife.[120] When the bird couple were away one day, leaving their children behind, the owl man saw his chance. The children were eaten by his dogs. Later, in revenge, the stonecurlew man killed the owl's dogs, but the owl escaped by hiding in the cave. After the creation, the owl continued to reside in dark places like caves, and the stonecurlews continued to mourn their young with their calls. In the Manning, Hastings and Macleay Rivers area of eastern New South Wales, it was Thoorkook the 'mopoke' (southern boobook) man who killed the sons of the two Byama brothers, whose wives later became bush stonecurlews.[121]

The mournful call of the bush stonecurlew, often associated by Aboriginal peoples with a lost child, is a central theme for many of the recorded mythological accounts of this bird. In the Lake Boga district of northern central Victoria, Will the stonecurlew started wailing mournfully in sympathy for a woman whose baby son had just been carried off by Nurrayil the wedge-tailed eagle.[122] The Tiwi people living on Bathurst and Melville Islands in the Northern Territory have a tradition that Wayayi the bush stonecurlew was once a woman who, due to an illicit affair, neglected her baby son who died as a result. She cried out *wayayi* when she saw her dead child, and became a bird that cries out mournfully for her son each night.[123] Among the Kungarakany people of the Finniss River, south of Darwin, Kiwililik the stonecurlew is much feared, and is said to have been 'named after one of its calls; at night, gives mournful, eerie cries; emits human-like screams'.[124] Kungarakany woman Kathy Mills explained:

> Whenever the curlew calls it is a reminder to take good care and protect the children. The cultural story relates to the estranged mother in search of her child that had been separated from her for disciplinary reasons. The screech of the Curlew as the abductor of neglected children was the reminder to those who might not take their responsibility seriously.[125]

The Jawoyn people in the Katherine area of the Northern Territory also associate the bush stonecurlew ancestor with a lost baby during the creation, and believe that during the night the bird cries out in sadness if it sees the footprints of toddlers who are learning how to walk.[126] In a related belief, the American golden plover, which also has a mournful call, is said by Aboriginal people in the south-west Kimberley to be heard calling out after the rains for her son who was lost during the creation.[127] Similarly, in the Kimberley, explorer/missionary James RB Love observed, with sadness, that:

> The curlews that wail so mournfully in the night are said to be the souls of babies that have died, and are now crying because they want to come and sit by the fire. Many a bereaved mother must have suffered anguish at night, hearing the sad wails of this bird, believing that her baby was crying and that it could neither come to her nor could she go to it.[128]

As an omen species, the hearing of a bush stonecurlew was considered a powerful portent. The Berndts said that, across Aboriginal Australia, people shared similar beliefs about stonecurlews. 'In much of Aboriginal mythology, Curlew is a taciturn and gloomy fellow. As a bird, his cries echo mournfully and are often taken as presaging death'.[129] This was still the case when Tindale conducted fieldwork in the south-east of South Australia in 1933, where local Aboriginal people were frightened of the bird.[130] Ronald M Berndt noted that on Yorke Peninsula in South Australia 'It was said by the Narunga [Narangga], that upon hearing a curlew's cry, a death would shortly occur, or that it is crying for its young. This was also said of the curlew in the Ngadjuri country [in the mid north of South Australia]'.[131] Berndt said that Lower Murray people believed that hearing the wailing call of this bird foretold the death of a close relative.[132]

During my own Lower Murray fieldwork during the 1980s, elderly Ngarrindjeri woman Leila Rankine reminisced about summer holiday camping with her family at Ngalang on Younghusband Peninsula in the northern Coorong area during the 1950s. They once heard the mournful call of a bush stonecurlew at night, which was an ill portent.[133] The species in the Lower Murray region was locally extinct by the 1980s.[134] In

Victoria, during my fieldwork in the 1980s, an Aboriginal man at Shepparton said that it was a local tradition that when a bush stonecurlew is heard to cry three times during the night, someone will die.

In the central Western Desert, the bush stonecurlew calls out to Aboriginal people as a 'Warning that there might be killers around'.[135] The Berndts worked with Antikirrinya people at Ooldea on the edge of the Great Victoria Desert in the 1930s, and recorded the same belief concerning the bush stonecurlew as a 'death bird'.[136] They reported that:

> The medicine-men are the only members of the tribe who have converse with spirits, *mam:u*. They cannot be seen, but make their presence known by the 'swish' of a current of air, or a long drawn howl, and sometimes are represented by a bird, the curlew, *wilu*. These spirits' chief attribute was to claim people and children for the grave.[137]

In the north central Kimberley, the Wunambal Gaambera people consider *wirndi* the bush stonecurlew to be the 'countryman' (has shared characteristics such as habitat) of *bilimbili* the beach stonecurlew, which cries out when the tide comes in.[138] Aboriginal accounts of the bush stonecurlew credit this bird with human-like qualities and needs. From the Wunambal Gaambera, it is recorded that:

> ... legend says that if you have a campfire at night out bush, this bird will come in steal some of the fire to keep warm. It will carry the hot coals in its beak and make its own fire, it then spreads [its] wings around the fire to keep warm.[139]

An explanation of the bush stonecurlew omen from Wardaman people south-west of Katherine in the Northern Territory provided additional details about the context in which bird calls are heard. It was said that:

> The distinctive night call of Wiliyuga [bush stonecurlew] can be a forewarning of bad luck or bad news. This can be especially powerful if the call is made when Wiliyuga is very close or if you are away from home and you hear the call persistently. This bird is locally known as the Night Curlew.[140]

Grey currawong

To balance the negative influence of many bird spirits, some bird species were helpful sometimes by warning people of the presence of danger. Tindale recorded a belief from Yaraldi-speakers of the Lower Murray that involved *kiling-kildi* the grey currawong:

> If Jarildekald [Yaraldi-speakers] men are hunting singly and hear a kelinkeli [*kiling-kildi*] bird (*Strepera*) calling they at once return to their companions for the cry of this bird is a warning of the presence of visible or invisible spirit

enemies. The miwi (mind) of a man (believed to be in his belly) cannot resist this warning, so strong is their fear of the sound.[141]

An ornithologist described the grey currawong as having a loud metallic 'clink' or ringing call,[142] which appears to be the basis for its Yaraldi and Tangani names. Tindale recorded from the Tangani-speaking clans of the Coorong in South Australia what appears to be a related tradition for the same species. He wrote that 'Kilindi [grey currawong] – bird was messenger going ahead and warning people of strange tribes, before he became a bird'.[143] The *kilindi* was described as a 'crow with white tail; makes lot noise' and a *ngaitji* species, although the associated clan is not known.[144] Tindale related an Aboriginal encounter with the grey currawong that took place during the early 20th century in the Encounter Bay area of the Lower Murray, 'Once when Karloan and Frank Blackmoor [Blackmore] were at Waitpinga [Beach] even in relatively recent times and were prospecting for metals they desisted and returned home abandoning their pick, as soon as they heard a jay [grey currawong] screaming at them'.[145]

The currawong is significant to other Aboriginal groups, such as the 'Wa-noo-rong [Woiwurrung] or Yarra tribe' of Victoria whose members believed that at the conclusion of the creation period '*Ballen-ballen* (the Jay [currawong]), who at that time was a man, had a great many bags full of wind, and being angry, he one day opened the bags, and made such a great wind that Bund-jel [Bunjil, Supreme Creator] and nearly all his family were carried up into the heavens'.[146]

Crow/raven

During my fieldwork in various parts of Aboriginal Australia, crows and ravens were occasionally said to be connected with sorcerers, in the same manner as a black cat is the avatar for a witch in Western European folklore. During the 1980s when I was working at Raukkan (Point McLeay) in the Lower Murray, 'crows' (Australian raven) seen in the proximity of the homes and acting strangely were sometimes said to be spirit familiars of a *kuratji* or sorcerer. Similarly, in 2016 when visiting the Aboriginal community at Cherbourg in south-east Queensland, I was told by members from one family of an occasion when several 'crows' flew to the verandah of their house while a particular Aboriginal man was staying there, and this was said to be due to a sorcerer looking for their visitor. Apparently, the man was being sought for punishment because of a 'tribal law' he had broken. Such beliefs are consistent with the generally bad opinion about crows and ravens that is reflected in the early mythologies, as described in Chapter 2.

Asian koel

The theme of bird spirits targeting babies in their killings is common across Aboriginal Australia, as shown with the above discussions of the brown goshawk, collared sparrowhawk, wedge-tailed eagle and *mingka*-bird. Another example comes from the north central Kimberley, where the Wunambal Gaambera people claim that they have 'a dreaming story associated with this bird [Asian koel] that is connected with its ability to kill small babies who cry too much'.[147] According to Wiynjorrotj and her fellow Jawoyn community members in the Katherine area of the Northern Territory:

> When *jowok* [Asian koel] gives a prolonged call close to someone it indicates bad news is coming, like a death in the family or of someone close to the family. This call makes people frightened and it is considered that *jowok* has special powers.[148]

The Asian koel is a nuisance species in the south-west Kimberley. Kim Akerman recorded that here:

> When *kurayikurayi* [Asian koel] are nearby, you have to hold your hands over your genitals and breasts and cover your mouth. If you don't, the birds may swoop down and steal your pubic hair or peck your nipples. They can also make your mouth, breasts or genitals grow enormously.[149]

Western gerygone

In comparison to many of the other avian spirits so far mentioned, the Wardaman people south-west of Katherine in the Northern Territory recognise that they have a more beneficial relationship to *nyorijban* the western gerygone, which is a species of peep-warbler. It is said that:

> Nyorijban has special powers and can pass messages to Wardaman people in language. He has the ability to indicate to Wardaman people that a relative has passed away, and is also involved in the [initiation] ceremony that changes boys into young men. Nyorijban can also indicate to a hunter that a Kangaroo is in the vicinity but still may be some distance away, or that someone else is approaching.[150]

Other spirit beings and birds

In Aboriginal belief, spirits in human form interact with birds. In the Lower Murray, it has been my experience that an unexplained loose feather found in a person's home may be taken as evidence for the presence of a *kuratji* ('feather foots', sorcerers), as it is widely believed by Aboriginal people that their 'magic slippers' are made from feathers and human hair string, as described in Chapter 9. Across Australia, the Aboriginal English term 'feather-foots' is often used to mean a 'traditional sorcerer'.[151] On one occasion in 1993, Ngarrindjeri man Henry Rankine and I were out on the Raukkan farm when we noticed two wedge-tailed eagles being chased by white-backed magpies. Henry, who was a keen duck shooter, remarked that one should never shoot at an eagle, because they might be a *kuratji* and this will anger them.

Anthropologist Yasmine Musharbash explained that the Warlpiri people of the Tanami Desert in the Northern Territory believe that the Jarnpa and Kurdaitcha revenge parties, which comprise spirit beings in human form that wear feather shoes, 'have companion birds who announce their presence to Warlpiri people versed in reading these signs'.[152] The belief that eagles might be sorcerers is widespread. In the Adnyamathanha language of the northern Flinders Ranges, *wildu ungi* refers to a 'witchdoctor in the form of an eagle'.[153] A belief among the Wardaman people in the Katherine area of the Northern Territory is that a 'black hawk' and a 'white hawk' are used by sorcerers who are magically taking kidney-fat out of their victims.[154]

Across Australia, spirit beings were often depicted in the records as having food tastes similar to those of people. For instance, in the Wimmera district of western Victoria, the much-feared Nargan spirit, who had only one and a half legs, was said to follow the tracks of people in order to get food. According to missionary John Bulmer, the Nargan would reach the Aboriginal camp:

> If they get up and give him all they have, he is content, he is very fond of eggs (*booyang*) and when they are thrown to him he will at once pick them up and devour them. He will then allow the blacks to proceed on their way.[155]

In the 1980s, I recorded many oral history accounts of the 'little men' or *kintji* (*kindja*) in the Lower Murray, associated with events taking place since the early 20th century.[156] It was Ngarrindjeri tradition that these anthropomorphic spirits lived in family groups and their camps were associated with hills and cliffs. The 'little men' were often said to be hunters, with a particular desire for ducks. This behaviour brought the 'little men' into occasional contact with Aboriginal hunters in the wetlands, who would leave a duck or two behind to appease them. Similarly, in the western Arnhem Land region of the Northern Territory, the human-like but thin Mimih spirits of the escarpment country were said to engage in hunting birds and mammals, although some of their plant foods were poisonous to people.[157]

Birds could protect people from the ravages of malevolent spirit beings. In the eastern Western Desert region of South Australia, anthropologist Ute Eickelkamp described birds as frequently being seen as protectors by the Pitjantjatjara and Yankunytjatjara peoples. She wrote that:

> One old man specified that it used to be *luunpa* (red-backed kingfishers) and *mirilyirilyi* (fairy wrens) that warned people about the approaching danger. As protectors, these small birds come in flocks, covering the hut of potential victims like a blanket.[158]

Totemic spirits

In many parts of Australia, it was an Aboriginal belief that an individual could gain special powers via the totemic animal, plant or object from which their clan derived descent, as described in Chapter 2.[159] The belief in such powers is not unique to Aboriginal peoples, as Frazer noted that in Samoa the totems, such as owls, 'gave omens to their clansmen'.[160] In Aboriginal Australia, it was commonly believed that everyone with a totemic familiar ('friend' or 'protector') could derive spiritual assistance from them, but only spiritually powerful people could fully take on the character of one of their spirit familiars.[161] This included an individual either shape-shifting or projecting their spirit into the form of their totem. For instance, Bates reported that in the south-west of Western Australia:

> The belief is also held in the southern districts that the natives were once animals or birds. For instance the Nagarnooks [one of the Nyungar marriage classes] are called Wejuk (emus), and are even supposed at the present time to be able to transform themselves from men to emus at will.[162]

Anthropologist Adolphus P Elkin recorded a 'medicine man' in the Dampierland of the south-west Kimberley in northern Western Australia who had three *rai* ('alter egos'), which were 'namely the wild turkey, a fish species and his own spirit-double'.[163] These *rai* were the spirit 'coats' described by Butcher Joe Nangan, who was from the same region.[164] The Australian pelican was Butcher Joe's *jalnga*, or totemic spirit guardian, who would come to him in dreams, and during the day a pelican flying past would give a sign recognising him.[165] Butcher Joe said that he had a dream during which his deceased aunt's spirit taught him how to perform the Mayarta Nurlu, the pelican spirit dance, as depicted in his art.[166] These and other spirit beings would look after the animals and people with who they are totemically connected.[167] Anthropologist Kim Akerman observed that Butcher Joe, as a *jalnganguru* or 'clever man' doctor, would speak with certain birds and animals in order to obtain the special powers they held, 'In the early morning, Nangan would often sit with a pannikin of tea, listening to the "news" brought by his bird familiars in their dawn chorus'.[168]

Tindale recorded that it was a Lower Murray tradition that a spiritually powerful man could transform himself into his *ngaitji* and 'He will sometimes remark that he has been about in the ngatji [*ngaitji*] form and you will recall seeing the strange bird'.[169] He recorded from Milerum that the:

> *Minka* [*mingka*] may travel as a *ngaitji* (totem being), the eagle, dog and *kurki* (hawk [collared sparrowhawk]), are worst to meet. In olden times one always had a waddy [club] handy. If you meet one in the form of a *ngaitji* you must give him a hit; the animal may be many yards away, never mind that, just strike downward, twisting the wrist. This breaks the *nunggi* or *kortui* (like a spider web) which holds the man spirit in the *ngaitji*, if you do this it goes back to the man. The *nunggi* remains coiled on your sacred waddy, it leads you to the man; you find him lying down; you strike him and kill him; there must be no blood; burn him to ashes.[170]

A person could send out their spirit in *ngaitji* form, but once the connection was broken the spirit returned to its human body. Tindale recorded that in the Lower Murray a woman would watch over the body of a man who, as a spirit, was away travelling as his *ngaitji* along a *ngildi*, a 'flying spider web'. As a sign:

> She will see perhaps an owl on a branch of a tree nearby. Then the bird disappears over the hill and she will know the spirit is returning and remark 'He will come

soon'. Then the man stirs; he breaks the *ngildi* and sitting up describes his journey and the events he has witnessed.[171]

In the Lower Murray, the Berndts recorded an event that took place in the early 20th century of a Ngarrindjeri man projecting his spirit as a bird. It involved a group of men who were at Marunggung near Wellington, collecting wood grubs for fishing bait, when a large black eagle (probably the wedge-tailed eagle) landed in a tree nearby.[172] Later, a man named Manuel Karpany who was not among them claimed to have been that eagle, and he was believed since he could list the people who had been present. Manuel was warned that he might have been shot, but he responded by saying that 'they could not have shot him; he himself had remained back in the camp and the eagle was his other self, his ngatji [*ngaitji*], a projection of his real self'.[173]

It was part of the 20th century Aboriginal folklore of the south-east of South Australia, that I and other anthropologists recorded, that Joe Lock and his brother Chalker were able to project their spirits into the form of their totem, which was a species of 'grey eaglehawk' not found in the region.[174] The foreign identity of the bird involved is significant, as these Aboriginal men were originally from the mid north of South Australia, but for decades had lived and worked on pastoral properties in the Lower Murray and south-east of South Australia. According to Lola Cameron-Bonney, Joe sometimes stayed at the Blackford Reserve in the south-east of South Australia, where he reportedly on occasion turned into an eagle that flew away. Lola explained that, to achieve this, Joe would lie on a blanket out in the open, and enter a trance. No one was allowed to move him, as in bird form he may not have been able to relocate his body. Once, when soaring high as an eagle, Joe spotted several people in a horse and buggy travelling towards Blackford, and later back in his human body was able to accurately describe them before they arrived.

Chalker was said to have been involved in a similar transformation event, which reportedly took place when a party was out gathering fruit on the Coorong Hummocks and a woman saw a large bird that resembled a Cape Barren goose.[175] Chalker later claimed that he had been that bird and, rather than looking for potential victims to attack, he was just out there protecting people from other 'flying sorcerers'. It was claimed by the Berndts that powerful Aboriginal 'doctors such as Chalker wore the feathers like a skin and were able to turn themselves into birds, even birds which were not their own *ngatji* [*ngaitji*]'.[176]

The living animals associated with each *ngaitji* could also act as messengers on their own.[177] When I was conducting fieldwork in the 1980s, there were several stories told of observed bird *ngaitji* behaviour having been used to notify family members of a death.[178] For example, Lola claimed that her mother's *ngaitji*, which was a welcome swallow, would fly up near to her family members to 'sob like a child' if someone close was dying, and that this had happened when her cousin's baby had died.[179] In 1933 it was recorded from the south-east of South Australia that in the past 'At Bordertown the natives waited, while burying a person, until a bird flew past; that was the spirit of the deceased going away'.[180] Totemic species and objects were considered to have the power to warn people of impending danger or even death.

The *ngaitji* of a deceased person would sometimes appear during events, such as funerals. This was said to have happened in 1993, when pelicans were observed congregating at the punt crossing the Narrung Narrows as mourners were coming into Raukkan for the funeral of a prominent woman, whose *ngaitji* was *nuri* the pelican. The identity of the deceased and the totemic familiar were seen by Ngarrindjeri observers as merged. Ngarrindjeri man Henry Rankine used an account of his uncle's death in the mid 1950s to explain the importance of reading signs. He said:

> We believe in different signs. My father [Hendel Rankine], who was, say, about eight kilometres away from our house, in our land one day, he was down there shooting ducks along the shore and his brother was here in Adelaide, the Royal Adelaide Hospital, sick. My father was walking along and, all of sudden, this pelican came, landed, settled on the dry land, walked towards my father with his wing dragging. My father put the gun up, was going to shoot the pelican and the pelican started talking to him and then my father put the gun down and the pelican got up, flew up, and headed straight from Point McLeay straight to Adelaide. When my father got home, he walked through the door and said to my mother, 'My brother died,' he said, 'You watch, you will see the boss coming in a minute.' Two minutes after that the boss walked down the hill and just before he got to the door my father called out and said, 'There's no need to come here, I know, I already got the message'. The superintendent looked at my father and said, 'What did you get?' He said, 'My brother died, about half an hour or an hour ago,' and the superintendent said, 'That's right. How did you know?' And he said, 'We get our signs. We get these things.'[181]

*****

In a worldwide review of birds as signs, anthropologist Felice S Wyndham and linguist Karen E Park remarked that 'The honed ecological awareness acquired by paying attention to birds translates more generally to sophisticated connective worldviews that extend our perceptive awareness in space and time'.[182] This is so in Aboriginal Australia, where birds are seen as mediating the connections between the lived world of humans and the spirit world. The flight ability of birds imparts to them the ability to act as messengers and carriers of deceased souls, when travelling to and from the Skyworld. While the specific behaviour of certain species, such as the willie-wagtail, appears to be a major factor in their designation as messengers, the peculiar behaviour of other birds, like the welcome swallow and Australian pelican, that is observed on rare occasions may be interpreted as a portent. A bird could also be seen as the spirit form of a deceased person or as the 'skin' of a sorcerer or another being. Spirit being lore contains many warnings about dangerous places and risky behaviour, with children being the intended audience. The environmental knowledge that Aboriginal peoples hold is structured in reference to Indigenous beliefs about spirit beings and their continued role in human affairs after the creation.

Aboriginal peoples recognised the presence of the spirits who were associated with death, and allowed their beliefs in them to modify and temper their own actions.[183] Any sighting of individual birds of certain species, particularly when they are acting peculiarly, is seen as a potentially harmful spirit in disguise. Bird species that are able to co-exist with people, such as magpies, crows/ravens and willie-wagtails, are prominent among the recognised spirit beings, as are other birds which have nocturnal activities, such as owls and stonecurlews. Aboriginal peoples consider that certain spirit beings are able to shift their form at will, between that of a human and a bird.

From an Aboriginal perspective, the observable displays of birds' innate intelligence makes them likely candidates as spirit beings. Western scientists have more recently begun to appreciate that birds possess cognitive and social abilities which earlier generations of researchers had thought were uniquely human. Australian biologist Gisela Kaplan recently remarked that:

> Birds and humans may have some common social rules and similarities in how much and for how long they can learn and live. This connection is possible despite the enormous evolutionary distance between humans and birds.[184]

Endnotes

[1] Clarke (1999a, 2007b, 2018b) and Brown & Naessan (2014).
[2] Pellis (1981) and Kaplan (2019).
[3] Elkin (1977).
[4] Moreman (2014) and Wyndham & Park (2018).
[5] G Róheim (cited in Berndt 1940c:462).
[6] Silberbauer (1981:72).
[7] Ichikawa (1998:106).
[8] Ichikawa (1998:105)
[9] Murphy (2002) and Bukowick (2004).
[10] G Róheim (cited in Berndt 1940c:462).
[11] Frazer (1890:124).
[12] Moreman (2014:484).
[13] Musharbash (2014:11–13).
[14] Tuan (1974).
[15] For overviews of the ethnographic literature relevant to Aboriginal spirit beings, refer to Buchler & Maddock (1978), Clarke (1999a, 2003d, 2007b, 2016b, 2018b, 2018c), Musharbash & Presterudstuen (2014) and Nicholls (2014).
[16] Healy & Cropper (2006:17,131,175–180).
[17] Clarke (2007b:148,150, 2014a:28).
[18] Elkin (1977:103), Clarke (1999a, 2007b, 2016b, 2018b) and Akerman (2020:43–65).
[19] Clarke (2016b).
[20] Bradley *et al.* (2006:93).
[21] Bradley *et al.* (2006:93).
[22] Tindale (1938–56:73).
[23] Berndt *et al.* (1993:253). A spelling variation for *muldhari* is *muldari*, for *ritjaruki* is *ritju-rukeri* and for *menmenengkuri* is *men-menangkuri*.
[24] Kloot & McCulloch (1980:176) and Warne & Jones (2003).

[25] Kloot & McCulloch (1980:82).

[26] Berndt *et al.* (1993:253).

[27] Clarke (2016b:751).

[28] Berndt *et al.* (1993:253).

[29] Gunn (1847:2), Blandowski (1855:73), Smyth (1878, 1:436) and Clarke (2018c:37–39).

[30] Anonymous (1845:2).

[31] CJ La Trobe (cited in Norman 1951:6).

[32] Wathen (1855:123), Massola (1957:83, 1969:80–81) and Clark (2010:566).

[33] Anonymous (1857:6).

[34] Hassell (1934a:244).

[35] Dawson (1881:92).

[36] Langloh Parker (1905:30,108,143). See also Radcliffe-Brown (1930:344), Whitley (1940:132) and Mountford (1978:59). Langloh Parker wrote the name of the spirit as Gowargay, while Mountford spelled it as *gouage*. Radcliffe-Brown and Whitley wrote it as *Gauarge*.

[37] Mathews (1898c:117). See also Ash *et al.* (2003:63) for the derivation of the name.

[38] Mathews (1898c:118). See also Mountford (1978:60). Mathews wrote the spirit's name as *gowarkee*, while Mountford spelled it as *gowarkee*.

[39] Bates (1901–14:231).

[40] BJ Nangan (cited in Akerman 2020:29).

[41] Mountford (1956:212, 1978:69–70) and McCarthy (1965:19).

[42] McEntee *et al.* (1986:21).

[43] McEntee *et al.* (1986:21–22).

[44] Clarke (2016b:752).

[45] Raymond *et al.* (1999:103).

[46] Karadada *et al.* (2011:76).

[47] Tindale (1936c:61,69).

[48] Teichelmann & Schürmann (1840:9). See discussion by Johnston (1943:281).

[49] Clarke (1996b:82–84).

[50] Schürmann (1846:234–235).

[51] Mathews (1898d:143).

[52] Moreman (2014:11).

[53] Clarke (1999a:161–162, 2009b:41).

[54] Thurman (2014:34).

[55] Akerman (2016:16).

[56] Berndt (1940c:461).

[57] Ramsay Smith (1930:322).

[58] Pine (1897:173). The Ti-ya-tinity name is probably a rendering of *tenitenidi*, which Tindale (c.1931–91b) identified, probably incorrectly, as the Tangani name for the 'boobook owl'.

[59] Penney (1842–43:52).

[60] Dawson (1881:49,52–53).

[61] Dawson (1881:49).

[62] Dawson (1881:52).

[63] Mathews (1904:370–373). The Dyuni-dyunity name appears to be a rendering of 'Ti-ya-tinity' recorded in Tatiara (Pine 1897:173) and of *tenitenidi*, recorded by Tindale (c.1931–91b).

[64] Mathews (1904:373–375).

[65] Langloh Parker (1953:35).

[66] McConnel (1957:55–56). The description of the bird in the myth suggests that the 'mopoke' referred to is the southern boobook rather than the tawny frogmouth, although both occur in this region.

[67] McConnel (1957:56).

[68] Clarke (1994:135–137, 1999a:159–161, 2016b:752,756,760) and Bell (1998:46,310,312–318,326–327).

[69] WE Harney (cited in Elkin 1977:132).

[70] Tindale (c.1931–91b), Berndt (1940c:461) and Clarke (2019a:132,136,144).

[71] Wiynjorrotj *et al.* (2005:141).

[72] Raymond *et al.* (1999:110).

[73] Slater (1970:396).

[74] Doonday *et al.* (2013:144).

[75] Bates (1981b:58).

[76] Akerman (2020:60).

[77] Dixon (1991:162).

[78] Tindale c.1931–91b (cited in Condon 1955a:84).

[79] Tindale (1955).

[80] Clarke (1996b:84).

[81] Mountford (1978:47–48).

[82] Massola (1968:96).

[83] Thomson (1935:37).

[84] Cane (2002:106).

[85] Mathews (1898b:92). See also Langloh Parker (1905:134) and Isaacs (1980:18). Ash *et al.* (2003:106) identified *maliyan* (i.e. Mullion, Mullyan) as the wedge-tailed eagle.

[86] Dawson (1881:93).

[87] Low (2014:139).

[88] BJ Nangan (cited in Akerman 2020:118–121,186–187).

[89] BJ Nangan (cited in Akerman 2020:118).

[90] Clarke (2014b:2226).

[91] Tindale (1960:63).

[92] Anonymous (1911b, 1922) and Ramson (1988:583).

[93] Doonday *et al.* (2013:145).

[94] Mathews (1909).

[95] Roth (1903:12).

[96] Fletcher (2007:83).

[97] Clarke (1999a:161). The name *ritjaruki* has also been spelled as 'richerookerie' (Anonymous 1922:6) and 'Richarookerie' (Anonymous 1930b:54).

[98] Brusnahan (1992:42).

[99] Angas (1847b:96).

[100] Bates (1992:157).

[101] Hassell (1934b:319).

[102] Tindale (2005:372).

[103] Berndt & Berndt (1989:188).

[104] Berndt & Berndt (1989:188).

[105] Brown & Naessan (2014:31,47,52–53).

[106] Mountford (1965:152).

[107] Ramsay Smith (1930:342).

[108] McEntee *et al.* (1986:18).

[109] Mountford (1965:152).

[110] Karadada *et al.* (2011:87).

[111] Raymond *et al.* (1999:105).

[112] Wiynjorrotj *et al.* (2005:132).

[113] Blakers *et al.* (1984:144,663).

[114] Cayley (1931:754) and Blakers *et al.* (1984:145).

[115] Slater (1970:314).

[116] Cayley (1931:754).

[117] Condon (1941:6–7).

[118] Hassell (1934b:324, 1975:115), Berndt (1940c), Gosford (2010) and Doonday *et al.* (2013:138).

[119] Schürmann (1846:241) and Johnston (1943:275–276).

[120] Berndt (1940c:458–459). Berndt simply described *winta* as 'owl', but Tindale (1936c:67) classified it as the barn owl (*Tyto alba*). It is assumed that Berndt's 'curlew' is a reference to the bush stonecurlew.

[121] Berndt (1940c:459–460).

[122] Massola (1968:96).

[123] Puruntatameri *et al.* (2001:92).

[124] Bishop (2000:10).

[125] Mills (2019:22).

[126] Wiynjorrotj *et al.* (2005:126).

[127] BJ Nangan (cited in Akerman 2020:88–91).

[128] Love (1936:151).

[129] Berndt & Berndt (1989:196–197).

[130] Tindale (1931–34:149).

[131] Berndt (1940c:460).

[132] Berndt (1940c:460–461).

[133] Clarke (2016b:752,755).

[134] Blakers *et al.* (1984:144) documented the decline of this species in south-eastern Australia.

[135] Brown & Naessan (2014:59).

[136] Berndt (1940c:461, see also 1940b:291).

[137] Berndt (1940c:461).

[138] Karadada *et al.* (2011:84).

[139] Karadada *et al.* (2011:71).

[140] Raymond *et al.* (1999:102).

[141] Tindale (1930–52:266). The grey currawong is described as the 'black magpie'.

[142] Slater (1974:282).

[143] Tindale (1934–37:256).

[144] Tindale (1931–34:223). The term *kilindi* is also written as *kelindi*.

[145] Tindale (1938–56:28).

[146] Smyth (1878:1:427).

[147] Karadada *et al.* (2011:77).

[148] Wiynjorrotj *et al.* (2005:129).

[149] Akerman (2020:150, see sketch on 151).

[150] Raymond *et al.* (1999:106).

[151] Arthur (1996:31–32).

[152] Musharbash (2014:43, see also 42).

[153] McEntee & McKenzie (1992:125).

[154] Harney (1957:21).

[155] Bulmer (1855–1908:50).

[156] Clarke (1994:127–129, 1999a:151–154, 2007b:150). See also Berndt *et al.* (1993:208).

[157] Berndt & Berndt (1970:18,51,192) and Taylor (1996:183–189).

[158] Eickelkamp (2014:68, see also 62). The identification of *mirilyirilyi* is uncertain, with Goddard (1992:69) translating *mirilyirilyi* as the 'fairy wren (*Malurus* species)'. In the neighbouring Antikirrinya language, *mirrilyirrilyi, mirrily-mirrily* is recorded by Brown & Naessan (2014:117) as the 'dusky grasswren (*Amytornis purnelli*)'. It is possible that the term was routinely applied to several species of small bird.

159 Petri (1952:84–99).
160 Frazer (1910:23).
161 Elkin (1977:20,22–23,25,39,74,92,106,108,119–120).
162 Bates (1906).
163 AP Elkin 1933 (cited in Petri 1952:85).
164 Akerman (2020:29).
165 Akerman (2020:29–41). .
166 Akerman (2020:31–41).
167 Akerman (2020:44–45).
168 Akerman (2020:133).
169 Tindale (1938–56:73).
170 Tindale (1931–34:228–229). Clarke (2019a:135) identified *kurki* as the collared sparrowhawk.
171 Tindale (1938–56:73).
172 Berndt *et al.* (1993:248–249).
173 Berndt *et al.* (1993:249).
174 Berndt *et al.* (1993:249–250) and Clarke (1999a:161).
175 Berndt *et al.* (1993:250).
176 Berndt *et al.* (1993:250).
177 Berndt *et al.* (1993:197–198).
178 Clarke (2016b:753–754).
179 Cameron-Bonney (1990) gave an account of Aboriginal folklore for the south-east of South Australia.
180 Tindale (1931–34:149).
181 Rankine (1991:115–116).
182 Wyndham & Park (2018:533).
183 Clarke (1994:126–139).
184 Kaplan (2019:xiii).

5

Bird nomenclature

What people observe in the environment is as much a product of the importance their culture places upon each type of object, as it is upon the acuity of their vision. For languages, the reading of the landscape creates a proliferation of words in categories that are deemed important by the speakers. A combination of utilitarian and other cultural factors combines to determine which organisms will attract a specific name, and how distinctive those names will be.[1] Anthropologist Eugene S Hunn explained that 'we humans recognise a diversity of kinds of birds, giving each a name, often a name descriptive of the birds' voices, colours, actions or haunts'.[2]

In Australia, linguists and amateur word collectors have compiled regional lists of Aboriginal bird names.[3] Reference to such studies is necessary when using historical records to investigate early Indigenous relationships with birds, but they are imperfect because on the frontier of British colonisation the first recorders as a group were inconsistent with both their use of English-derived names and the spelling systems they used for Indigenous languages.[4] Since some of these early sources concern languages that are no longer spoken today, it is now often difficult to directly assess the accuracy of their work. According to Western Australian government biologist Ian Abbott, the 'scope for error in recording Aboriginal names is very great'.[5] In my own experience, there are errors in many of the recorded Aboriginal vocabularies, both in terms of the sourcing of words and the identification of the avian species concerned.[6] Many other bird species have been described in the ethnographic literature, but are not properly named to the extent of identifying species.

Ornithologists have synthesised this historical data on avian names, in recognition of the potentially valuable insights it provides into past bird species distributions. For instance, in 1933, T Theodor Webb, the missionary/ornithologist based at the Methodist Mission Station at Milingimbi, published a list of Aboriginal bird names for the eastern Arnhem Land region.[7] In 1943, zoologist T Harvey Johnston at the University of Adelaide published the results from a detailed ethnozoological study in north-eastern South Australia, based on extensive fieldwork and historical research, which included lists of Aboriginal bird names and a catalogue of the uses for each species.[8] In 1955, Herbert T Condon, the curator of birds at the South Australian Museum, listed Aboriginal names for birds from across South Australia, many of them gained from curator of ethnology Norman B Tindale, as a key for identifying the birds mentioned in the early historical and ethnographic literature.[9]

In 1981, biologist Ian Mansergh and linguist Luise Hercus argued that by compiling Aboriginal names for animals from earlier recordings, maps could be generated of the past distributions of locally extinct species in the Gippsland region of Victoria.[10] Historical geographer Sue Wesson suggested in 2001 that the former ranges of birds, such as magpie geese and bustards in southern Australia, might be determined through the analysis of early recorded Aboriginal vocabularies.[11] In 2018, a research team comprising ornithologist Andrew

Black and the linguists John McEntee, Peter Sutton and Gavan Breen produced its findings from a project that drew upon data from Aboriginal ethno-linguistic sources to help determine the likely pre-European distribution of the galah before its southern expansion after European settlement, due to an altered environment.[12] In 2019, linguist Aung Si published the results of an investigation into the processes by which terms for animals were exchanged between different Aboriginal language groups across northern Australia. He established that there were several loanword 'corridors' along which borrowing frequently occurred.[13]

A national catalogue of Aboriginal avian names was produced in 2006;[14] however, it has the same problem as many of the early to mid 20th century compiled word lists,[15] in that the source languages from which the listed names are drawn were not identified.[16] In 2009, Abbott produced a detailed list of Aboriginal names for bird species from the south-west of Western Australia, and made insightful suggestions concerning their adoption into common English usage.[17] Overall, though, trained linguists have fared much better in striving to accurately and more fully record Indigenous bird names.[18] In particular, the involvement of linguists with language speakers in the building of encyclopaedic dictionaries for Aboriginal languages in the arid and tropical parts of Australia has provided a wealth of ethno-ornithological data.[19]

In many parts of Australia, where the Aboriginal languages are no longer fluently spoken, some Indigenous bird names are nevertheless still known and in use within the local variety of Aboriginal English.[20] During my own fieldwork in the Lower Murray in the 1980s, Indigenous bird names were recorded from elderly people who in some cases were uncertain about the common English name for the species involved. For instance, Ngarrindjeri woman Fran Kernot (née Rankine) described the *kripari* bird as being 'like a snipe, if not a snipe', that 'runs along the sand' and 'looks like a sparrow with long legs and has its eggs in the sand'. By viewing published photographs, it was later identified by Aboriginal peoples as the Australian painted-snipe.[21] In an example from my fieldwork in the south-east of South Australia during the same period, elderly Aboriginal people Ron Bonney and Lola Cameron-Bonney described *nyrukin* as a bird they knew well – a black, white and brown-coloured 'stilt' with a 'turned up beak', which I concluded may have been a reference to the bar-tailed godwit.[22] Similarly, Ngarrindjeri man Henry Rankine at Raukkan did not know the English name for *turi* but described it as 'like [blue] bald coot, but swims out in the lake'. Other sources assisted in identifying it as the Eurasian coot.[23]

Indigenous classifications

All of the world's languages contain encoded information about how their speakers interact with their environment.[24] European scholars of the colonial period largely ignored the existence of complex classification systems in the cultures of the hunter-gatherers they encountered, while the modern field of anthropological linguistics uses the study of language as a tool for investigating how specific cultures order the universe.[25] An evolutionary psychological approach to the study of ethnoclassification is the search for formal hierarchies and covert categories within the Indigenous classification of organisms.[26] This book takes an ethnoecological approach to ethnoclassification,[27] which strives to understand and explain ecology as it is experienced and imagined by Aboriginal peoples.

In Australia, the study of plant and animal classifications provides deep insights into Aboriginal views of their world.[28] At the time of first European settlement there were at least 200 distinct languages in Aboriginal Australia, and within those were numerous speech varieties or dialects.[29] The ways in which Aboriginal languages reflect how people see their environment is far too vast for full treatment here, although examples of language terms are given to show how birds are classified in ways that are different from formal scientific classification systems. In my experience with many parts of Australia, contemporary Aboriginal peoples are well aware that their own ways of looking at plants and animals in the environment are unique.

It is apparent that the common English definition of 'bird' does not have an equivalent term in most of the better-recorded Aboriginal languages. For example, in the Kaurna language spoken on the Adelaide Plains in South Australia, *walta* was recorded as a term for a 'large bird, turkey [Australian bustard], eagle, &c'.[30] In the Lower Murray group of languages, all small flying birds were classed as *pulyeri* but this term excluded all larger species.[31] Similarly, *yuta* was recorded in the Wailpi language of the northern Flinders Ranges of South Australia as a 'general term for small bird' and 'any little bird'.[32] In the Walmajarri language of the Great Sandy Desert of Western Australia, *tuurru* is a term generally used just for various small bird species.[33] In the Kunwinjku language spoken in western Arnhem Land in the Northern Territory, the term *mayhmayh* is used for a group of small birds while many, but not all, larger birds are identified as *mayh*.[34] Some recently published Aboriginal dictionaries do contain a single Aboriginal term for a bird,[35] although I suggest that in most cases this is either due to a misunderstanding by the recorder or because only a single term for one group of birds was remembered by Aboriginal speakers, due to language loss.

Among the classification systems of the Aboriginal languages of Australia, there is an absence of a broad category of 'birds' that would be equivalent to the class of Aves in the Linnean system.[36] Aboriginal peoples interacted with groups of birds independently and in different ways. For instance, in Gupapuyngu (Kopapingo) language spoken at Milingimbi in Arnhem Land, Webb claimed:

> It may be worth noting that among the aborigines I have discovered no evidence of interest in birds as such, and, particularly among the smaller varieties, such as Honeyeaters, Flycatchers, etc., there is apparently no definite name for many species. The usual practice is to refer to all these small members of the feathered world as 'tchikai', practically no attempt being made to separate them into genera or species.[37]

The lack of a perfect fit between Aboriginal and scientific classifications of the biota has, in the past, been an obstacle for biologists wishing to engage with Aboriginal biocultural knowledge as a potential source of 'new' data.[38] For instance, in 1928 ornithologist C Sullivan provided Aboriginal names for a list of birds from the west coast of South Australia. He wrote:

> With some species, such as the *Amytornis* [grasswrens] and the Hawks, one may doubtful as to the coupling of the native name with the bird in question; with

others, e.g., the Pink Cockatoo [Major Mitchell's cockatoo] or Little Penguin, there would be no doubt. I give the native names for what they are worth, without vouching for the accuracy of my transcription or the correctness of identification on the part of the native.[39]

In Aboriginal languages, the clusters of species are not only based on similar bird size and shape; other factors such as habitat and behaviour are also important. In Yolngu-matha spoken in north-east Arnhem Land, *djarrak* is a generic term for 'gull'. It includes the crested tern, little tern, silver gull and whiskered tern.[40] The term is presumably based on the related ecology of this group of different-looking birds, as here the individual species also have their own names. Other Aboriginal languages also have birds clustered into named groups. For instance, in the Wik Mungkan language spoken in western Cape York Peninsula, there are some distinct subcategories within the birds. The following list of avian names was extracted from the Wik Mungkan dictionary, which is used at Aurukun.

minh – edible animals, including birds that are edible, such as:
minh achamp – emu,
minh keech – all large white wading birds which have long necks and legs, with straight pointed bills: cattle egret, great egret, little egret, intermediate egret, eastern reef egret (white-phase only),
minh kuunth – Australian bustard,
minh miwan – green pygmy-goose.[41]
panch – most birds that are not edible, such as:
panch aay – black-faced cuckooshrike, white-bellied cuckooshrike,
panch pak – all small honeyeaters: black-chinned honeyeater, white-throated honeyeater, brown honeyeater, white-streaked honeyeater, bar-breasted honeyeater, rufous-banded honeyeater, rufous-throated honeyeater, dusky myzomela and olive-backed sunbird,
panch thomp – all small beach birds, often seen in groups: red-kneed dotterel, red-capped dotterel, black-fronted dotterel, grey plover, American golden plover, lesser sand-plover, greater sand-plover and oriental plover.[42]

As shown by the Wik Mungkan system of identifying what Europeans call birds, there is an underlying logic to which species are grouped together, but this is not a system built upon likely evolutionary relationships. In the past, Europeans who have looked for terms relating to what ornithologists would regard as a genus have generally been disappointed. This fact remains a warning for contemporary scientists who may want to uncritically import into their work a single piece of Aboriginal environmental knowledge without understanding its original context. In New South Wales, a former colonist ALP Cameron writing about Aboriginal terms for animals and plants remarked that:

Metaphorical expressions are conspicuous by their absence, and we meet with very few generic terms. They have a name for every tree that grows, every bird that flies, and indeed for everything that exists, but very seldom a word signifying a whole genus.[43]

In Aboriginal languages, the emu is not classed among the other birds, probably because it is large and flightless.[44] Even in the Wik Mungkan example given above, the emu is in a category of edible animals, not just edible birds. The separation of emus from most other birds is perhaps not unexpected, given that its plumage is atypical; an early scientist remarked upon the emu's 'strange almost hair-like feathers'.[45] In the mythology of the Wiilman people in the south-west of Western Australia, it was considered that 'Waitch [Emu] has the best feathers of any bird because they are a mixture of ordinary plumage with Omer's [western brush wallaby's] fur'.[46] In the vocabulary of the Arrernte language spoken in the Macdonnell Ranges of Central Australia, as recorded by missionary Carl Strehlow in 1909, the unusual structure of some emu feathers is apparent. The short emu feathers are called *ilia punga*, literally 'emu-hair', and the long emu tail feathers are termed *iliapa*.[47] Doctor and Aboriginal protector, Herbert Basedow, observed that among the Aboriginal groups living along the Nullarbor Plain 'It might be explained that the emu, because it cannot fly, is not regarded as a "bird" in the generally recognised sense, and consequently the wings are looked upon as "fingers"'.[48]

The exception to the observation that emus are generally separated from other birds is the southern cassowary. On Cape York Peninsula, both birds were called *yirrkungarr* on Flinders Island at the eastern end of Princess Charlotte Bay, and *urrkungkurru* nearby at Barrow Point.[49] Linguist/anthropologist Peter Sutton remarked:

> I know of at least one language where both cassowaries and emus were classified together under a single name. In Wurriima aka The Flinders Island Language, eastern Cape York Peninsula. My teacher was a fully fluent speaker, Wodhyethi (aka Johnny Flinders). I thought it odd so double checked … I just also checked the language on the immediate south of Wurriima (Flinders Is Language), namely The Barrow Point Language, and it also only a single term for both cassowaries and emus.[50]

The situation of the emu and southern cassowary being classed separately from other birds also occurs in relation to some of the world's other large flightless birds known as Ratites, such as the common ostrich for the San in the Central Kalahari Desert in southern Africa[51] and the northern cassowary for the Karam people of the upper Kaironk Valley in the Schrader Mountains of New Guinea.[52] If a zoologist working in Australia was to adopt an Aboriginal classification for their taxonomic work, then the emu and southern cassowary would probably be treated as belonging in the same category as macropods like kangaroos and wallabies, as they all have much in common in terms of Aboriginal hunting and cooking techniques (see Chapters 6 and 8).

Possession of the flight ability has led to various non-avian animals being grouped with birds in various languages. In some Western Desert dialects, bats and even certain flying insects are classified with the flying birds.[53] In other Aboriginal languages only the birds and bats are grouped together, and insects are excluded. For instance, in the Kaytetye language of Central Australia, *thangkerne* is a general term for 'fleshy flying creatures including birds and bats but not emus'.[54] In the Kayardild language spoken on Bentinck Island of the Wellesley Islands of the southern Gulf of Carpentaria in Queensland, all flying game animals are termed *kalanda yarbuda*; this includes most birds, but also bats and flying foxes.[55] This was also the case in the Yanyuwa language at Borroloola in the south-western Gulf of Carpentaria of the Northern Territory, where the same animals were classified as *julaki*.[56] Similarly, in the Anindilyakawa of Groote Eylandt in the north-western Gulf of Carpentaria of the Northern Territory, *wurrajija* is a category for creatures with wings. It includes many insects, flying foxes and bats.[57]

In the Kabi Kabi language spoken south of Bundaberg in south-east Queensland, *dhip'pi* for bird was also recorded as 'a generic name applied to winged creatures generally'.[58] By definition, the classification of winged animals excludes some avian species. For instance, in the dictionary for Eastern and Central Arrernte, spoken in the Macdonnell Ranges of Central Australia, the term *thipe* (*uthipe*) refers to:

> Fleshy flying creatures including birds and bats, but usually translated as just 'bird'. This does not include the emu (*arleye* / *ankerre*) because it does not fly, or for some speakers the bush turkey [Australian bustard] (*artewe*), presumably because it does not fly much.[59]

Feathers and a bird shape alone do not combine to define a 'bird', as is illustrated by the description of the bunyip, which is an Aboriginal water spirit originally termed *banib* in Wergaia dialect of Wemba in south-west Victoria.[60] In south-eastern Australia, these much-feared Aboriginal river spirit beings were described as feathered and having a shape resembling that of an emu, particularly the head and neck area.[61] The perceived call of the *banib* was apparently that of a bird, as early naturalists proposed that the eerie booming call of the Australasian bittern was behind the numerous bunyip 'sightings' in southern and eastern Australia.[62] This bittern's secretive behaviour added to its identification with the bunyip, as geographer/writer Charles Fenner wrote, 'The mysterious booming sound made by the bittern (*Botaurus poiciloptilus*), a very shy bird, has become associated with the bunyip, but actual observers have usually described the sound of the latter as a roar or a bellow'.[63] Concerning the bittern, a settler remarked that 'I think to this bird, more than any other cause, we can attribute the longevity of the "bunyip" myth'.[64] A former Victorian settler stated that:

> Gippsland was supposed to be the home of the bunyip. The 'booming' of bitterns from the Moe Swamp gave rise to the supposition that the noise was made by some large amphibious animal. The aboriginals kidded the whites that it was so, and they said the animal's name was Buneep … Since then the curl and ebb and flow of our flexible Anglo-Saxon tongue has transformed Buneep into Bunyip. Observe, too, in the first syllable of this word an illustration of the principle known to philologists as Onomatopoeia – i.e., an attempt to make

the pronunciation conform to the sound. The 'booming' of the bittern is to be conveyed by a long drawn-out Bunyip! Something intended to be decidedly uncanny.[65]

The form of a large emu was a characteristic of other spirit beings. There were beliefs in giant 'birds', such as the *meeheeruung parrinmall* ('big emu') in the south-west of Victoria.[66] Similarly, the *ngindyal* in the north-west of Victoria was described as 'a bird-like animal, having the shape and feathers of an emu, but of enormous proportions, and was moreover, a great magician'.[67] Elsewhere, other Aboriginal spirit beings were described as having the power of flight. For instance, in the south-west of Western Australia it was reported that Aboriginal peoples believed in a 'huge winged serpent' known as Waugal.[68]

As discussed in the previous chapter, Aboriginal spirit beings sometimes took on the form of birds like eagles, owls, crows, willie-wagtails and grey currawongs. Birds can therefore have dual identities: animal and spirit. A symbolic classification system exists separately from the ethnotaxonomy of the group, meaning that animals like birds can be grouped in categories according to their symbolic meanings.[69] This situation is similar to what has been described by ethno-ornithological studies outside of Australia, such as at Flores in eastern Indonesia where anthropologist Gregory Forth observed that the:

> Nage [people] identify Falconiformes (eagles, hawks, and falcons), Strigiformes (owls), and several kinds of dark-colored, nocturnal, or scavenging birds (crows, drongos) as members of a symbolic category of 'witch birds.' As the name suggests, the category is defined by the several ways each of these birds is connected in Nage cosmology with 'witches' (polo).[70]

At the species level, there appears to be superficial agreement between the classification systems used by Aboriginal peoples and ornithologists. The extent of this agreement is not always clear in the available historical records, which are imperfect and often fail to distinguish between species of birds in certain groups.[71] This may be a shortcoming on the recorders' part, but in at least some cases it is known that there were Indigenous terms covering several species of birds from the same habitat which possessed similar looks and behaviours. For instance, Aboriginal peoples in many areas may not have distinguished between the *Corvus* species that in Australian English are referred to as 'crows' and 'ravens'.[72] This appears to be the case with these birds in some languages of Central Australia,[73] in the south-west of Western Australia[74] and in the Lower Murray.[75] This should be expected, since even ornithologists have difficulties when trying to distinguish between various Corvids in the field.[76] There are also examples of the lumping of species together. For instance, many early European observers considered the sarus crane and brolga to be the same species,[77] while in the Wik Ngathan language of western Cape York Peninsula in northern Queensland they were not: the former were *yoompnham* and the latter were *thuul*.[78] Other groups, such as the Yanyuwa people of the south-western Gulf of Carpentaria and the Mawng speakers of coastal western Arnhem Land, however, referred to them by the same name.[79]

In the Lower Murray region, the evidence suggests that the apparent lack of recorded individual names for various small hawks/falcons was probably not a failing of the recorder, but more likely due to local Aboriginal peoples grouping them under a single term, *kiriki*.[80] This was noted by other scholars for seven species of hawk in the Antakirrinya language of the Western Desert,[81] and for the four 'fierce falcons' in Adnyamathanha spoken across the northern Flinders Ranges of South Australia.[82] Similarly, in the Anmatyerr language of Central Australia, the brown falcon and grey falcon are both called *irrkerlanty*.[83] In the Yidiny language from the rainforest region inland from Cairns in northern Queensland, *bijuu* refers to the grey goshawk, but also probably to the collared sparrowhawk and brown goshawk.[84] From the data assembled from the Lower Murray, it appears that in the Ngarrindjeri language the welcome swallow and martins were classed together with a single name, possibly along with swifts,[85] which is also a folk term grouping noted in certain languages outside of Australia.[86]

Australian raven. Its Aboriginal name is generally derived from its call, such as *waa* for central Victoria languages. This species often shares a name with various crows. Philip A Clarke, Laratinga Wetlands, Mount Barker, South Australia, 2015.

In some Aboriginal languages, however, a distinction is made between swallows and swifts. In the southern Kimberley of northern Western Australia they are classified separately as moiety totem birds, as well as being 'rainbirds' that are harbingers of the wet season.[87]

For Aboriginal languages, a combination of similar physical form, related calls and shared preferred habitat are all important characteristics for determining which bird species – as an ornithologist would classify them – might be grouped under a single term without any other distinction. In the Adnyamathanha language spoken in the northern Flinders Ranges, the galah and little corella are both called *warrandu*,[88] possibly due to their similar habits and occasional cohabitation. It is possible that the name was broadened to include the galah, when the bird extended its range there during historical times.[89] In the Tiwi language spoken on Melville and Bathurst Islands in the Northern Territory, *arntirringarika* is the name used for the red-winged parrot but it is also used as a generic term for all parrots, and *kurlutuki* is a general name for doves and pigeons, due to the shared call of some common species.[90] Similarly in Tiwi, *arlipunyika* refers to all honeyeaters found on the islands.[91] There is evidence for a hierarchy of bird names for some avian groups. In Tiwi, raptors such as the black-shouldered kite, black kite, brahminy kite, little eagle, western osprey, wedge-tailed eagle and white-bellied sea eagle are all known by individual names but are also grouped together as *murtati*.[92] A general Tiwi term for all owl-like species is *pinjoma*. This includes the barn owl, boobook owl, rufous owl and tawny frogmouth, although some species also have separate names.[93]

In the Jawoyn language of Katherine in the Northern Territory, the black kite, square-tailed kite and brown goshawk have the same name, *karrkkayn*, which is derived from their call.[94] The grey shrikethrush and black-faced cuckoo-shrike have the identical Jawoyn name of *jawayakwayak*,[95] possibly due to a shared habitat and common physical features. Examples from the neighbouring Wardaman language, which is spoken to the south-west of Katherine, include all finches that are generically referred to as *nilngman*, all ducks and the green pygmy-goose that are called *jibilyuman*, cockatoos collectively referred to as *ngeleleg*, and the peaceful and diamond doves that are both known as *golorog*, the last being a representation of their call.[96] In Wardaman, a collective term for owls and nightjars is *mugmug*, which is said to be based on the call of some species.[97]

Regardless of name sharing, Aboriginal peoples recognise that different types of birds can have close relationships with each other. In some Aboriginal folk systems, organisms may be described as kinsmen, or called 'countrymen' or 'mates' to each other, if they occupy a similar habitat niche or have a shared mythological origin.[98] For example, Adnyamathanha people in the northern Flinders Ranges considered the zebra finch to be the 'eldest sister' of the Australian bustard and the emu.[99] Similarly, finches in the south-west Kimberley are said to possess 'brother-in-law' and maternal 'uncle' relationships with two of the grass species which produce the seeds they chiefly eat.[100] Here also, the cockatiel, possum and cicada, who each possess homes in trees, are said to be 'brothers'.[101] Among the Walmajarri people of the Great Sandy Desert in Western Australia, the barn owl is considered to be a 'close countryman' to the southern boobook,[102] probably due to their shared nocturnal habits like hunting. In the Wik Ngathan language spoken on western Cape York Peninsula in Queensland, the palm cockatoo is considered to be the 'mate' of the red-tailed black cockatoo.[103] Aboriginal people who travel to a different Country sometimes use kinship terms when comparing animal species. For instance, Bobby Brown from the Antakirrinya said that 'The Grey Butcherbird is the big brother of the Kookaburra',[104] which he had seen when visiting Adelaide.

Aboriginal peoples express the close connections between birds by highlighting their shared habitats. In the Jawoyn language of Katherine in the Northern Territory, the black-faced cuckoo-shrike is said to be a 'countryman' of the white-bellied cuckoo-shrike.[105] The same is said of couples such as the northern fantail and willie wagtail,[106] the northern rosella and red-collared lorikeet,[107] and a larger group containing the Australian magpie, magpie-lark, grey butcherbird and pied butcherbird.[108]

There are associations between birds for which the reasons are less apparent, possibly due to more complex ecological and cultural links. In the Adnyamathanha language, *anda-anda* is both the singing honeyeater and the grey shrikethrush,[109] while *awi-ita-na* covers the rufous whistler and mistletoebird.[110] In the Antakirrinya language of the Western Desert, *ruurl* and *luurn* are terms used for the sacred kingfisher, red-backed kingfisher and rainbow bee-eater, all of which nest in burrows dug into riverbanks.[111] Similarly, in the Walmajarri language of the Great Sandy Desert, the rainbow bee-eater and the red-backed kingfisher are both known as *luurn*.[112] A cluster of small bird types known as *nyilynyil* in Walmajarri includes the red-backed fairy-wren, purple-crowned fairy-wren, white-winged fairy-wren, crimson chat and yellow chat.[113]

At the other end of the spectrum of names are birds that are differentiated according to gender or age. Ornithologists William S Peckover and LWC Filewood, when working with Papua New Guinean people, noted that many 'have sufficient knowledge of their birds to name almost all of them at species level, sometimes giving two or more names for different colour phases or sexes'.[114] From Aboriginal Australia, a Lower Murray example of sexual differentiation is the musk duck, as the drake was called *nanawuli* while the hen recorded as *tilmuri*.[115] Similarly, in the Ramindjeri speech spoken at Encounter Bay in South Australia, the hen emu was known as *preki*[116] while the cock was *yarli*.[117] In the Mawng language spoken in coastal western Arnhem Land, the female magpie goose is called *manimunak* while the male is *parrngarnaki*.[118] Linguist Jeffery Heath noted that in the Nunggubuyu (Wubuy) language of south-eastern Arnhem Land, 'Many birds have special juvenile terms and some distinguish male from female'.[119] For example, here the adult whistling kite is called *gurujujurg* while the juvenile of the same species is *damirnmirn*.[120] The silver gull in adult form is called *jarrak* while the immature form, which is grey, is generally called *kurumburra*.[121] In terms of gender separation in names, Heath's examples were from the Australian pelican, brown goshawk, white-bellied sea-eagle, great-billed heron and emu.[122] In the Thaynakwith language of north-west Cape York Peninsula in northern Queensland, the male of both the brolga and eastern sarus crane is called *ndrril*, the female is *kurrungganh* and the chick is *njichan*.[123]

Some derivations

Having considered the Indigenous classification of birds, the focus in this section is on the varied meanings behind individual bird names. Condon remarked that 'Some of the names are descriptive or refer to habits and call-notes, and a few indicate the suitability of the species simply as "tucker".[124] It has been noted across the world that the bird names in many non-Western European languages are onomatopoeic,[125] a fact which contributes to bird names changing little through time.[126] This is particularly so in Aboriginal Australia.[127] An early ornithologist, WM Sherrie, stated:

> The blacks – as is common among wild things – had a singularly acute sense of hearing, and they were very exact in reproducing the calls of birds. As a general thing they named birds after their calls. Thus the wood-duck became g'naroo, and the black swan gunyick or gunyuck.[128]

Calls

When a linguist is building a dictionary for a language and compiling names for different types of bird, it is possible to record three things: the bird name; a word that represents its call; and a customary imitation of its call.[129] The second and third words can be very similar, but done in a different voice. By way of example, in English the following four expressions are associated with pet canines – 'dog', 'bark', 'woof woof' and 'rrraow'. Bearing this in mind, we should expect a large variation with the recording of names for the same bird species.

In Aboriginal Australia, the analysis of names recorded for different types of bird indicates that many are derived from the bird calls.[130] Examples from the Lower Murray include *kukaki* for the laughing kookaburra,[131] *tjipeteruku* for the magpie-lark[132] and *wak* for raven or crow.[133] Here, a bird call that is used in modified form for the brolga is *prolgi*, which apparently derived from the call *prolg! prolg!*[134] This imitation of the bird call has implications for other subjects, through the extension of word meanings. In the Yaraldi dialect of the Ngarrindjeri language of the Lower Murray, the springs and swamps where the brolga lived were termed *prolginali* while the south-east winds were *prolgi-mayi*, the latter from *prolgi* (brolga) and *-mayi* ('wind').[135] The significance of the meaning of *prolginali* may have been based upon an association between seasonal wind changes and bird migration.

The term *prolggi* appears to be related to the Australian-English term brolga, which was a borrowing by Europeans from an Aboriginal language – possibly the Gamilaraay language in eastern New South Wales, where the bird was called *burralga*.[136] The word source may have been yet another language – in the various languages of the Lake Eyre Basin of eastern Central Australia there are related recorded brolga terms, such as *pooralkoo, puralku, bouralko* and *bralgo*.[137] In spite of the likely onomatopoeic origin of the name, a popular explanation expressed in an early 20th century newspaper was that 'The aboriginal name for the bird is brolga – said to mean dancing bird – and the black name has therefore much to recommend it'.[138] The 'native companion' is an early European term for the brolga.[139]

While the bird call was widely used as the basis for the Aboriginal name for the brolga, there appears to be some variation in the chosen name due to the existence of a wide range of complex and varied calls from which to choose. For instance, in addition to the examples given above, in the Wadeye area south-west of Darwin the brolga is known as *ku kulerrkkurrk*, which is said to be based on its call.[140] In the Jawoyn language spoken round Katherine in the

Northern Territory, the brolga is called *pornorrong*, also apparently based on one of its calls.[141] As with early recordings of Aboriginal languages, the writing of ideophones is highly variable. It was noted by colonist and Aboriginal Protector Archibald Meston in Queensland that:

> The native companion had more names than any other bird or animal in Australia, many having the 'coor' sound, taken from the voice of the bird, as cooroor, cooradooc, cooralbin, cooroora, and coorcorl; the last of which an aboriginal could sound exactly like one of the calls of the bird … Coorahlba, coorahlga, and coorahlbong were among the names of the native companion in South Queensland.[142]

It is sometimes difficult to tell to what extent the apparent difference in the Indigenous name recorded for various languages is due to the phonetic system used. Taking the names for the willie-wagtail as an example, in the Paakantyi language spoken in the Darling River area of western New South Wales the name is *tirrygirryka*,[143] in western Victoria it was *dyirri-dyirritch*[144] and in the Great Victoria Desert of Western Australia it is *tjintirrtjintirrpa*.[145] Allowing for the fact that *dy* is equivalent to *tj*, these names are similar in spite of coming from widely separated places, probably because they are all derived from the same call of the species.

Aboriginal languages from across temperate Australia are rich in onomatopoeic bird names. In the Ngaiawung language spoken along the Murray River north of Murray Bridge, *tildabki* was recorded as the magpie-lark.[146] A recorded name for the white-browed babbler from the northern Flinders Ranges region of South Australia was *unulunula*, which is thought to be onomatopoeic.[147] In the Barngarla language of Eyre Peninsula in South Australia, the term *ngarkarko* for 'species of wattlebird' appears to refer to the red wattlebird, due to the similarity with its call.[148] The Albert's lyrebird in the mountains in south-eastern New South Wales had the Aboriginal name of *bullen-bullen*, which according to early naturalist George Bennett 'is an imitation of the cry of the adult'.[149] In some of the Nyungar languages spoken in the south-west of Western Australia, the black-winged stilt was called *djandjarok*, possibly based on its high-pitched call.[150] Similar sounds from different bird species have been the basis of a shared name. For instance, in Nyungar, both the willie-wagtail and the restless flycatcher were called *tyitti-tyitti*.[151]

There are many onomatopoeic bird names in the languages still spoken today in arid and tropical Australia. In the Walmajarri language of the Great Sandy Desert of Western Australia, the willie-wagtail is *jirntipirriny* due to its constant twittering, the magpie-lark is *tiyatiya* based on its call, the masked lapwing is known as *tintirrpari* with its night-time call being 'tintirr … tintirr … tintirr', the silver gull is *kirtkirtpunta* based on the call 'kirt … kirt … kirt' and the southern boobook is *kurrkurr* based on the 'kurrkurr … kurkurr' noise it emits when hunting.[152] Examples from the Wunambal Gaambera language spoken in the north central Kimberley are *gurirririrri* the magpie-lark, *jonjonon* the striated pardalote and *gorragorra* the blue-winged kookaburra.[153] In the Endeavour River area of Cape York Peninsula in Queensland, the laughing kookaburra is called *gugaa* in the Guugu-Yimithirr language. The origin of this onomatopoeic name is enshrined in a creation myth concerning the kookaburra ancestor laughing at the carpet snake man, whose wife had swum away as a fish.[154]

In the Jawoyn language of the Katherine area in the Northern Territory, the pheasant coucal has the name *pukpuk*, which is based on its call 'buk-buk-buk-buk', that is said to make when it is 'singing' the bush cucumber to make it fat and ripe.[155] Similarly, the Jawoyn name for the grey-crowned babbler is *ngakngak*,[156] and for the black-tailed treecreeper it is *wikwik*,[157] both of which are renderings of their calls. Other onomatopoeic examples from this language include the weebill as *miynpiyu*, the red-backed fairy-wren as *wirrmi*, the rainbow bee-eater as *wirrirtwirrit*, the bar-shouldered dove as *kolototok*, the zebra dove as *kottowkottowk*, the yellow-throated miner as *wijwij*, the barking owl as *mukmuk* and the southern boobook as *korrkporok*.[158] In the cockatoo/parrot group the Jawoyn names are predominantly derived from calls, such as the red-tailed black cockatoo as *karrak*, galah as *pilkpilk*, red-collared lorikeet as *tettet*, red-winged parrot as *weley* and hooded parrot as *jikkilirrij*.[159]

There was much similarity between the various Indigenous names recorded across Australia for ravens and crows, which modern Western Europeans recognise as a cluster of related species. For instance, the Australian raven was *wa* or *wak* in the languages of south-eastern South Australia[160] and *waa* in central northern Victoria.[161] Similarly, from further afield in the northern Flinders Ranges of South Australia the recorded Adnyamathanha name for the Australian raven is *wakala*.[162] In the Gamilaraay/Yuwaalaraay languages of inland northern New South Wales, the Australian raven is called *waan*, *waagaan* and *waaruu*.[163] The Australian crow is called *wanggura* in the Wunambal Gaambera language of the north central Kimberley,[164] and is *wakwak* and *wagwag* respectively in the Jawoyn and Wardaman languages spoken in the Katherine region of the Northern Territory.[165] The Australian crow is known as *warrk* in eastern Arnhem Land.[166] In the Walmajarri language spoken in the Great Sandy Desert, the Australian raven and little crow are collectively known by the terms *wankurta*, *waakwaak* and *kaarnka*.[167] It is possible that some languages formerly had different Aboriginal bird names for the various Corvids.[168] Evidence for this is the recording of the Western Desert term *kaarnka* for 'crow', which appears to be onomatopoeic for the little crow found in this region.[169]

Some common bird names used in contemporary Australian English are derived from Indigenous words that were also onomatopoeic.[170] Examples include the galah from *gilaa* in Yuwaalaraay and neighbouring languages of northern New South Wales,[171] corella probably from *garila* in the Wiradhuri language of south-west New South Wales,[172] boobook owl from *bug-bug* in the Dharuk language of Sydney,[173] kookaburra from *kukuburra* of the Wiradhuri language,[174] currawong probably from *garrawang* of the Yagara language of Brisbane[175] and weelo (welu, wheelow, willaroo) the bush stonecurlew from *wirlu/wirul widluru* in languages spoken from the south-west of Western Australia to southern South Australia.[176] In the Mawng language spoken in coastal western Arnhem Land, *kurrwirlu* is the onomatopoeic name for both the bush stonecurlew and the beach stonecurlew.[177] In the case of mopoke (mopehawk, morepork), which in Australian English is confusingly used for both the southern boobook and the tawny frogmouth, the name is probably an English representation of their calls.[178]

The adoption of Aboriginal words for animals and plants into Australian English has not been uniform in time and space. It largely occurred during the early period of British colonisation, when varieties of creole and then Aboriginal English were, through necessity, more extensively used.[179] Australian English language scholar William Ramson has determined that about one-tenth of the Aboriginal vocabulary of ~250 Dharuk words at Sydney recorded by colonist Captain John Hunter in 1790 remains current in Australian English. This is about five times as many as those eventually borrowed from any other Aboriginal language.[180]

Behaviour

Bird names were occasionally taken from behavioural characteristics. For instance, in the Nyungar language spoken at Esperance in the south-west of Western Australia, the Pacific gull was called *yorringup*, literally meaning 'standing (in/at) water'.[181] The name *yi'lai bodhau'man* was identified as a 'crane' in the Kabi Kabi language spoken south of Bundaberg in south-east Queensland, and was said to literally mean 'cray-fish eater'.[182] Similarly, in the Ngarrindjeri language of the Lower Murray, an unidentified species of 'crane' was called *moik-amaldi*, from *moiki* ('crab') and *-amaldi* ('fellow'), meaning a bird that eats lots of crabs.[183] In this region, the name of the blue-billed duck, *pulki-nyeri*, referred to *pulki* ('well' or 'trench') and *-inyeri* ('belonging to').[184] The *wate-eri-on*, grey currawong, was said by Lower Murray people to have a name that meant 'one who follows in another's tracks', as this species was believed to warn other birds of the approach of people in the distance.[185] It is possible that *waranggaiperi*, which is an alternative Lower Murray name for the *pungkari*, hardhead duck, has the etymological root of *wurangi* ('mad') and *-raiperi* ('pertaining to').[186] A recorded Lower Murray name for the crested pigeon is *kurei wuni*, which may be a reference to the noise the wings make in flight.[187] In the case of the Australian raven, *marangani* was said here to refer to the 'walk of birds'.[188] Similarly, in the Guugu-Yimithirr language spoken at Hopevale in northern Queensland there is *bilu-warra*, meaning 'hip-crooked', for the Australian pelican, based on its walk.[189]

In the Diyari language of eastern Central Australia, the spotted crake was called *tampatampana*, with *tampana* referring to the jerky progressive movements typical of rails.[190] On Eyre Peninsula in South Australia, the Barngarla name for the white-faced heron is *wadna*,[191] possibly in reference to a throwing stick and the stance of the bird when looking for its prey.[192] The flocking behaviour of the budgerigar appears to be behind its association by Aboriginal people in the northern Flinders Ranges with the act of being very nervous or excited about something. It has been noted that 'Whilst in English we speak of having 'Butterflies' in the stomach the Adnamatana [Adnyamathanha] speak of 'Budgerigars' [*uliri*]'.[193]

Among the Wardaman people who live south-west of Katherine in the Northern Territory, the grey-crowned babbler is sometimes called the 'happy family' bird in Aboriginal English, due to its 'gregarious and chatty habits'.[194] The same description is used for babblers in the Walmajarri language spoken in the Great Sandy Desert.[195] In the Tiwi language spoken on Bathurst and Melville Islands in the Northern Territory, all quails are known as *puwarrirr*, which is apparently based on the sound that these energetic birds make when they are startled and fly quickly away.[196] In the Wunambal Gaambera language spoken in the north central

Kimberley, the blue-faced honeyeater is called *meme*, due to its habit of eating much fruit, *me*.[197] In Walmajarri, the name of the black-necked stork is *kunyarrlanujuwal*, which is said to refer to its ability to kill dogs,[198] and the great bowerbird or 'stealing bird' is *ngurriji*, which means 'robber' in reference to the bright shiny objects it takes for its bower.[199] In relation to the great bowerbird, the reference to their stealing behaviour is apt. Biologist/anthropologist Donald Thomson noted that on Cape York Peninsula in northern Queensland their bowers are often decorated with 'odds and ends gathered from the native camps'.[200] In the Wik Ngathan language spoken on western Cape York Peninsula, the white-browed crake is called *May–umpiyel-nhee'enh*, with *May–umpiyel* being the lotus lily and *nhee'enh* the verb 'to stand on', so literally the 'lotus-lily-stander'.[201]

Habitat

The type of country associated with a bird species was also referred to in their Indigenous descriptors. For example, in the Lower Murray *wetjungali*, which is the name for the superb fairy-wren in the Yaraldi dialect, referred to *wetji*, lignum bushes, which is a link explained by myth.[202] During the creation, the supreme male ancestor Ngurunderi landed at Nguldemalang on the south-eastern shore of Lake Alexandrina, where he came across people hiding in the reeds and 'whispering' among themselves.[203] When Ngurunderi spoke, they turned into superb fairy-wrens, who are now always seen among the lignum.[204] In another creation account from this region, the lignum bird people, known as *watji pulyeri* (from *watji*, 'lignum bush' and *pulyeri*, 'small bird'), were out gathering yams and putting them in their baskets when they were chased by other people – the fish hawk, sparrow hawk and night owl men who were cannibals.[205] To escape, they turned into the small birds and disappeared into the dense lignum thickets.

Some bird terms were used for groups of birds that were regarded as closely related, due to the habitat they prefer. In the Lower Murray, the Yaraldi word *tloperi* was used for the 'black-and-white ibis; also the white ibis and spoon-billed ibis. No distinguishing term used for these three kinds of ibis, because they fed in the same way'.[206] In spite of this, a search for bird terminology has found other names for these birds, except for the white ibis that was specifically recorded as *tloperi*.[207] It appears, therefore, that there was some flexibility with use of this broad term. Similarly, in the Antikirrinya language of the Western Desert there are non-mutually exclusive avian categories described as camp followers ('our relations'), messengers, hollow log nesters, ground nesters, scrub birds, waterfowl ('long-neck family') and little birds.[208] For instance, the 'long-neck family' waterfowl group, which includes the Australasian darter, white-faced heron and white-necked heron, birds are collectively called *ngurntiwarlarta*, from *ngurnti*, 'neck' and *warlarta*, 'long'.[209]

There are other examples of waterfowl species who occupy a similar niche and share an Aboriginal name. In the Jawoyn language spoken around Katherine in the Northern Territory, the yellow-billed spoonbill and royal spoonbill are both called *kem'mu*.[210] Another shared name for other waterfowl in this language is *pamparnja* for the great egret and similar birds such as the white-necked heron.[211] Similarly, in the Wardaman language spoken south-west of Katherine, the glossy ibis, sacred ibis and straw-necked ibis are collectively called *gurndirndin*.[212] The Mawng language, spoken in coastal western Arnhem Land, has the generic term *yampurlpurl* for spoonbills and ibis.[213] Also in Mawng, the word *kawagawa* refers to all the white egrets: the great egret, intermediate egret, little egret and cattle egret.[214] In Yolngu-matha, spoken in north-east Arnhem Land, *gananhdharr* refers to a variety of waterfowl who occupy a similar ecological niche. These include the white-necked heron, pied heron, cattle egret and great egret.[215]

Appearance

A few names, often used as alternatives, are highly descriptive of the bird's physical form. For instance, in the Yaraldi dialect of the Lower Murray it was claimed that the bird name *talkuri*, the Australasian bittern, was derived from *talgi* (thistle) and *kuri* (neck), because when it is not gathering food it holds its head up 'like a thistle'.[216] An alternative Lower Murray name for *nuri* the pelican was *rorika*, said to mean 'empty head' in reference to the bird's large bill which is used for holding fish.[217] The pied oystercatcher in the Narangga language spoken on the Yorke Peninsula in South Australia was *tiarti*, literally 'sharp',[218] presumably in reference to its pointed beak. In the same language, the name for the Australian bustard, *waltja*, was said to refer to a 'long neck' or 'throat'.[219] The appearance of birds can be related to people. Sutton observed that among the Wik peoples on Cape York Peninsula a man in a recognised joking relationship with another in a certain kinship category might say to him 'Oh brother in-law, you have a penis like the head of an ibis!'[220]

In the Diyari language of Lake Eyre, the name for the cockatiel was *koornyawillawilla*, and it is suggested that this is a reference to the rainbow, *kuranya*, due to the coloration of the bird's head.[221] Here, the welcome swallow was known as *mulyamulyayapunie*, which is thought to be drawing attention to the appearance of the bird's mud nest, because *mulya* referred to mud or mourning cap.[222] In the Wunambal Gaambera language spoken in the north central Kimberley, the name for the variegated fairy-wren is *marrirri*, which refers to the blue colouring on its head.[223]

Extending bird names

Aboriginal people have used bird-related names to describe themselves, as well as a range of other phenomena. Studying the ways in which people chose to name themselves and identify with their Country can be intellectual windows to investigating the wider aspects of each culture.

Birds as people names

In Aboriginal Australia, individuals receive their first names as part of a process that provides them a position within the society. Often a wide range of kin are consulted about appropriate names. Significant events at the time of perceived conception, along with the animals and objects associated with them, are taken into consideration for naming rituals.[224] Early colonist David Collins in the Sydney region remarked that children were given names at about four to six weeks of age, and that the name was 'taken from some of the objects constantly before their eyes, such as a bird, a beast, or a fish, and is given without any ceremony'.[225] This is supported by an account from colonial surveyor John Frederick Mann, who stated that in New South Wales 'permanent names are given to the youths' during the Bora ceremony, which is held at the conclusion of their 'minority' (younger years).[226] He recorded that here:

> The names of children often partake of an abbreviation of that of the parent, unless something occurs immediately at their birth of sufficient importance to record by adopting it as a name. The sight of a bird, insect, or animal supplies a name during minority.[227]

Anthropologist Isobel White stated that 'This gift is presented as if it were a valuable object since the name becomes part of the child's personality, not to be treated lightly'.[228] In some Aboriginal cultures, the name may not be given for a considerable time after birth. In the southern Western Desert, Ronald M Berndt remarked that:

> ... a child is never named, in this region, until it is able to walk, as if it were, one would run a grave risk of one's child dying. If one of these evil spirits [*mamu*] were to hear that a very young child had been named, it would visit it at night as it slept and cause sickness.[229]

In the Lower Murray region, the Berndts recorded that children did not receive a personal name until they could sit up. This was reportedly 'the time when a child received its personal name, which referred to the *ngatji* [*ngaitji*] that belonged to it through its patrilineal line'.[230] Later, 'When a child reached the next stage, on being weaned, that name would be dropped and another substituted, together with further reference to the *ngatji* [*ngaitji*]'.[231] The child, when able to talk, would be given yet another personal name, which was generally transmitted through the descent group from one generation to the next, as well as a public name that was generally used in camp.[232] A similar naming system was described for groups further upstream on the banks of the Murray River.[233] Life experience shaped the chosen names that people were called. Among the Ramindjeri people a man who hunted emus was referred to as *prekinyeri-orn*, from *preki* ('female emu'), *-nyeri* ('pertaining to') and *-orn* ('man').[234] An individual's name could change several times through life.

Indigenous bird vocabularies were a rich source of words for naming individuals at birth, as well as later when new names were given to those who had been initiated. For instance, explorer/artist George French Angas recorded in the Lower Murray during the 1840s that Aboriginal birth names were derived:

> ... either from the spot where they were born, from some trivial occurrence, or from a natural object seen by the mother soon after the birth of the child; for example ... *Rolcoorolca* (the noise of the emu) ... *Ungoontah-ungoontah* (stamping of the emu); *Peetpeerim* (the whistle of a bird).[235]

Indigenous terms for birds, as with other animals, were consciously changed due to Aboriginal avoidance customs, whereby the name of a recently deceased individual was not publicly spoken out of respect and through the fear of the person's spirit lingering.[236] South Australian colonist and writer James Dominick Woods explained:

> … the practice of never uttering the names of dead persons. Now as the names of the natives are mostly the names of some natural features of the places in which the bearers are born, or after some creature or some special occurrence, the death of a person necessitates a change in the name of these also, and the new names in time become fixed terms in the language instead of serving as a temporary expedient.[237]

There is a record of both the name avoidance and the alteration of a bird's name in operation. South Australian settler and overlander James C Hawker noted in his journal a conversation he had with King Tenbery of the 'Moorunde tribe' on the Murray River, who spoke Ngaiawung, about his bird shooting. He recorded that:

> In the course of our talk about shooting I happened to mention the word 'torpoul' a teal. Almost all the natives have sobriquets and one of Tenbery's sons was called the teal who died a short time since; amongst the natives, who are exceedingly superstitious, it is counted unlucky to name any person who had died, otherwise than as an individual, in my using the word 'torpoul' he immediately corrected me and told me to call it the 'touyoum' which is the new name of the bird: in this manner they are continually altering the names of plants, birds, animals etc. as almost every native is named after some of these.[238]

In Aboriginal Australia, birds were sometimes used for names, both formally and informally. Formal names were often tied to associations people had with particular parts of the Country. During my 1980s fieldwork in south-eastern South Australia, Lola Cameron-Bonney claimed that while she had a 'fly' as her *ngaitji* (clan ancestor), she had been ritually given the 'Aboriginal name' of 'Thrunkinyum' that meant 'place of the mountain duck [*tarankinyi*]'.[239] Informal names, which are typically descriptive of the person's form or a reference to an unusual life event, are widely used in Aboriginal communities across Australia.[240] Partly due to the frequent re-use of given names, many Ngarrindjeri people in the Lower Murray also had nicknames. There were people within the community who were known as Nuri (pelican), Pungk (short for *pungkari*, hardhead duck) and Big Eagle. Birds had a modern symbolic significance for the community as well. In the 1980s, Aboriginal people in the Lower Murray adopted *kungari* the black swan as the Ngarrindjeri emblem, for use on sporting uniforms for example. Often the design included a frame of spears and boomerangs.

Contemporary Aboriginal peoples have surnames derived from Indigenous names related to birds. Examples include the Karadada family of the Wunambal Gaambera community in the north central Kimberley, whose surname is a rendering of *gurrudoda*, a local term for the pied butcherbird and grey butcherbird.[241] In this way, the Karadada family claims to have a close personal association with what they see as one type of bird. In the same community, a family bears the name of *marrnga* the brahminy kite.[242] In the Wadeye area south-west of Darwin, local Aboriginal people remember a ceremonial leader known as Old Wagon who had the Indigenous name of Kianoo Thimarri, after *ku thimerri* the black-necked stork.[243] Here, an elderly man renowned for his ability to climb trees to collect eggs had gained the nickname of Tumtum (eggs).[244] Aboriginal people also have English names derived from their totemic birds. In the case of the Wik peoples on Cape York Peninsula, there was once a man called Cockatoo and another called Peret (his pronunciation of Parrot), while the surname Yunkaporta comes from *Yangk* and *poot*, meaning the 'lower leg tendon [of the brolga]'.[245]

Placenames

In Aboriginal Australia, birds are prominent elements in the Indigenous landscape iconography.[246] Placenames are sometimes based on bird associations. For instance, the name of the town of Curramulka on the Yorke Peninsula in South Australia is said to be derived from the Narangga placename Garimalka, meaning 'emu white', in reference to 'a limestone waterhole where emus come to drink'.[247] The placename, in the form of *Garrdimalga*, was used by palaeontologists to describe a new genus of fossil megapode birds, since Curramulka was where the type material was located.[248] Other 'emu' (*garil kari*) related placenames recorded here include Carribie Station Well, which was apparently derived from the placename Karibi, 'where emus drink', and Kulkari, 'place where emus made a noise'.[249] In most cases, the link to birds probably stems from an event involving the ancestor or a place where the particular bird species was frequently caught after the creation, or perhaps in many instances to both.

The iconography of the Lower Murray cultural landscape has particularly strong avian influences. Barn Hill, which is north of the Inman River on the Fleurieu Peninsula, was called Towarangk and was said to mean 'place of the *towari* [*teiwuri*, brown treecreeper]', with relevance to a Ramindjeri myth.[250] The name of Yaltung (Bald Hills in Encounter Bay) related to Yalti the great cormorant/black-faced cormorant ancestor, which has totemic connections to a local clan known as Yaltalindjera.[251] The Lower Murray topography contained different types of places that were classified according to their use by birds. For instance, a perching place for *karipanyi* the black-shouldered kite was called *karipa ngola*, with *ngola* referring to 'camp'.[252] Similarly, the swamps and springs where brolgas congregated were called *prolginali*.[253] Certain places were aesthetically linked with a bird species. For instance, along the Murray River, Naberuwa (Wood Point, north of Tailem Bend) reportedly represented the beak of the black duck.[254]

Places where birds were often hunted were named as such. At Tailem Bend on the Murray River, Peindjalang (Pintjalang, Penjulung) is a site on the cliff edge with a 'name based on Peindjali [*pinyali*] – emus went down cliff here to drink hence the name',[255] and here hunters would spear them in the late afternoon.[256] The Point McLeay lagoon was known as Tenaityeri, which was recorded as meaning 'The lagoon of gulls',[257] in an apparent reference to the Caspian tern (*tenatjeri*). To the north of the Lower Murray at the bend of the river between Murray Bridge and Mannum is a lagoon known locally as Neeta – *neetha* was said in the

Ngaiawung language to refer to the 'bittern' (probably the Australasian bittern) that was once common there.[258]

The important fishing and bird hunting site of Monteith Swamp near Murray Bridge was Piwingang (Pewingang), meaning 'place of the hawks [*piwi*, swamp harriers]'.[259] This place appears to have been economically important for Aboriginal hunter-gathering, as an archaeological investigation here in 1911 found 'very numerous and, sometimes, very large heaps of broken Unio [river mussel] shells and many blackened cooking-stones, indicating long occupancy by the natives'.[260] It is claimed that the name of the Tanggarangangal (Tangkeri, Tangkerangk) islands near the north-eastern shore of Lake Albert, which was where the Tangka ancestor and his companions froze to death, was derived from a combination of the words Tangka and *ranari* (*raneri*, duck snare).[261]

There are many other avian-based placenames across Australia. For instance, in the Katherine region of the Northern Territory, Manbulloo Station received its name from the Wardaman language term *manburlu*, for the pallid cuckoo.[262] Although it is not always recorded, many of the bird-related placenames are based on mythological connections. In the northern Flinders Ranges, the Adnyamathanha people call a particular large cave on the side of a hill Wakarla Adpaindanha, which means 'the painting of the crows' in reference to the wedge-tailed eagle ancestor having badly burnt the Australian ravens there during the creation, giving them a distinct blackness.[263] Another hill is known by them as Warturlipinha, from *warturli-vipi* meaning 'Australian ringneck's egg', in apparent reference to its shape.[264]

The placename of Pimpama, located between Brisbane and the Gold Coast in Queensland, is recorded as having been derived from the Yugambeh (Yugumbir) name *pim'pimba* or *bim'bimbam*, and said to translate as 'place of soldier birds [noisy miners]' from *bim'bim* for 'soldier bird' and *-ba* for 'of or belonging to'.[265] Analysis of the examples given above shows that avian-based placenames may often be directly related to the creation traditions, but in some cases they also encoded Aboriginal environmental knowledge. Among the Wik peoples of Cape York Peninsula, anthropologist John von Sturmer found that there was a category of placenames that could be translated, such as:

> Names which record events, e.g., where certain birds were observed (e.g., *wewe* – 'crow place', *madhan* – 'sea pelican place', etc.) ... Sometimes these actions are considered habitual, e.g., *(agu) minha ngakanga*, 'the place where the brolgas come for water' (in the dry season).[266]

Names for other things

Aboriginal peoples used bird names to describe other organisms and objects that at first glance would appear to be unrelated to avian species. For instance, in the Ngameni and Wangkangurru languages of the lower Diamantina River region, the serrated goodenia was called *kalla-toora-milkie*, due to similarity of the flower to the eye of a bustard (*kallatoora*).[267] In Yidiny, spoken in the rainforest region inland from Cairns in northern Queensland, *murrgu* means both 'the incubating mound of a scrub-hen or scrub-turkey [orange-footed scrubfowl or Australian brush-turkey]' and the 'deep hole for kapamari cooking', both of which involve digging in mounds.[268] An example from the Ngarrindjeri language of the Lower Murray region is the term *ngauwandi*, which refers to the stakes for an Aboriginal brush shelter and to the nest of a bird – the two meanings converge on a related structure involving similar materials.[269]

On the colonial frontier, the meaning and use of some Aboriginal words was extended by the speakers, due to a similarity in form. An example is the Lower Murray term for 'money', *nganhari*, which appears to have been derived from the word for 'egg shell', *nanhari*.[270] Similarly, in Yidiny spoken near Cairns in northern Queensland, *binirr* means both 'shell (of egg, turtle, mollusc, nut); money (in coins)'.[271] According to linguist John M Black, the word for emu, *pinyali*, in the Lower Murray was borrowed in the form of *pingyali* to mean 'horse'.[272] In central New South Wales, *maliyan*, for the wedge-tailed eagle, was on occasion used as a name for a 'policeman'.[273] Aboriginal terms have often been reapplied to exotic avians, based upon their similarity to indigenous species. For instance, in the Wardaman language of the Northern Territory, the term *girribug* for the pheasant coucal is also used for the introduced Indian peafowl.[274]

Europeans in Australia have also taken local Aboriginal words and applied them to different things. In the late 19th century, a schooner named *Punkari* was involved in trade across the lakes of the Lower Murray.[275] Australian English had appropriated the word *punkari* (*punkary*) from *pungkari* used for the hardhead duck in the Lower Murray languages.[276] Further upstream along the Murray River in the early 20th century, the river barge *Crowie* had a name derived from the Ngarrindjeri term *krauli*, the white-faced heron.[277] Such word borrowings were intended to impart a local flavour.

Early British colonists utilised a description of Aboriginal ceremonial performances, as 'corroborees', to described the behaviour of the Albert's lyrebird.[278] For instance, in 1860 George Bennett wrote that 'Each bird forms for itself three or four "Corroboring places", as the sawyers call them; they consist of holes scratched in the sandy ground, ~2 1/2 feet [76 cm] in diameter, by 16, 18, or 20 inches [41, 46 or 51 cm] in depth, and about three or four hundred yards [274 or 366 m] apart, or even more'.[279] The noisy miner is also said to exhibit 'corroboree' behaviour, which is defined as a ritualised group display, usually performed by several birds on adjacent branches.[280] With such human-like avian behaviours, it is little wonder that Aboriginal peoples saw birds as kinsmen.

As discussed in Chapter 2, palaeontologists have on occasion utilised Aboriginal terms for ancestral beings, as recorded in ethnographic and linguistic texts, to scientifically name newly discovered extinct species. Examples include the Ankotarinja man of the Arrernte people in Central Australia, who was used to describe a genus of tiny marsupial carnivores, in 'An allusion to *Ankotarinja*, a dreamtime ancestor'.[281] In the case of reptiles, the ancestral serpents known as the Wonambi of the Western Desert and the Yurlunggur of eastern Arnhem Land have been used to name vanished species of large snakes, such as *Wonambi naracoortensis*[282] and *Yurlunggur camfieldensis*.[283] The Kadimakara ancestors of the Diyari people living in the Cooper Creek area of eastern Central Australia inspired the naming of the *Kadimakara* genus of lizard-like reptiles,[284] as did the Quinkin spirits from eastern Cape York Peninsula in northern Queensland for the *Quinkana* genus of extinct crocodiles.[285] There was never any suggestion by any of the authors that Aboriginal peoples had co-existed with the reptilian species concerned.

A case for possible Aboriginal connections with extinct species has been made for *Genyornis*, which was a genus of large flightless ground birds present when the ancestors of Aboriginal peoples first arrived in Australia.[286] Ancient Aboriginal rock engravings and paintings potentially record *Genyornis*, either as footprints or in body shape.[287] As Dromornithids, palaeontologists termed them 'thunder birds' and 'mihirungs'. The reference by palaeontologists to mihirung was based on the giant bird-like spirit being of that name in the mythology of south-west Victoria.[288] This link with Aboriginal creation narratives was based upon Tindale's speculations concerning the identity of Aboriginal mythical beings as long-extinct species.[289] Given the composite nature of Aboriginal ancestor and spirit being descriptions, as discussed in Chapters 3 and 4, such a conclusion appears tenuous in my opinion. In 1979, palaeontologist Pat Vickers Rich noted that Tindale and his contemporaries had drawn a link between myth and fossil species and 'Based on that legend, the common name "Mihirung Birds" is here given to this group, referring to their gargantuan proportions'.[290] Rich was not convinced that there was a demonstrable link, as she and Peter F Murray later wrote that 'Whether the traditions are based on actual contact with living animals or are based on bones found preserved around local waterholes and swamps is difficult to substantiate'.[291]

Rather than relying on Greek and Latin as word sources, palaeontologists have on occasion actively used Aboriginal vocabularies to compose names that enhance the heritage value of the fossil species being described for the first time in the scientific literature. For instance, in 1979 Rich used the name *Barawertornis* for a new genus of Dromornithids, from fossils of the Late Oligocene to Early Miocene sub-epochs (~26.5–16.3 million years ago) found at Riversleigh in north-west Queensland.[292] The name is derived from '*Barawerti*, Aboriginal word for ground; *ornis*, G., bird',[293] although the source language is not given. Similarly, in 1979 Rich named the *Ilbandornis* genus of extinct birds from the Late Miocene or Early Pliocene sub-epochs (~11.6–3.6 million years ago) using '*Ibanda*, Aboriginal word for ground; *Ornis*, G. [Greek], bird';[294] the source Aboriginal language was again not given. In 1987, C Paterson and Rich named a Miocene (~23–5.3 million years ago) species *Dromaius gidju* (later renamed *Emuarius gidju*), with the species epithet derived from 'From an Aboriginal word meaning "small" … language not specified'.[295] In the case of *gidju*, it is elsewhere listed as a word for 'small' in a language spoken along the Warrego River in central New South Wales.[296]

A more recent Aboriginal naming example by palaeontologists is a bird species from the late Miocene (~11.6–5.3 million years ago) discovered in Central Australia, which Adam Yates and Trevor Worthy named *Dromaius arleyekweke*,[297] with the species name taken from *arleye* (emu) and *akweke* (small) from the Eastern and Central Arrernte languages of the region.[298] As the *Dromaius* species concerned is so old, there was no intention to suggest that Aboriginal peoples and the species concerned were ever co-occupants of the landscape.[299] With so many Indigenous Australian languages now having published dictionaries, the use of Aboriginal terminology in the naming of fossil species is likely to continue.

From the above examples, it can be concluded that the lack of correspondence between an Aboriginal word chosen for the scientific name and the region associated with the fossil species concerned was not an issue for past palaeontologists. If the intention was to use an appropriate Indigenous name from a language relevant to the area where the fossil species once roamed, then the former linguistic diversity of Australia poses a problem. For example, a senior member of the Aboriginal community in south-eastern South Australia correctly pointed out to me a few years ago that the Aboriginal part of the name for the Wonambi Fossil Centre at the Naracoorte Caves[300] was not actually an Indigenous term from his Country. In general, though, in my opinion such naming practices, when applied thoughtfully with respect to the choice, have the potential to rightfully acknowledge that the ancestors of Aboriginal peoples have existed on this continent for many millennia, during which they have withstood major changes in climate and the environment. When these ancestors first arrived some 50 000 years ago, they would have encountered a range of animal species on the continent that became extinct relatively quickly, possibly due to Aboriginal predation as argued by palaeontologist Tim Flannery.[301]

*****

The names that were used in the past for Australian birds came about in a variety of ways, in part due to Australia's relatively recent colonial history. In 1906, English ornithologist and sportsman Collingwood 'Cherry' Ingram wrote about his bird shooting trip to the Murray River area of South Australia. He remarked that 'In their vernacular the Australians have adopted a very loose nomenclature for the natural objects which surround them … from a vague outward resemblance to things that were once familiar to their forefathers in the "old country", as they still call England'.[302] Ingram went on to reason that, because of this, 'the Australian Bustard (*Eupodotis australis*) will for all time be wrongly known as the Wild Turkey, and it would be futile therefore to write of the bird by any other appellation'. While the attraction of the bustard as a gamebird for shooters undoubtedly influenced its description as a 'turkey',[303] this chapter has also given examples of bird species, such as the kookaburras and currawongs, that have Australian English names based on words appropriated from Aboriginal languages during the early phases of colonial contact when an Aboriginal creole was spoken on the frontier of British settlement.

The recorded vocabularies of Aboriginal languages are a potentially rich source of words for birds, if there is an attempt to introduce new terms into Australian English. It has long been recognised that the use of Indigenous terms for birds found only in Australia makes more sense than the over-extended use of existing common names for Northern Hemisphere species, to which in some cases they bear only a partial resemblance.[304] Bates observed that many Indigenous bird names are 'infinitely more agreeable to ear and eye' than the Australian English terms.[305] Zoologist John MacPherson remarked that 'It seems to me a pity not to preserve as far as possible the native names for the various animals and plants, and, where practicable, popularise them as preferable to the clumsily manufactured ones accepted in authoritative works'.[306] Condon observed that 'Many of the native words are pleasant-sounding and could be used to replace the cumbersome English vernacular names in the Official Checklist of Australian Birds, published by the Royal Australasian Ornithologists' Union'.[307]

A recognised problem with the future active borrowing of Aboriginal terms for use as Australian English names for plants and animals is the large number of recorded languages from which to choose.[308] This means that in some cases the deliberate adoption of an Indigenous bird name will result in a word being used outside the area where its parent language was originally spoken. Furthermore, the adoption of local terminology for species that range across other parts of the world, where they already have long-established names, may increase confusion among international birdwatchers.[309] However, colonists' adoption of an existing bird term from their homeland for newly encountered bird species, based solely on superficial similarities, has also led to misunderstandings. Ornithologist Mark Bonta warned that where bird names from one area have been misapplied to different species in another, the 'transmutation of knowledge produced a riotous jumble of terminology'.[310]

This chapter has shown the varying ways in which Aboriginal environmental knowledge concerning birds is ordered and maintained by speakers of local Aboriginal languages. The decline of Australia's Indigenous languages represents a serious threat to the maintenance of this knowledge, and this adds to the ongoing need for language maintenance and revitalisation.[311] The differences between Indigenous and scientific Linnean systems of nomenclature have sometimes resulted in the incorrect matching of local and scientific bird names in scholarly publications.[312] In Australia, the value of ethno-ornithological research into the proper identification of bird names mentioned in early historical records is that it leads to additional sources of data for zoologists when reconstructing the former ranges of bird species.

The high diversity of languages spoken across the world favours the recognition of multiple names for what scientists would regard as the same species. A zoologist observed that 'In Europe, for example, each nation has its own vernacular name for each species of bird, and no one common name is regarded as more correct or fundamental than another'.[313] The lay and scientific uses of English-based names have little or no appeal in non-English speaking countries. In Japan, a locally based nomenclatorial system continues to develop and is in common use by a broad range of scholars.[314] It has been argued that this proliferation of common names does not pose a problem, as long as the names are consistently used and can be cross-referenced.[315]

Endnotes
[1] Bulmer (1967), Hays (1982), Hunn (1982) and Macindoe (2012).

[2] Hunn (2010:xi).

[3] For example, Bates (1928) collected bird names for the Western Desert, Stone (1911:447–448) gathered names from Victoria, and Gason (cited in Smyth 1878:Vol.1:224–226) provided a list of bird names from the Cooper Creek area in north-east South Australia.

[4] Condon (1955a:74).

[5] Abbott (2009:223).

[6] Clarke (2019a:125–126).

[7] Webb (1933).

[8] Johnston (1943:270–287).

[9] Condon (1955a, 1955b).

[10] Mansergh & Hercus (1981:111).

[11] Wesson (2001:7).

[12] Black *et al.* (2018).

[13] Si (2019).

[14] Peter (2006).

[15] Sorenson (1920), Endacott (1944) and Cooper (1962).

[16] Healey (2007).

[17] Abbott (2009).

[18] For example, Hercus (1966) and Mansergh & Hercus (1981) for Victoria, Sutton (1978b) for Wik Ngathan in northern Cape York Peninsula, Turpin (2013) for Kaytetye in northern Central Australia and Zorc (1996) for Yolngu-matha in north-east Arnhem Land.

[19] For example, see Goddard (1992), Henderson & Dobson (1994), Glass & Hackett (2003), Green (2010), Turpin & Ross (2012), Green *et al.* (2019) and Verstraete (2020).

[20] Clarke (1994: Appendix 2, 2003b: Table 1).

[21] Clarke (2019a:135).

[22] Clarke (2019a:137).

[23] Gale (2009:155) and Clarke (2019a:124,141). The 'blue bald coot' is the purple swamphen.

[24] Mühlhäusler & Fill (2001) and Mühlhäusler (2003).

[25] Brown (1986), Lakoff (1987), Berlin (1992), Foley (1997) and Duranti (2001).

[26] Berlin (1992).

[27] Bulmer (1967) and Dwyer (2005).

[28] Heath (1978), Waddy (1979, 1982, 1988), Walsh (1993) and McKnight (1999).

[29] Blake (1981), Yallop (1982), Schmidt (1993), Thieberger & McGregor (1994) and Henderson & Nash (2002).

[30] Stephens (1889:498).

[31] Gale (2009:119).

[32] Tindale (1924:Appendix). Note that today the Wailpi are generally treated as part of the Adnyamathanha group.

[33] Doonday *et al.* (2013:147).

[34] Taylor (1996:161).

[35] For example, *dhigaraa* as a 'general name for any bird' in the Gamilaraay-related languages of central New South Wales (Ash *et al.* 2003:60). Similarly, *dyirda* in Wirangu spoken in western South Australia was recorded as just 'bird' (Miller *et al.* 2010:25).

[36] Bulmer (1978:4), McKnight (1999:154–155) and Dwyer (2005:120).

[37] Webb (1933:18).

[38] Clarke (2019a:125–126).

[39] Sullivan (1928:165).

[40] Zorc (1996:97).

[41] Kilham *et al.* (1986:53,80,114–115,168).

[42] Kilham *et al.* (1986:168–169). Details from this difficult-to-obtain source are available online (http://ausil.org/Dictionary/Wik-Mungkan/lexicon/mainintro.htm).

[43] Cameron (1893).

[44] Maddock (1975:102–105), Green (2010:521), Brown & Naessan (2014:15–16) and Naessan (2017:348–349).

[45] Mjöberg (1915:112).

[46] Hassell (1934a:248).

[47] Strehlow (1909:202).

[48] Basedow (1925:251).

[49] P Sutton (pers. comm.).

[50] P Sutton (pers. comm.).

[51] Silberbauer (1981:69).

[52] Bulmer (1967).

[53] Brown & Naessan (2014:17–22).

[54] Turpin & Ross (2012:589).

[55] Evans (1992:278–279).

[56] Bradley *et al.* (2006:81,Table 8).

[57] Waddy (1982:73, 1988:73–76,98,101).

[58] Watson (1946:11).

[59] Henderson & Dobson (1994:549).

[60] Dixon *et al.* (1992:109) and Ramson (1988:109).

[61] Smyth (1878:Vol. 1:435–444), Worsnop (1897:167,Plate 83), Norman (1951), Mulvaney (1994:36), Clarke (1999a:157–159, 2018c:37–42) and Holden & Holden (2001:175–178).

[62] Fabian (1931), Fleay (1940), Anonymous (1944), Meadow Argus (1947) and Fields (1951).

[63] Fenner (1946:107).

[64] SENEX (1923).

[65] Anonymous (1914).

[66] Dawson (1881:92–93).

[67] Mathews (1904:365).

[68] SENEX (1927). The Waugal (Wawgal, Woggal) from the south-west of Western Australia is normally described as a giant snake (Bates 1901–14:219–221, 1925).

[69] Forth (2010b).

[70] Forth (2010b:140).

[71] Tidemann & Whiteside (2010:156).

[72] Condon (1995b:95).

[73] Brown & Naessan (2014:26,46–47).

[74] Von Brandenstein (1977:171–172).

[75] Clarke (2019a:133,135,141).

[76] Horton *et al.* (2013:1).

[77] Blakers *et al.* (1984:141).

[78] P Sutton (pers. comm.). See also Kilham *et al.* (1986:225–226,273).

[79] Refer to Bradley *et al.* (2006:89) for Yanyuwa, and Singer *et al.* (2021:160) for Mawng.

[80] Clarke (2019a:122,133).

[81] Brown & Naessan (2014:99).

[82] McEntee *et al.* (1986:4,21,24).

[83] Green (2010:348).

[84] Dixon (1991:158).

[85] Clarke (2019a:132–133,136).

[86] Ichikawa (1998:108).

[87] K Akerman (pers. comm.).

[88] McEntee *et al.* (1986:19,22).

[89] Black *et al.* (2018:46).

[90] Puruntatameri *et al.* (2001:97–98).

[91] Puruntatameri *et al.* (2001:99).

[92] Puruntatameri *et al.* (2001:101–102).

[93] Puruntatameri *et al.* (2001:100).

[94] Wiynjorrotj *et al.* (2005:142). Similarly, in Mawng, which is spoken in the neighbouring western Arnhem Land region, *karrkany* is a generic term for the black kite and possibly also the red goshawk (Singer *et al.* 2021:73).

[95] Wiynjorrotj *et al.* (2005:127,131).

[96] Raymond *et al.* (1999:106,108–109,112).

[97] Raymond *et al.* (1999:109–110).

[98] Peile (1980:59) and Clarke (2012:100).

[99] McEntee *et al.* (1986:18,22–23).

[100] BJ Nangan (cited in Akerman 2020:140–141).

[101] BJ Nangan (cited in Akerman 2020:152–153).

[102] Doonday *et al.* (2013:143).

[103] Sutton (1978b:17).

[104] Brown & Naessan (2014:57).

[105] Wiynjorrotj *et al.* (2005:127–128).

[106] Wiynjorrotj *et al.* (2005:128).

[107] Wiynjorrotj *et al.* (2005:137).

[108] Wiynjorrotj *et al.* (2005:129).

[109] McEntee *et al.* (1986:18).

[110] McEntee *et al.* (1986:18).

[111] Brown & Naessan (2014:43,93).

[112] Doonday *et al.* (2013:145).

[113] Doonday *et al.* (2013:146).

[114] Peckover & Filewood (1976:10).

[115] Tindale (c.1931–91b).

[116] Meyer (1843:56). The term *preki* was written as *breke* (Clarke 2019a:138).

[117] Meyer (1843:65). The term *yarli* was written as *yarle* (Clarke 2019a:143).

[118] Singer *et al.* (2021:146).

[119] Heath (1978:53).

[120] Heath (1978:54).

[121] Van Egmond (2012:420).

[122] Heath (1978:53).

[123] Fletcher (2007:84).

[124] Condon (1955a:75).

[125] Berlin & O'Neill (1981), Hunn & Thornton (2010) and Ibarra *et al.* (2020).

[126] Forth (2010a:232–234).

[127] Dawson (1881:lvii), Bates (1928), Worms (1938) and Tidemann & Whiteside (2010:158–159).

[128] Sherrie (1918).

[129] P Sutton (pers. comm.).

[130] Dawson (1881:lvii), Bates (1928), Worms (1938), Johnston (1943:270), Giacon (2010:251,257–258), Brown & Naessan (2014:25), Naessan (2017:353–355) and Clarke (2019a:122).

[131] Gale (2009:26).

[132] Berndt *et al.* (1993:561).

[133] Gale (2009:159).

[134] Berndt *et al.* (1993:239).

[135] Berndt *et al.* (1993:239) and Gale (2009:116). A spelling variation of *prolgi-mayi* is *prolgimaii*.

[136] Dixon *et al.* (1992:31,87–88,218) and Ramson (1988:95).

[137] Johnston (1943:276–277).

[138] Anonymous (1913b).

[139] Ramson (1966:122, 1988:95).

[140] Hardwick (2019b:62).

[141] Wiynjorrotj *et al.* (2005:145).

[142] Meston (1921).

[143] Teulon (1886:213).

[144] Massola (1968:18).

[145] Chapman *et al.* (1995:359).

[146] Tindale (1964).

[147] Johnston (1943:285).

[148] Johnston (1943:286), in reference to Schürmann (1844:2:45).

[149] Bennett (1860:181).

[150] Abbott (2009:249). The call has been described as a barking 'ke-yak' (http://www.oiseaux-birds.com/card-black-winged-stilt.html).

[151] Von Brandenstein (1988:112).

[152] Doonday et al. (2013:139,143,145,149).

[153] Karadada et al. (2011:77,86,88).

[154] Gordon (1979:21–23).

[155] Wiynjorrotj et al. (2005:124).

[156] Wiynjorrotj et al. (2005:124).

[157] Wiynjorrotj et al. (2005:131).

[158] Wiynjorrotj et al. (2005:132,134,137,140–141).

[159] Wiynjorrotj et al. (2005:135–137).

[160] Smith (1880:131) and Tindale (1931–34:168).

[161] Stone (1911:448).

[162] McEntee et al. (1986:22).

[163] Ash et al. (2003:134–135,171).

[164] Karadada et al. (2011:76).

[165] Raymond et al. (1999:101) and Wiynjorrotj et al. (2005:127).

[166] Webb (1933:21).

[167] Doonday et al. (2013:149).

[168] K Akerman (pers. comm.).

[169] Glass & Hackett (2003:22).

[170] Ramson (1966:89,121–123).

[171] Johnston (1943:283), Dixon et al. (1992:90–91) and Black et al. (2018:55).

[172] Dixon et al. (1992:89–90).

[173] MacPherson (1931:370), Troy (1994:55) and Dixon et al. (1992:20,27,87). Historically, the bug-bug (bubuk) name has also been associated with the Australian owlet nightjar, probably by mistake (Troy 1994:54).

[174] MacPherson (1931:371), Dixon et al. (1992:27,91–92) and Troy (1994:55).

[175] Dixon et al. (1992:90).

[176] Schürmann (1844:Pt2:71), Bates (1928), Tindale (1936c:68), McEntee et al. (1986:23) and Dixon et al. (1992:94).

[177] Singer et al. (2021:137).

[178] Ramson (1966:88–89,122, 1988:401–402), Turner (1966:61–62) and Dixon et al. (1992:87). According to Slater (1970:392,395), the southern boobook call is 'mo-poke' or 'more-pork', while the tawny frogmouth call is 'oo-oo-oo' and repeated many times with breaks.

[179] Troy (1994:12–18) and Clarke (2008a:49).

[180] Ramson (1966:102). See also Turner (1966:199–200).

[181] Von Brandenstein (1988:141). The term yorringup was also written as yuagang-qaip.

[182] Watson (1946:12).

[183] Gale (2009:55). The term moik-amaldi was also written as moikemalde.

[184] Gale (2009:118–119).

[185] Berndt et al. (1993:124,234). The wate-eri-on was also known in the Lower Murray as killing-kildi ('black magpie').

[186] Gale (2009:163).

[187] Condon (1955a:77).

[188] Condon (1955b:95). The term marangani was also written as marangane.

[189] Roth (1901b:5). This author wrote the name as *belu-warra*, which P Sutton (pers. comm.) rewrote in the form given here.

[190] Johnston (1943:273).

[191] Schürmann (1846:64).

[192] Johnston (1943:277).

[193] McEntee *et al.* (1986:19).

[194] Raymond *et al.* (1999:99).

[195] Doonday *et al.* (2013:147).

[196] Puruntatameri *et al.* (2002:99).

[197] Karadada *et al.* (2011:87).

[198] Doonday *et al.* (2013:133).

[199] Doonday *et al.* (2013:148).

[200] Thomson (1935:76).

[201] Sutton (1978b:10).

[202] Berndt (1940a:173). See also Gale (2009:164).

[203] Berndt *et al.* (1993:224,227,243). Nguldemalang is also written as Ngulunmalang.

[204] Mountford & Roberts (1969:20–21).

[205] Berndt *et al.* (1993:237). The term *watji pulyeri* is also written as *watji-puldjeri*.

[206] Berndt *et al.* (1993:559). The term *tloperi* was also recorded as *tloppere, tolopori, troperi*.

[207] Clarke (2019a:141).

[208] Brown & Naessan (2014:2,25,30–44) and Naessan (2017:355–365).

[209] Brown & Naessan (2014:38). The *ngurntiwarlarta* group includes the black swan, Australian pelican, Australasian grebe, great egret, hoary-headed grebe and Australasian pied cormorant.

[210] Wiynjorrotj *et al.* (2005:149).

[211] Wiynjorrotj *et al.* (2005:146–147).

[212] Raymond *et al.* (1999:103).

[213] Singer *et al.* (2021:231).

[214] Singer *et al.* (2021:75).

[215] Zorc (1999:118).

[216] Berndt *et al.* (1993:556).

[217] Berndt *et al.* (1993:215).

[218] Tindale (1936c:66).

[219] Tindale (1936c:67).

[220] P Sutton (pers. comm.).

[221] Johnston (1943:283).

[222] Gason (1879:286) and Johnston (1943:285).

[223] Karadada *et al.* (2011:86).

[224] Spencer & Gillen (1904:Ch.20), Thomson (1946), McConnel (1957:139–141), Elkin (1964:154–155,202), Dousset (1997) and Simpson (1998).

[225] Collins (1798–1802:465).

[226] Mann (1883:640).

[227] Mann (1883:640).

[228] White (1983:5).

[229] Berndt (1940b:291).

[230] Berndt *et al.* (1993:146).

[231] Berndt *et al.* (1993:146).

[232] Berndt *et al.* (1993:147–148).

[233] Eyre (1845:2:324–330).

[234] Gale (2009:114).

[235] Angas (1847b:92).

[236] Bonwick (1870:145–146), Fison & Howitt (1880:249–250), Roth (1903:20), Spencer & Gillen (1927:2:432–433) and Thomson (1946:163).

[237] Woods (1879:4).

[238] Hawker (1841–45:12 May 1844).

[239] For the identity of *tarankinyi*, refer to Clarke (2019a:140).

[240] Love (1936:217), Stanner (1937) and Sutton (1982:9,187,192–194).

[241] Karadada *et al.* (2011:76).

[242] Karadada *et al.* (2011:89).

[243] Hardwick (2019b:58).

[244] Hardwick (2019b:59).

[245] P Sutton (pers. comm.).

[246] For examples of avian placenames, refer to Hercus & Potezny (1999) for northern South Australia, Mattingley (1905) for the Sydney area and Meston (1921) for the eastern states.

[247] Tindale (1936c:69).

[248] Shute *et al.* (2017:43).

[249] Tindale (1936c:69).

[250] Tindale (c.1931–91a).

[251] Berndt *et al.* (1993:292,311). *Yalti* is sometimes written as *yoldi* (Clarke 2019a:124–125,132,143).

[252] Gale (2009:82).

[253] Berndt *et al.* (1993:239).

[254] Berndt *et al.* (1993:413).

[255] Tindale (1987b, see also c.1931–91a).

[256] Berndt (1940a:168) and Berndt *et al.* (1993:14).

[257] Taplin (1874:130).

[258] Bellchambers (1931:104).

[259] Berndt (1940a:168) and Berndt *et al.* (1993:14,22,223).

[260] Stirling (1911:14).

[261] Tindale (1987b).

[262] Raymond *et al.* (1999:101,105).

[263] Tunbridge (1988:25,28–29).

[264] Tunbridge (1988:124).

[265] Watson (1946:104,109).

[266] Von Sturmer (1978:254).

[267] Johnston & Cleland (1943:156).

[268] Dixon (1991:158).

[269] Gale (2009:78).

[270] Gale (2009:68–69).

[271] Dixon (1991:157).

[272] Black (1917:11).

[273] Ash *et al.* (2003:106).

[274] Raymond *et al.* (1999:101).

[275] Padman (1987:21).

[276] Condon (1955a:83), Ramson (1988:508) and Dixon *et al.* (1992:93–94).

[277] Roberts *et al.* (2017:6). These authors incorrectly identified 'Crowie' as the brolga, which Clarke (2019a:134) has reidentified as the white-faced heron.

[278] Ramson (1988:173). The Australian English word 'corroboree' was derived from the Dharuk word, *garaabara*, in the Sydney region (Ramson 1988:172).

[279] Bennett (1860:184).

[280] Dow (1975).

[281] Archer (1976:54). For a description of Ankotarinja, refer to Strehlow (1933).

282 Smith (1976:41–44) first described the *Wonambi* genus, and Palci *et al.* (2018) provided a palaeontological account of these snake species. For descriptions of Wonambi refer to Berndt & Berndt (1989:118–125) and Clarke (2014a:76–77).

283 Scanlon (1992, pers. comm.) described the *Yurlunggur* genus, and Palci *et al.* (2018) provided a palaeontological account of these snake species. For descriptions of Yurlunggur refer to Mountford (1978) and Warner (1937:32,250,25–256,259–261,264–267,273–275,280,408).

284 Bartholomai (2008). For description of the Kadimakara refer to Gregory (1906:Ch.1, see also 116,353).

285 Molnar (1981) described the *Quinkana* genus. For descriptions of the Quinkan spirits and the associated rock paintings refer to Morwood (2002:47,59,126,130–131,256–275) and Trezise (1969).

286 For descriptions of *Genyornis*, see Rich (1985), Mayor & Sarjeant (2001:159–161), Murray & Vickers-Rich (2004) and Low (2014:142–143).

287 Hall *et al.* (1951), Tindale (1951) and Gunn *et al.* (2020).

288 Dawson (1881:92–93).

289 Tindale (1951:381). See also Rich (1979:16).

290 Rich (1979:1).

291 Murray & Vickers-Rich (2004:8–9).

292 Rich (1979:23–26).

293 Rich (1979:23).

294 Rich (1979:36).

295 Patterson & Rich (1987:97). Taxonomy revised by Bowles (1992).

296 Mathew (1899:233).

297 Yates & Worthy (2019).

298 Henderson & Dobson (1994:76,208).

299 T Worthy (pers. comm.).

300 See the Naracoorte Caves website (https://www.naracoortecaves.sa.gov.au/discover/above-the-ground/wonambi-fossil-centre).

301 Flannery (1994).

302 Ingram (1906:334).

303 Ramson (1966:88–89, 1988:693).

304 Anonymous (1917) and Sherrie (1918).

305 Bates (1928).

306 MacPherson (1931:371).

307 Condon (1955a:75).

308 Sherrie (1918), Clarke (2008a:Ch.3) and Abbott (2013:111).

309 Schodde *et al.* (1978).

310 Bonta (2010b:95).

311 Schmidt (1993), Tsunoda (2005), McConvell & Thieberger (2006), Monaghan & Mühlhäusler (2015) and Amery (2016).

312 Ng'weno (2010:104) and Clarke (2019a:125–126).

313 Abbott (2009:225).

314 Callomon (2016:8).

315 Doran (1903:38), Eisenmann & Poor (1946) and Callomon (2016:6).

6

Early hunting and gathering

European observers formerly considered that the world's hunter-gatherer societies possessed passive relationships with the environment, meaning that they responded in an *ad hoc* manner to whatever natural resources were found, aimlessly wandering about the land waiting to find what each season would deliver by way of sustenance. As the result of more detailed studies of Indigenous lifestyles since the mid 20th century,[1] it is known that hunter-gatherers took a more active role in managing their environment, through their food selection, use of fire and hunting pressure. Prior to European settlement, Aboriginal hunter-gatherers were able to restrict their impact upon the landscape by dispersing thinly and by constantly moving according to their own calendars.[2] For them, birds as highly seasonal animals were excellent sources of food and artefact-making material.

What gamebird is that?

In Aboriginal Australia, there are two main categories of recognised food: animal and vegetable.[3] By way of example, in the Lower Murray of South Australia all animal food was generally classed as *mami*, while vegetable food was *nguni*.[4] Here, food was also classified according to foraging zones, with coastal bird species considered to be part of *ngamataro*, 'shore food' – a category that included birds, fish and shellfish that are all found along the beach.[5] In the Western Desert language, the two main food categories are *kuka* for meat and *mai* for vegetable food.[6] A similar separation exists in many other Aboriginal languages, such as in Gugu Gulunggur spoken in Bloomfield River in northern Queensland, where meat is *minya* and vegetable food is *maiyi*.[7] In the Wik Ngathan language of western Cape York Peninsula, all animal foods, including eggs, are termed *minh*, while vegetable food is *kuthel*.[8]

Contrary to recent non-anthropological opinions, as expressed by popular writer Bruce Pascoe,[9] Aboriginal peoples at the time of European settlement were not 'farmers' in the Western sense of the term, but highly skilful nomadic foragers who possessed a wealth of knowledge about the environments with which they interacted.[10] In Aboriginal society, men and women had different but complementary roles in foraging for food.[11] The division of labour was such that men fished and hunted more mobile animals, such as birds and large marsupials, while women generally focused on collecting more static resources, such as plant foods, shellfish, eggs, ground insects, reptiles and small burrowing mammals. Nevertheless, members of either sex and of any age would collect most foods, including birds, if presented with an opportunity.

The number of food species used by Aboriginal peoples across Australia is immense in comparison to the foods typically used by agricultural societies, such as the sources brought out by the first European settlers.[12] Food sources were seasonal: the proportion of animal food in the overall diet varied throughout the year from being a major source to a minor one. Not all potential foods were fully utilised during more benign times.[13] Most animal foods were generally highly desirable and therefore when available were selected over more reliable plant food sources, which in many cases were either less palatable or more time-consuming to prepare. Not all bird species were hunted under normal circumstances because, as discussed in Chapter 8, a few of them were regarded as unpalatable. Others were avoided due to the spirit being associations described in Chapter 4.

Birds would have been a major food source for most groups across Aboriginal Australia. In riverine regions, such as in south-eastern Australia where there are extensive wetlands with large waterfowl populations, birds were recorded as a significant category of meat for subsistence. In some seasons they may have even outranked fish and mammal sources.[14] For drier inland regions, emus were often described as being a highly favoured source.[15] In general, Australian hunter-gatherers would have consumed the flesh and eggs of most bird species if the chance arose,[16] as did European settlers on the frontier.[17] Eggs, in some regions of uncertain climate, were a highly opportunistic food source. Doctor and anthropologist Herbert Basedow explained:

> The eggs and fledglings of all birds yield abundant food supplies during favourable seasons. In central Australia such seasons are dependent entirely upon the rains. Birds breed usually after the setting in of rain, which might be once or twice a year, but in the driest regions, like the Victoria Desert [in Western Australia], perhaps only once every few years. There is no doubt that emu, black swan, and native goose are amongst the biggest suppliers of eggs. Of the two last-named birds, in particular, enormous harvests of eggs are occasionally wrested during exceptional seasons. At these times the tribes who have been so bounteously favoured carry on a regular trade with neighbouring tribes, who have perhaps not had the same opportunity or good fortune.[18]

Generally, birds are under-represented in the archaeological record, which is biased towards stony materials and the larger bones from humans and big animals. Avian materials comprising feathers, skins and small bones are unlikely to persist long in the ground. Aboriginal middens do sometimes contain the archaeological remains of birds, particularly emu egg shell fragments and bones of large species that are dense enough to persist.[19] For instance, small pieces of eggshell from the now-extinct Kangaroo Island emu were found on the surface of a former Aboriginal campsite on Kangaroo Island in South Australia.[20] At an archaeological excavation at Swanport along the Murray River in South Australia, Aboriginal middens were found to contain a wide range of animal remains, including large bones from the Australian bustard and Australian pelican, along with human burial material.[21]

'Muttonbird' (short-tailed shearwater) bones have been found in several archaeological deposits in Tasmania[22] and in the Sydney area of New South Wales.[23] This is testimony to the reliance of some coastal Aboriginal groups upon this species as a food source. Among the bird remains excavated from Aboriginal sites in the Sydney area are those of the Australian brush-turkey, Pacific baza, short-tailed shearwater and little penguin.[24] Palaeontologists have suggested that recovered fragments of egg shell, identified as belonging to the extinct *Genyornis* bird, had been burnt in Aboriginal cooking fires.[25] *Genyornis* was a large flightless Dromornithid, which was an extinct family endemic to Australia.[26] It is also possible that some of these egg shells were from contemporaneous extinct Megapodiid birds, such as *Latagallina, Progura* and *Garrdimalga*, which were biologically related to the Australian brush-turkey, orange-footed scrubfowl and malleefowl.[27] All of these extinct species therefore appear to have co-existed with the ancestors of Australian Aboriginal peoples. When describing Dromornithid material found at Penola in the south-east of South Australia, naturalist Rev Julian Edmund Tenison-Woods claimed that 'In 1866 I found the remains of a Struthious bird, much larger than the Emu, in one of the kitchen-middens of the natives in South Australia. The bones were marked by the scrapings and cutting of flint knives of the blacks'.[28] Later palaeontologists were unable to find Tenison-Woods' specimens, so his assertion remains unverified by scientists with better analytical tools and techniques.[29]

In my own field experience, when working along the eastern escarpment of the Mount Lofty Ranges in South Australia, small emu shell fragments that are weathered white are occasionally found in the soil of the floors within rock shelters that, from the smoke-darkened roofs, are likely to have been Aboriginal camps. Fragments of emu egg shells are regularly found in archaeological excavations.[30] It is worth noting that birds can also create accumulations of faunal remains, composed of shell, bone and crustacean exoskeletons, that could be mistaken for Aboriginal middens.[31] In southern Australia, the Pacific gull is one such coastal species that produces middens of shellfish remains.[32] Similarly, owls and small carnivorous mammals leave bone deposits from their kills in caves where Aboriginal peoples seasonally sheltered.[33] The old mounds made by the orange-footed scrubfowl in eastern Australia can take on the appearance of a midden.[34]

Rock engravings and paintings that feature birds, identified by their prints and occasionally by the outline of their bodies, are another source of data concerning Aboriginal perceptions of what ornithologists classify as birds. For instance, footprints of the emu and Australian bustard are common motifs in the ancient Panaramittee tradition of rock engravings in temperate Australia, dated from recent times to over 10 000 years in age.[35] While it will never be known exactly why the artists chose to depict birds in their work, it is possible to draw some conclusions. In western Arnhem Land in the Northern Territory, there are paintings of birds such as the magpie goose, black-necked stork, purple swamphen and little egret, within rock shelters. The paintings are linked to the period from 1500 years ago to the present, when the freshwater swamps in the region built up, replacing the existing saltwater environment.[36] The detail in some of the artworks, such as a broken neck presumably caused through hunting, and the depiction of internal organs and fat reserves, make it clear that these birds were favoured game.

It is probable that the ancestors of contemporary Aboriginal peoples hunted the large Dromornithid, *Genyornis newtoni* or 'thunder bird', that widely occurred in Australian forests until going extinct some thousands of years ago, possibly due to human predation.[37] The Panaramittee rock engraving tradition described above has depictions of what Norman B Tindale believed to be the footprints of this species,[38] along with what his South Australian Museum colleagues Charles P Mountford and Robert Edwards thought were possibly those of the diprotodon and a saltwater crocodile;[39] however, more recently a rock art expert has disputed this.[40] In 1925, Basedow observed that:

> ... there are some exceptionally large bird tracks carved into the rocks at Balparana, in the Flinders Ranges, which seem too big to be intended for those of an emu; the question might reasonably be asked whether they could not have been made by a primitive hunter at a time when the now extinct 'moa' or Genyornis still lived in Australia.[41]

It is tempting to postulate a connection between possible engravings of extinct megafauna and the mythic recordings of large bird-like ancestral entities in the creation. Writer Jennifer Isaacs claimed that 'References to giant animals who lived in the Dreamtime are found abundantly in myths from many tribes. The giant emu (Genyornis) appears in several stories'.[42] This statement is at odds with my own observation that in Aboriginal tradition the creation ancestors and spirit beings were often seen as much larger than contemporary forms, as a reflection of their cultural significance. There is no direct evidence that recorded myths with large bird ancestors are ancient memories of long-extinct birds like *Genyornis* or perhaps a megapode. This is in spite of the fact that it is reasonable to assume that such species, whose presence overlapped the arrival of the ancestors of Aboriginal peoples some 50 000 years ago,[43] were hunted.

In order to determine whether the knowledge of an extinct animal is likely to survive long-term within the corpus of environmental knowledge held by an Indigenous group, we can investigate an example outside Australia. In the neighbouring landmass of New Zealand, several species of large birds called 'moa' (such as *Dinornis robustus* and *D. novaezelandiae*) became extinct within the last 1000 years. Ornithologist Mark Cocker noted that 'It seems remarkable how quickly the Māori's collective memory of the giants was largely erased once the birds were extinct'.[44] The disappearance of all knowledge was probably not instantaneous, as linguistic research has demonstrated that Indigenous Māori oral traditions include ancestral sayings that explicitly refer to a few extinct species, and most likely the moa.[45] The Māori names for most other bird species that vanished before European colonisation are now forgotten.[46] Archaeologist Atholl Anderson summarised what early scholars had reported as Māori knowledge of the moas, and concluded that overall they were probably not genuine recollections.[47] He wrote that the 'Alleged traditional evidence is thoroughly contradictory and clearly confused by naïve [European] interpretation of a complex ethno-ornithology',[48] meaning that there were misunderstandings over the bird species involved. The circumstances that would allow for the long-term survival of knowledge specific to the Australian megafauna, such as *Genyornis*, are far more challenging, since the extinction events were at least tens of thousands of years earlier than those of the various species of New Zealand moa.[49]

As discussed in Chapter 2, the nature and content of environmental knowledge used in myths is reflective of the speaker's world view in an often rapidly changing environment, with new elements eventually replacing or altering any redundant information. Given this, anthropologically it seems very unlikely that Australian Aboriginal mythology can be used today as scientific evidence for the existence of certain animal species tens of thousands of years ago.[50] This is the finding of linguist Dorothy Tunbridge, who in the mid 1980s recorded knowledge of mammals held by Adnyamathanha people in the northern Flinders Ranges of South Australia. She was able to record some details concerning species that had gone extinct during historic times, but stated that:

> The extinction of native mammals and other fauna of the Flinders Ranges has been a significant cause of the 'death' of some Dreaming histories. As knowledge concerning a mammal was lost sometime after the mammal itself had vanished, so the Dreaming history became irrelevant and people could no longer hold onto it.[51]

As discussed in Chapter 2, important totemic bird species were avoided by particular hunters most of the time due to their personal connection. In some cases the whole community would suspend hunting of a particular species at certain times, such as after the death of an individual associated with it. In Chapter 4, it was described how particular birds, such as the bowerbird group, were not hunted due to their perceived spirit associations. In relation to the satin bowerbird, the early New South Wales naturalist George Bennett remarked that 'at the Murrumbidgee that the natives have a veneration for this bird, never killing it'.[52] In central New South Wales, the colonist and writer Katherine Langloh Parker described how male initiates were taken to the 'playground' or bower of 'weedah' (*wiidhaa*) the spotted bowerbird in order to gain enlightenment. 'The reason given for taking him to a weedah's playground is, that before the weedah was changed into a bird, he was a great wirreenun ['wizard']; that is why, as a bird he makes such a collection of pebbles and bones at his playground'.[53]

On Cape York Peninsula in northern Queensland, biologist and anthropologist Donald Thomson noted that the bowers of the great bowerbird are often built near Aboriginal camps, but the birds are left unmolested because they are not eaten and the 'aboriginal does not destroy bird life wantonly'.[54] The collecting prowess of the members of this bird group is widely recognised. For instance, Anmatyerr people in Central Australia say that the western bowerbird 'goes along stealing native tomatoes', which are a favourite Aboriginal food source.[55] I have observed a western bowerbird in Central Australia that plucked from a quandong tree its entire crop of fruit, while still unripe, in order to place them as decorations in its bower.[56] Aboriginal peoples considered the bowerbird's act of collecting special objects to decorate the bowers within which they perform, along with the practice of building the bowers near camps, to be the characteristics of a sorcerer who should therefore be left alone and not hunted.

Foraging

The movements of Aboriginal hunter-gatherers were highly seasonal, and this helped to minimise their impact upon the populations of game animals.[57] The avoidance of certain areas surrounding important myth sites had the benefit, possibly unintended, of providing a refuge for game species.[58] For instance, in 2002 anthropologist Marcus Barber was on a hunting trip with Yolngu people in north-east Arnhem Land and observed that a prominent man refrained from killing a brolga because the bird, when sighted, was at its ancestral site.[59] Other foraging areas may have been left alone when people living there had moved away due to ceremonial commitments elsewhere, such as for mourning the death of a senior member of the land-owning clan.[60]

In well-watered regions, such as southern South Australia, the regular seasonal pattern was for Aboriginal foraging bands to leave their summer camping sites close to the coast or lakes in autumn and progressively move towards the more sheltered inland regions for winter and spring.[61] In desert regions the seasonal circles were less predictable, but after widespread rainfalls Aboriginal bands moved away from the major waterholes into more distant areas in order to find populations of animals and plants which had not been utilised for some time, and thereby allowed the Country to recover by spreading themselves more thinly.[62] In tropical northern Australia, the climate heavily shaped local Aboriginal movement patterns.[63] Here, most of monsoonal rains falls during the monsoon 'wet' generally beginning in late December, with the 'dry' commencing around May. Most of the yearly rainfall occurs in the few months often referred to as 'the wet'.

The Australian Aboriginal practice of deliberately burning the vegetation, described by archaeologist Rhys Jones as 'fire-stick farming', kept parts of the landscape open for ease of travel and foraging.[64] The presence of fires also attracted birds who caught grasshoppers and small lizards exposed by the flames.[65] Martu people in the northern Western Desert regularly burnt the spinifex that dominated parts of their Country, resulting in a green patchwork after rains that brought in Australian bustards, emus and other game.[66] Regular burning would probably have led to the increased abundance of certain bird groups, such as raptors and wood-swallows that forage around fires, and granivores that are attracted to recently burnt areas dominated by grasses and forbs.[67] In relation to the timing for setting fires, so as not to cause harm to gamebirds, colonist Ethel Hassell in the south-west of Western Australia remarked that:

> All the young of the birds that build their nests on the ground were hatched and the young ground rats old enough to run about before these were made. When the time was held ripe for the bush fires (*man carl*) the *man carl* corrobboree was held. This dance was done at night.[68]

The frequency and timing of fires was important, not just for the Aboriginal inhabitants but for the environment in general.[69] It is likely that animal species which are closely associated with fire-loving forms of vegetation would have struggled to maintain large and dispersed populations under totally natural regimes of infrequent and unpredictable fires. The most significant impact of the Aboriginal-controlled fire regime in many parts of Australia would have been maintaining the dominance of the sclerophyllous flora, along with its resident bird populations.[70] Pockets of fire-sensitive vegetation, such as forests of conifers and rainforest species, would have been kept in check along with their associated fauna.[71]

Aboriginal insights into where to forage were sometimes perceived as having come from the spirit realm.[72] For instance, in western Victoria, colonist James Dawson noted that 'If a man dreams he will find a swan's nest in some particular spot, he visits the place with the expectation of finding it'.[73] Hunters had intimate knowledge of the plant foods that were favoured by game animals.[74] For instance, in the Murray River region of South Australia, the explorer/overlander Edward John Eyre recorded 'a vegetable food called war-itch (being that the emu feeds upon)'.[75] Similarly, in western South Australia it was noted by anthropologist Daisy M Bates in 1918 that:

> One special. plant, popularly called 'lady's finger' or 'bull's eye' (*Anguillaria dioica* [early-nancy]), but known to the natives as 'wardring ma' (turkey [Australian bustard] food) grows profusely in good seasons, and flocks of turkeys come to feed on the sweet-tasting flowers during their brief period [in spring], departing as soon as they have exhausted the supply of this succulent food.[76]

To hunt such a diverse range of gamebirds, Aboriginal peoples utilised their detailed environmental knowledge in order to select capturing methods that were designed for the target species.[77] While some birds could be brought down by throwing-stones,[78] in other cases hunters used a wide variety of clubs, throwing-sticks, spears, snares, traps and nets[79] – the selection of which depended upon factors such as the species, time of year and location. The weapons that Aboriginal peoples use today now include weapons such as shotguns and rifles, but the hunting of gamebirds remains a symbolically important element for the local economy.[80] For instance, in my experience bustards are considered excellent food by many groups in northern Australia, and while they were previously caught on the open plains using light spears, traps, boomerangs, throwing-sticks and clubs, today they are generally hunted with guns by people in motor vehicles.[81] This is also the case with other large birds such as black-necked storks[82] and emus.[83] I have also seen Aboriginal children bringing down birds using throwing-stones and slingshots.[84] As described in Chapter 10, these are modern relationships that Aboriginal peoples have with birds, although they are derived from traditional origins. The following is an account of the bird-hunting techniques that Aboriginal peoples utilised when Europeans first arrived in their Country.

Egg and nestling collecting

As stated earlier in this chapter, in Aboriginal Australia all bird eggs would have been eaten if the need and opportunity existed. The eggs of certain bird species, however, were targeted during foraging expeditions. Tasmanian Aboriginal people were recorded as gathering and consuming the eggs of ducks, gulls, Caspian terns and penguins.[85] In the Western Desert of Western Australia, eggs of all kinds of bird were eaten, particularly from the larger birds such as the emu, eagle and bustard.[86] In the tropics, a wide variety of eggs, such as from various ducks, were consumed, but those from the magpie goose were so highly favoured that special expeditions were organised for their gathering.[87] In this region, emu eggs are still eaten and are considered very tasty.[88] In the seas off coastal Northern Territory, greater crested terns lay their eggs on sandy islands during the 'knock-em down' rains of the late wet season, and Tiwi people collect them for cooking and eating.[89] In the Wadeye area south-west of Darwin, hunters seasonally visited major waterbird colonies comprising Australasian darters, little pied cormorants, little black cormorants and straw-necked ibises, in order to harvest eggs and nestlings.[90] On the Wellesley Islands in the Gulf of Carpentaria, eggs for eating were chiefly gathered from nests on exposed parts of reefs, rocky islets and cliffs.[91] In times of plenty, only a few types of eggs would have been avoided, such as those of the pelican which western Victorian people considered too fishy to normally eat.[92]

Bird eggs are highly nutritious foods,[93] but in the wild they can only be collected seasonally. For instance, across south-eastern Australia eggs and nestlings were chiefly available from spring into summer.[94] In the Coorong area, museum researcher Norman B Tindale claimed that:

> In September eggs of swans, pelicans and ducks became important and excursions were made to places where emus had begun to lay their giant eggs. It was also the beginning of the season for exploiting mallee fowl eggs, their incubating mounds being robbed but the birds left unmolested.[95]

Malleefowl eggs were universally highly desired as a food source.[96] The importance of them in the Lower Murray was such that, according to the Yaraldi-speaking people who Ronald M Berndt and Catherine H Berndt worked with, 'These birds were rarely caught, although considered good meat. In preference to killing them, their eggs were collected from large heaped up nests in September and October'.[97] The mounds made by malleefowls were sometimes large enough for Europeans to mistake them for Aboriginal graves.[98] The eggs are light pink and have a matt surface.[99] It was claimed that the Ngarkat people in the mallee east of the Lower Lakes in South Australia relied heavily upon malleefowl (*lawan*) eggs.[100] A Murray River settler remarked that 'On account of their size and delicious flavour the [malleefowl] eggs are keenly hunted by the white and black populations of Australia'.[101] He stated that they were fragile and would break if boiled like fowl eggs, although they could be cooked in the coals of a fire and when sucked raw were good for relieving thirst. During my 1980s fieldwork in the south-east of South Australia, Ron Bonney claimed that, unlike some other bird eggs, those from the malleefowl cannot be kept once collected as the shell is too thin.[102]

Swan eggs are large and greenish-white with a coarse surface.[103] They are equivalent in volume to at least three average farm chicken eggs and although their taste is much the same, that of the swan is richer. Swan egg-collecting places were important for hunter-gatherers. For instance, in the Lower Murray a camping and foraging area for the Manangka clan was Pelbarangalang at Reedy Point on the western shore of Lake Albert, where swans were hunted and their eggs gathered annually.[104] Here, the rights to collect eggs would have been determined by the clan ownership of foraging territories, as was recorded for the Kurnai of the Gippsland region in eastern Victoria where the permission of local clan leaders was required before taking swan eggs at particular places.[105] Similarly, in south-west Victoria Dawson claimed that 'The penalty for robbing a swan's nest in a marsh belonging to a neighbouring tribe is a severe beating'.[106]

In the upper south-east of South Australia, Bonney claimed that his adopted father, Moandik man Alf Watson, and other 'old people' always knew when the local swan eggs were in season, due to the commencement of flowering for a yellow daisy (probably *Senecio* species) in the scrub surrounding the lagoon where the nests were located.[107] As discussed in Chapter 7, Aboriginal foragers often took the flowering of particular plants as a sign that certain birds would be laying eggs. The use of gathered eggs from a few species could be extended. Bonney stated that Aboriginal peoples along the Coorong and in the south-east formerly stored swan and emu eggs, which have comparatively thick shells, for months by burying them in damp soil.[108] Swan egging was an activity that persists to the present in places such as the Lower Murray.[109] Aboriginal artist Ian Abdulla, who grew up along the Murray River, wrote that when swan egging as a child with his family they would 'test the eggs to see which ones to eat by putting the eggs in the water … the ones that didn't float were good ones for eating'.[110]

In the north central Kimberley, the Wunambal Gaambera people cooked and ate a wide variety of eggs of species including the emu, orange-footed scrubfowl, pied imperial pigeon, bar-shouldered dove, peaceful dove, diamond dove, terns and brown booby.[111] Aboriginal people in the Kimberley regard the eggs of the orange-footed scrubfowl as very tasty, but claim that it takes a lot of digging to extract them from the nesting mound.[112] For Aboriginal people living along the Finniss River south of Darwin, it is 'goose-egg time' in February and the middle of 'wet-weather-time', which is when the magpie geese fly in honking to make their nests on reeds in the wetlands.[113] In the Wadeye area south-west of Darwin, during the wet season hunters used watercraft in order to collect magpie goose eggs from nests in the swamps.[114] Here, hunters constructed raised large wooden platforms in trees within the wet season swamps as a basecamp when using dugout canoes to collect magpie goose eggs.[115] Mud was gathered to make a fireplace on the paperbark sheeting, in order to cook their bounty and to generate the smoke necessary to drive away mosquitoes. Arnhem Land people living in the Arafura Swamp region had similar practices for collecting magpie goose eggs.[116] Here, they used special shallow bark canoes to access nests, then filled their string bags with eggs which they took back to the swamp camp – a raised platform of sticks covered with paperbark, and with a mud-lined fireplace.

In northern Queensland, the Wik peoples on the western side of Cape York Peninsula ate the eggs of many types of bird, in particular those of the bustard, orange-footed scrubfowl and magpie goose, when the wet season broke around March.[117] Here, the giving of magpie goose eggs was a formal way of welcoming visitors to Country.[118] The eggs of the orange-footed scrubfowl and other large birds were recorded as being eaten at Princess Charlotte Bay, on the eastern side of Cape York Peninsula.[119] The collecting of magpie goose eggs can be a dangerous activity, as the bird tends to nest in the same habitat as saltwater crocodiles.[120]

In the Torres Strait, Islanders went on expeditions in the late wet around October to collect the eggs of the greater crested tern and silver gull.[121] In earlier times, the foraging impact was minor due to limited opportunities, but Thomson recorded that after British colonisation here:

> Each year, upwards of 100 boats, some of them carrying from 16 to 20 natives, go out from Thursday Island alone, and the devastation proceeds without restraint throughout the season. It is true that in former days the aborigines visited the sand banks and islands for birds, their eggs, and young, but the cruising range

of their outrigger canoes was necessarily restricted, and the prolonged periods of bad weather, and the necessity for obtaining fresh water, combined to make negligible the effect of these raids.[122]

Across Aboriginal Australia, hunter-gatherers were expert climbers, often cutting steps into trees and using devices such as ladders and slings to ascend trunks without branches in order to collect highly desired foods, such as eggs and nestlings.[123] According to Adnyamathanha woman Annie Coulthard from the northern Flinders Ranges, steps called *yathinda* were cut into a tree trunk using a 'tommy axe', and 'It appears that in earlier times these steps, which were made by an individual to suit his own step, were not used by another, and that the way the steps were cut indicated who had made them'.[124] A settler in the Yankalilla area on the Fleurieu Peninsula of South Australia noted that 'native men were expert at climbing trees for birds and opossums. They cut notches in the trees for their toes with an axe, and could climb the tallest gum trees this way'.[125] In the late 19th century, Basedow noted that:

> ... notches are cut into the butt of the tree in step-like manner to allow the hunters to ascend for the purpose of chopping out their prey from the hollows. Whilst some are thus busying themselves aloft, others are waiting below in readiness to secure any which might attempt to escape.[126]

Among Aboriginal peoples, there appeared to be a marked preference for consuming nestlings rather than the undeveloped eggs. This was the case along the Murray River in South Australia, where nestlings would often be taken and the eggs were left behind.[127] On the southern edge of the Kimberley in northern Western Australia, naturalist George A Keartland from the Calvert Scientific Exploring Expedition of 1896 noted with regards to the red-tailed black cockatoo that 'Young birds were taken by the natives from the spouts [broken off hollow branches] of the eucalyptus on the Margaret River in early November'.[128] He also noted for the little corella that 'During November the natives secured immense numbers of nestlings, which they regard as excellent food'.[129] During my own fieldwork in the Pilbara in northern Western Australia during 2015, local Aboriginal people actively searched for the nestlings of various cockatoos and parrots in hollow trees for cooking and eating, using long sticks from the surrounding scrub to drag them out. Elsewhere, such as in the Wadeye area south-west of Darwin, nestlings are chopped out of trees using hatchets.[130]

The choice of whether to collect eggs or nestlings was based upon whether the young birds naturally left the nest soon after hatching. On the Murray River in South Australia, Eyre explained that:

> The eggs of birds are extensively eaten by the natives, being chiefly confined to those kinds that leave the nest at birth, as the Leipoa [malleefowl], the emu, the swan, the [magpie] goose, the duck, &c. But of others, where the young remain some time in the nest after being hatched, the eggs are usually left, and the young taken before they can fly.[131]

Stalking

When stalking animals, hunters wishing to reduce their noise would use a sign language to communicate among themselves,[132] which I have observed is still used by Aboriginal peoples in areas such as the Lower Murray.[133] In Central Australia, the hand movement to designate an emu was 'Bends the fingers of one hand at right angles to palm and sways the arm to and fro'.[134] Stalking can take place at any time of the year, although there are distinct seasons for particular birds. For instance, during the cold weather of the dry (around July) in the Katherine area south-east of Darwin, Jawoyn men would smear their bodies with mud to cover their scent when emu hunting, then either stalk the birds or hide among vegetation near the waterholes where they come to drink.[135] Similarly, in north-east Arnhem Land, from the late dry season to the build-up to the wet season (September to November) it is a good time to hunt magpie geese, because the birds are forced to congregate at the remaining water in the billabongs and therefore are more easily caught by hunters hiding among the reeds.[136]

During the late wet season (about March to April), central Arnhem Land hunters used specialised bark canoes to enter the Arafura Swamp to hunt magpie geese, which they cooked on mud platforms placed on raised platforms lined with paperbark.[137] Similarly, in the Wadeye area south-west of Darwin, platforms with paperbark sheets on top were built in trees on the wet season floodplains as a base for hunters in dugout canoes, who used their fishing spears to kill nesting magpie geese.[138] There are rock painting sites in western Arnhem Land that show the specialised dart-like 'goose' spears. These are associated with the Freshwater Period (1500 years ago to present), when lowland parts of the region became less saline and therefore a more suitable habitat for magpie geese.[139] The 'goose' spears consisted of a light-weight reed shaft with a thin hardwood head, and were thrown with special sabre-spearthrowers.[140]

On hunting expeditions, much knowledge of local game animals would have been gained through the reading of tracks. To catch some birds, it was simply a matter of the hunter hiding with a club, either along a pathway used by the game or behind a screen of reeds growing in the swamp.[141] In the Lower Murray, bird hunting was generally a focus for males. Groups of the *ratharathi* (masked lapwing) were rounded up by boys, who then clubbed them.[142] It was also recorded by the Berndts that:

> Young boys would ambush *pilokuri* (water hens [dusky moorhens]) by hiding in the lignum bushes and rushing out with a club when the birds approached. Cape Barren geese (*lalwari*) would require greater strength and dexterity, since a club had to be thrown for at least five yards [4–5 m] to be effective.[143]

Understanding bird behaviour was crucial when hunting. For instance, certain species, such as the pied imperial pigeon in northern Australia,[144] will return to the same tree each night to roost, making hunting them much easier than if they behaved more randomly. Flocking behaviour is also important. In 1988, a Ngarrindjeri bird shooter from the Lower Murray remarked to me that when a *ngulkani* (banded stilt) was taken down it was initially left there, as the birds' companions would soon flock round – providing the opportunity to shoot more of them. Ngarrindjeri people regarded the *ngulkani* as good eating. Here also, men with clubs hunted the great egret, which was cunning and would warn other birds the hunters approached.[145] In the Lower Murray, brolgas were stalked by hunters armed with clubs and spears.[146] Similarly, colonist R Brough Smyth recorded that in Victoria:

> The natives kill this bird [brolga] with a stick, a boomerang, or a waddy. When a flock is flying low at evening, they come within range, and a skilful man will easily secure at least one out of a flock. The flesh is said to be very good.[147]

In the material culture of Aboriginal Australia, the distribution of the returning boomerang, made from either wood or bark, may well have been tied to regions with extensive wetlands where there was much hunting of waterfowl and it was an apparent advantage to have a thrown object return to its owner.[148] In south-eastern Australia, a boomerang was thrown above flying ducks in order to force them back onto the water – the noise as it cuts through the air resembles that of a hawk, which is a predator the ducks fear more than they do humans.[149] After the first throw, the hunter would launch another boomerang much lower, hoping to break the neck of any bird it hits.

Aboriginal hunters were extremely accurate when throwing their weapons at gamebirds. For instance, Basedow claimed that 'It is astounding how adroitly an aboriginal can project the light reed spears; to fell a dove at a distance of from forty to fifty paces is child's play for an experienced thrower'.[150] In the Lower Murray, duck spears, *kuyudhi*, had a long shaft and were thrown into the flight path of flying birds.[151] Here, spears were used to bring down the silver gull, great skua, Caspian tern and sharp-tailed sandpiper along the coast and species of pigeons in the scrub.[152] Upstream on the Murray River, moulting ducks were speared using reed spears with a hardwood head.[153] In the Lower Murray, when hunting the white-faced heron, a group of boys would simultaneously throw their spears in order to confuse the bird and thereby make it more easily hit by men armed with clubs.[154] At Encounter Bay, little penguins on West Island and at Victor Harbor were speared or clubbed.[155]

Ducks were more easily caught when they had lost most of their feathers through moulting. In the Lower Murray, large hunting expeditions were organised at this time. Missionary George Taplin at Point McLeay noted in his journal on 31 December 1859 that 'The men are gone on a great duck hunt for black moulting ducks'.[156] Here, the hardhead and blue-winged shoveler were mainly caught in autumn when moulting.[157] In the Gippsland region of eastern Victoria, waterfowl were often caught when moulting, sitting on their eggs or when just fledged, often without the use of a noose or net.[158] It is also likely that waterfowl that need to regularly dry their wings, such as various cormorants and the Australasian darter, were more easily caught at this time. While some birds were the focus of hunting activities in particular seasons, at least some waterfowl, such as ducks, would be taken throughout the year if the opportunity arose. Smyth noted that for these birds 'As far as I can gather, they did not have a 'close season' in Victoria'.[159]

For Ngarrindjeri people, the *lawari* (Cape Barren goose) is a highly desired food. The large bird has a migratory presence in the Lower Murray, and is generally only seen here from October to February when young birds arrive from the Bass Strait islands to feed on herbage growing on the flats surrounding the lakes.[160] During my fieldwork, one widely spoken account of the 'old' method of stalking the *lawari* was told with much humour within the community by senior Ngarrindjeri men Lindsay Wilson and Henry Rankine.[161] Upon sighting a flock, the hunter takes off his clothes and walks backwards towards a group of birds, which at first are inquisitive and do not fly away. When he is close, he turns around and rapidly shoots one or more of the birds. In pre-European times, men used clubs to catch the *lawari* in the water.[162] In Bass Strait during the early 19th century, the sealers and their Aboriginal companions relied heavily upon the Cape Barren goose as a source of meat when the seal population declined.[163] Aboriginal hunters in the Wadeye area south-west of Darwin also took advantage of the curiosity of their game – a red handkerchief being waved while the hunter whistled and danced would lure an emu to within shotgun range.[164]

Cockatoos and parrots were widely hunted in the temperate Australian forests, with the use of boomerangs and throwing-clubs. Colonist George Grey provided a detailed account of the use of a *kiley* or boomerang, along with a wounded bird as a decoy, for cockatoo hunting in the south-west of Western Australia. Grey claimed that the upward-spinning boomerang confuses the game in flight, and after initial success the hunter:

> … avails himself of the extraordinary attachment which these birds have for one another, and fastening a wounded one to a tree, so that its cries may induce its companions to return, he watches his opportunity, by throwing his kiley or spear, to add another bird or two to the booty he has already obtained.[165]

Early South Australian colonists were greatly impressed with Aboriginal bird hunting skills. In 1868, colonial artist William Anderson Cawthorne wrote an article on 'Natives killing parrots', that accompanied his sketch on the topic in the *Illustrated Adelaide Post* newspaper.[166] Cawthorne wrote:

> The ease, the grace, and the sure aim with which a native throws his wirri, or waddy [hardwood club], as it is generally termed, are matters of surprise to the civilised white man. A group of natives stealthily and softly, with the footfall of a cat, engaged in killing parrots in the forest, is a study worthy the genius of the highest order, whether painter or sculptor. It is not an uncommon thing for a native to bring down four and five parrots at a height of 80 to 100 feet [24–30 m], at one throw of his waddy, and with such force is it thrown, it will often be seen spinning its flight far above the topmost branches of a tall gumtree, accompanied by the dead birds it has killed … The waddy used for killing parrots is generally of a light and handy description. In the season a native will in the course of a short time kill a very large number of birds, bringing down his two or three at each throw, with the facility of powder and shot, with the advantage of little or no noise, and so not alarming his prey.[167]

Aboriginal hunters in Victoria also used the boomerang, along with throwing-sticks, when hunting cockatoos and parrots. Smyth stated that:

> Parrots of many kinds are very numerous in the forests of Australia, and the natives are practised in killing them with the short heavy sticks they carry and with the boomerang. The cockatoo-parrots fly in large flocks. Sometimes at evening one may see hundreds of them high in the air, on the borders of the swamps, flying hither and thither and screaming loudly. They are wary birds, and a sportsman must use great caution in approaching them. In Gippsland the [sulphur-crested] cockatoo (*Braak*) and parrots of other kinds were not often killed by the boomerang.[168]

In Tasmania, along the coast and islands of Bass Strait, Aboriginal people seasonally hunted the 'muttonbird' (short-tailed shearwater), so called by Europeans settlers after its taste.[169] It was this food source that was reportedly being gathered and eaten by Aboriginal people at Cape Grim in 1828, when shepherds massacred 30 members of the group at the cliffs.[170] In the 19th century, sealers utilised Aboriginal labour to procure meat and copious oil from these birds, which became economically important when local seal populations had declined due to over-hunting.[171] During the same period on Kangaroo Island in South Australia, the sealing community, which included Aboriginal women taken from Tasmania, ate 'muttonbird' eggs.[172] The Cape Barren Islanders have maintained the muttonbird industry to the present.[173] Colonist James Bonwick observed that the 'muttonbird' is:

> Smaller than a duck, but somewhat larger than a pigeon, it accumulates fat to an enormous extent, and furnishes by pressure alone a considerable amount of oil. The time of incubation is toward the end of the year. The female comes to land, burrows in the sand of the shore, or the decomposed granite of the islands, often to the depth of four feet [1.2 m], and deposits its eggs. These were diligently procured by the black women, and carried by the sealers with their seal oil to Launceston and other markets.[174]

In the Cooper Creek district of Central Australia, a hunter would follow an emu during hot weather until it was tired and then kill it.[175] In arid areas, where bird populations relied heavily upon visits to waterholes, hunters armed with throwing-sticks and throwing-stones would lie in wait to bring down birds such as the galah, budgerigar, crested pigeon, various finches and honeyeaters.[176] Protector of Aborigines/writer WE (Bill) Harney described seeing Aboriginal hunters on the Barkly Tableland in the Northern Territory 'trilling like the kestrel-hawk [nankeen kestrel]' in order to make a flock of budgerigars bunch together when flying, so that boomerangs thrown at them would produce a large kill.[177] These boomerangs would not have been of the returning variety.[178] During the hottest part of the day, the budgerigars shelter in dense foliage and can be approached close enough to be knocked down with throwing-sticks.[179] Thomson described Pintupi people in the Western Desert capturing budgerigars:

> Great flocks come in to the desert rockholes to drink at evening and are killed by throwing sticks. The birds are gathered in armfuls by the hunters and the little boys watch a flock keenly as it circles low within range and rush to pick up dead or wounded birds.[180]

In the Wadeye area south-west of Darwin, during the dry season hunters would climb to the top of large trees, cutting off branches if necessary, in order to 'block the road of the [magpie] geese' as the birds returned from the floodplains each night to roost in the paperbark trees.[181] From their arboreal vantage point, the hunters deployed their throwing-sticks, which were made heavy by soaking them in water. In this region, flocks of radjah shelduck are stalked from behind sand dunes and mangroves, and in earlier times were brought down using throwing-sticks rather than shotguns.[182] Knowledge of avian food sources was crucial for hunting success. In the Wadeye area, it was recorded that:

> Hunters would wait near trees such as Emu Berry *mi memem* (*Grewia retusifolia*) when they had fruit in the dry season to spear the emus or near waterholes. The fruit of the Sand Palm are also favourite food for emus.[183]

During the dry season in the Arafura Swamp area of central Arnhem Land, when the water has receded to the deeper tree-lined waterholes, the hunters used special throwing-sticks to catch magpie geese. In 1937, Thomson recorded that here:

> They take a slender sapling, about six feet [1.8 m] long, leaving small protuberances along its length and the bark is stripped at the thin end to act as a counterpoise. The hunter climbs out to the end of a thick, leafy branch overhanging the water and waits until a flock of geese passes, flying low. Then he throws his stick; the protuberances prevent it glancing off the feathers, in which they tangle, breaking the wings of the bird and bringing it down. It sounds an improbable method, but it works.[184]

When stalking an emu, a method was to lie in wait next to an emu pad leading to water.[185] When the bird approached in late afternoon, the hunter would thrust out his hand and catch the bird by one of its legs then club it to death. Emus were hunted in many of the same ways as kangaroos,[186] making their shared identity with most other bird species problematic from an Aboriginal point of view, as discussed in Chapter 5. The hunting techniques utilised by Tasmanian Aboriginal women, who were brought by sealers to southern South Australia during the early 19th century, were so efficient that they led to the extinction of the Kangaroo Island emu.[187] Across Australia, emu hunters smeared themselves with mud and plant material to disguise their scent.[188] According to anthropologist Frederick McCarthy at the Australian Museum in Sydney:

> Innate curiosity enticed emus to investigate a bunch of feathers waved above a rock or low bush, enabling the hunters to spear them at close range. In central Australia the bird's call was made through a short hollow bough by a man in a pit to draw an emu within spear range or cause it to fall into one of a circle of pits around him.[189]

Aboriginal peoples took advantage of the emu's inquisitive nature. According to the Berndts, the Lower Murray hunter would seek out an emu in the mallee scrub:

> When he saw one, he would stand in such a way as to be clearly observed by this bird. Then he would lie on his back, moving his arms and legs: the emu, attracted to this display, would come towards him and circle around him. Jumping up, he would grab his club that lay beside him and hit the bird on the legs and then on the head.[190]

The characteristic curiosity of the emu has been noted elsewhere, such in the Kimberley where it was reported that 'The emu is extremely curious and this often leads to his own destruction. It is said that if one wraps oneself into a red blanket, one can attract the curious emus to come up close'.[191] Care was needed when hunting emus as they are capable of delivering a savage kick to the hunter and his dogs; emus later caused injuries to colonists' horses.[192] There is a recording of a Tangani song from the Coorong in South Australia that describes 'When an emu falls down on its back and strikes out with its feet it can hit with considerable effect'.[193]

Swimming

Across Australia, catching birds that had settled on an open section of water away from cover involved the hunters swimming quietly towards them.[194] Typically, a hunter, wearing a hat composed of mud and water plants like various reeds or bulrushes, would swim to the waterfowl and pull them under.[195] An anonymous historical account from the Lower Murray described methods of duck hunting:

> A common method is for the native, when stalking duck on the water, to cover his head with a bundle of long grass tied close to its extremities into something like a horse-collar, the extreme ends falling over on to his shoulders. He holds his spear in one hand and a bunch of leafy switches in the other. Viewed from the front the individual thus accoutred looks for all the world like a tussock of grass floating lazily along, so slow and silent are his movements. Sometimes, without troubling about a spear, he may, with his head thus masked, gradually steal upon the unsuspecting birds and catch them by hand. Another common

method of capture is for the native, clad in Nature's garb, to dive under a flock and pull them down from underneath, one after the other, without alarming the main flock.[196]

During Lower Murray fieldwork in the 1980s, several elderly Aboriginal people recalled the 'old time' techniques for catching waterfowl.[197] Ngarrindjeri man Oscar Kartinyeri remembered that as a boy he and his companions would go to the lagoons at Poltalloch Station on the southern shore of Lake Alexandrina to hunt ducks.[198] Here they would locate birds resting on the water surface, then swim underneath to pull them down and 'ring' (break) their necks. Ngarrindjeri woman Dulcie Wilson provided an account of duck hunting from her husband, Lindsay Wilson, who as a boy in the 1930s had camped on the Coorong Peninsula with Jacob Karruck Harris from the Wutaltinyerar clan. She recorded that:

> Lindsay was taught the habits of birds and animals in and around his area. He used to say 'If you watch a mob of ducks on a spit (mudbank) you will see that they have at least two birds on guard.' While others might be preening their feathers or sleeping, there are two who will be watching out for any danger. They always watch towards the land or sky, as they don't expect any danger from the water. When the boys were young, they would cover themselves in mud, put water-weeds on their heads and slide into the water near the ducks. Gradually, they would move closer and closer until they could grab a duck or two.[199]

Across northern Australia, magpie geese were also caught using this technique. In the Daly River area south of Darwin in the Northern Territory, Basedow observed that:

> At times the hunter plucks a large water-lily leaf [probably lotus], into which he cuts two holes for his eyes to look through. Holding this leaf over his face, he swims out to some [magpie] geese he has observed on a lagoon, and, when within grasping length of the prey, he simply pulls a bird under by its legs and strangles it.[200]

A modified method as used by Tommy Mungulung, involving a shotgun, was recorded from the Wadeye area south-west of Darwin. Writer Jeff Hardwick recorded:

> Duck or [magpie] geese on a billabong would appear undisturbed by the slow approach of a patch of water-lilies [*Nymphaea*] shrouding Tommy's head and shotgun. Taken completely by surprise, there was always a maximum number of ducks or geese per cartridge. Leaving the dead birds floating, Tommy would quickly secure those only slightly wounded, and ready to take off, by wringing their necks. Others that had fluttered off wounded into surrounding scrub were carefully noted and later retrieved.[201]

Emu drives

Drives of large animals, such as emus, wallabies and kangaroos, involved the highly coordinated activities of the whole foraging band.[202] Special techniques are required as emus are fast, running up to 50 km/hour across open country.[203] Drives were often used in conjunction with fenced enclosures, such as those described in 1798 by New South Wales colonist David Collins in the wooded river country inland from Sydney.[204] Emus habitually walked to waterholes along the same pad, where hunters could set up for them.[205] In Central Australia, yards were made near waterholes in order to capture emus.[206]

Long and robust game nets, composed of cord manufactured from plant fibres and animal sinews, were sometimes placed at the termination of the drive in order to entangle the emus,[207] particularly in the Murray and Murrumbidgee River regions.[208] The Adnyamathanha people of the northern Flinders Ranges in South Australia laid game nets, *mindi*, upon the ground on pads near waterholes frequented by kangaroos and emus, then raised the nets when the game was deliberately disturbed by a thrown stone. The entangled animals were dispatched with a hit from a club, *wirri*.[209] The Adnyamathanha used similar techniques involving nets and fences to catch yellow-footed rock wallabies.[210]

In central New South Wales, emu nets were made of cord 'as thick as a clothes-line' from either 'kurrajong' (bottle tree) or 'burraungah' grass fibres.[211] Here, brush fences were also often used in conjunction with nets.[212] According to McCarthy:

> Brush fences were built to form alleyways along which emus walked into a pit dug in a narrow opening. In western Queensland and New South Wales cord nets 60 feet [18 m] long and seven feet [2 m] high, with a mesh about a foot [30 cm] square, were secured to poles and trees to make a V-shaped corral into which the emus were driven and killed. While emus were drinking at a pool, a net was set across their pad; as they returned the hunters appeared suddenly behind them, and in the resulting panic one usually became entangled in the net.[213]

The layout of the country provided a backdrop for each game drive. Along the Coorong of South Australia, emus and kangaroos were driven towards parts of the landscape where they could be more easily trapped by water, with Monokoru Point on the eastern side of the main lagoon being a favoured place.[214] Tindale recorded that in the Coorong area:

> Kangaroos and emus were hunted by organised parties, which went out into the scrub country behind the Coorong and drove the animals to ambush, using fire to aid the hunt. The animals were usually driven on to one of several narrow-necked peninsulas jutting out into the Coorong [Lagoon], where they were cut off and easily dispatched. The whole clan, or perhaps several adjoining clans, possibly up to 100 men, would take part in the kill. Hunting was a man's game.[215]

The Aboriginal use of fire to drive game animals was recorded by the first European settlers in New South Wales.[216] The setting of fires was done not randomly but to a plan, and it involved a highly coordinated group of hunters. An early settler, Captain Dirk Meinertz Hahn, was in the Hahndorf area in the Mount Lofty Ranges east of Adelaide when he observed the Aboriginal use of fire for hunting large animals, such as kangaroos and presumably emus. He noted:

> They practice another peculiar and shameful kind of hunt in summer, when the grass is fully grown in the hills. However, this occupies a whole tribe. They form a circle of about twenty English miles [32 km] in diameter and light a fire all around this level area and direct it nearer and nearer to the centre of this circle; the long dry grass, brushes and young trees burn furiously, all the wild animals inhabiting this area flee from the fire in their fright and draw nearer and nearer to the centre, where the savages then capture them. This kind of hunt took place during our stay there and burnt for several days.[217]

Emus would be hunted at any time of year in the Coorong area, although they moved about the country seasonally, visiting water sources during the dry summer months and dispersing into the hinterland upon the arrival of autumn rains.[218] A Murray River settler remarked that emus 'in olden times feasted on the abundance of native currants in their season'.[219] The identity of the fruit mentioned here is possibly indicated by Bonney's observation during my fieldwork that emus in the Lower Murray and south-east districts had once moved in large numbers towards the coastal dunes during the spring fruiting season of the *nguli* (wild currant bush), then stayed there for the *mantari* (monterry) berries in order to 'get fat on them'.[220] Aboriginal peoples may have recognised other associations with emus and seasonal weather. A newspaper correspondent in the early 20th century remarked that 'Some bushmen regard the "booming" noise made by the emu as a reliable signal that rain is approaching'.[221]

Hides

The simplest hides were naturally dense branches in trees or bushes that would provide cover for hunters, who disguised their scent by smearing their bodies with mud.[222] Basedow recorded that in the Darwin area in the Northern Territory:

> The Larrekiya [Larrakia] and Wogait [Wadjiginy] tribes conceal themselves in the branches of a tree, the seeds of which are known to attract the emu. The hunters ascend the tree in the early hours of the morning and remain there perfectly quiet until the prey arrives. At an opportune moment, the bird is speared with a specially heavy spear known as 'nimmerima'.[223]

In more open areas, hunters used strategically placed hides in order to have game approach within reach of their arm or spear jab. It is recorded in Tasmania that structures made from sticks and grass were used as hunting hides, with bait placed on the ground nearby to attract ducks and ravens.[224] In the Lower Murray, it was reported that bird hides were 'carefully concealed by she-oak branches, interwoven with grass' and positioned near waterholes where emus could be speared as they approached to drink at sunset.[225] Along the Coorong during summer, emus were generally ambushed when coming to waterholes to drink.[226] In this area, the tops of limestone cliffs overlooking the main lagoon were good lookouts, 'The natives here construct elevated seats of platforms in bushy she-oak trees, for the purpose of watching and spearing the emu and kangaroo as they pass towards the water to drink'.[227] For night hunting in the northern Flinders Ranges, the Adnyamathanha people made stone walls 60–80 cm high, alongside an animal pad near a water source.[228]

In the Darwin hinterland region of the Northern Territory, a hide was constructed in a pit for catching magpie geese in places where they were known to seasonally forage; these were made with a roof of twigs, paperbark, grass and soil with one or two lookout holes.[229] The birds prefer to eat the corms of the spike-rush growing in the wetlands.[230] Hardwick described this technique as being used in the Wadeye area south-west of Darwin:

> During the late dry season when the spike rush bulbs were scarce and water on the plain dried up, traps were built. These were a hole in the ground big enough for two men to stand in. The roof was of cane *nanhthi pirn pirn* (*Flagellaria indica*) and sheets of paperbark with grass on top with four holes around the edge. Spike rush bulbs *mi walangka* were scattered around the trap. The men had to be in before about 4 o'clock when the geese came to feed. Four other men would stand about a 100 metres away to scare the geese towards the trap once they were feeding. The men inside would seize them quickly by the leg and pull them in to break their necks. They would end up standing on a large pile of dead geese inside. It was hot and dusty work. Brolgas were also caught in the same way.[231]

Hunters in the northern deserts of Western Australia covered themselves with tussocks of porcupine grass, with a small grass fire placed nearby to attract hawks.[232] In the 1890s, government geologist HYL Browne found hawk hunting hides in the area between Little Gregory and Big Gregory Creeks in the northern Kimberley while on an expedition:

> Some curious erections, made of flat stones, built up in the form of a horseshoe, were seen here, which are said to be erected by the blacks with a view to enabling them to catch hawks, which are plentiful in this region. The native covers one of these stone places with boughs, and ensconces himself inside, at the same time making a small fire, the smoke of which attracts the bird's notice. He is provided with a pigeon or other small bird which he exposes to view on the top of the stone 'trap,' and the hawk, seeing this, swoops down upon it and is grabbed by the man beneath.[233]

Hawk and eagle hunting hides in the Victoria River district of the north-west of the Northern Territory were substantial structures made from large stones, so much so that early explorers who found them in the absence of Aboriginal people thought that they might have been huts with religious functions.[234] The hides used by Wardaman people south-west of Katherine in the Northern Territory were similar. Hunters used them by pushing through the roof a long stick with feathers, skewered with small birds or mice, and twirling it to attract hawks and the wedge-tailed eagle.[235] The gamebirds most often caught from these hides were raptors with the behaviour of congregating round the edge of bushfires in order to prey upon small mammals, lizards and other creatures fleeing the flames. The birds of prey caught using the stone hides in the Wadeye area south-west of Darwin provided food and their feathers were used for decorations in ceremonies,[236] as described in Chapters 8 and 9.

Lures, calls and decoys

Lures are objects used to arouse the curiosity of birds, in order to encourage them to move closer to the hunter and thereby be more easily caught.[237] Aboriginal hunters also used calls for the same purpose. Decoys of live or imitation animals were often used to entice gamebirds to come within hunting range. For lure and call techniques to be effective, the hunters must possess deep knowledge of predictable bird behaviours.

The use of lures is recorded from across Australia, and their operation generally relied heavily upon the quick and timely actions of the hunter. For instance, colonist David Collins described how Aboriginal hunters in the Sydney area caught fish-eating birds:

> … a native will stretch himself on a rock as if asleep in the sun, holding a piece of fish in his open hand; the bird, be it a hawk or crow, seeing the prey, and not observing any motion in the native, pounces on the fish, and, in the instant of seizing it, is caught by the native, who soon throws him on the fire and makes a meal of him.[238]

Across eastern Australia, Aboriginal duck hunters would throw a boomerang, or sometimes a piece of bark or wood, into the air to imitate a hawk and thereby force the game to take to the water where they could be netted.[239] In comparison to fishing nets used in the same region, the game nets used for duck hunting would have been of large mesh, less strong and less water-resistant.[240] Along the Coorong, a basketry duck decoy was used to entice ducks to swim closer to the bank where hunters were hiding.[241] A former Lower Murray settler recalled that 'Out of emu feathers they made decoy forms, which enabled them to get close to and catch the bush game'.[242] A method to entice grey teal ducks towards the bank involved the hunter, hidden from view, waving a bunch of common reed flower heads.[243] In 1991, at a favourite duck hunting place on the shore of Lake Alexandrina, Ngarrindjeri man Henry Rankine demonstrated for me the same technique, although he used his beret rather than reed heads.[244] According to another account from a European observer:

> The blacks on the Coorong used to make an imitation wild-dog's tail of grass-plumes, stand behind a bush on the edge of the water, and gently wave and wag the grass about, keeping themselves strictly out of sight. A blackfellow once

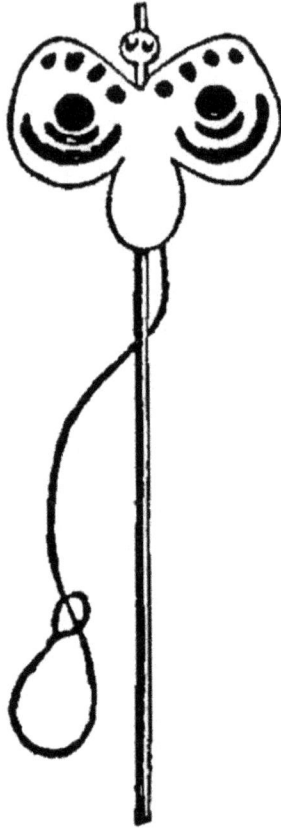

Butterfly lure, intended for catching the inquisitive Australian bustard. Smyth RB (1878) *The Aborigines of Victoria*, Vol. 1, Fig. 18.

explained the matter to me thus:- 'You see, plenty duck sit down 'long water – too long way blackfellow knock 'em down 'long a waddy. Blackfellow make 'em dog's tail 'long a grass, and make it all same dog sit down 'long a bush. One duck, he see him, and he say, 'long another duck, 'Ki! You see him wild dog, sit down 'long a shore!' 'No, that one bit o' grass.' All time duck he swim that way and that way (right and left) – all time he come close, and by'm-by he come plenty close. Then blackfellow he jump up and throw um waddy [throwing-club] – whir-r-r. My word! He kill um three, four, five duck.'[245]

The waving reeds apparently looked like a dog tail, and this attracted inquisitive ducks already on the water. When the birds come in close, the hunter stands up and throws clubs as they take flight. The duck lure technique was refined after European settlement, as described by settler John P Bellchambers:

After the admixture with other breeds, the natives, either by accident or design, evolved a wonderfully clever decoy dog. It was small, rather long in the body, of a creamy colour, with a bushy tail, which was carried erect. By skilfully working these dogs in the shallows, the ducks were enticed in to the shore, the yellow wagging plume of the animal attracting the curiosity of the birds.[246]

Black swans were also killed using the lure method. Ian Abdulla wrote that along the Murray River:

> Swans are very inquisitive birds. If you're on one side of the swamp and they're on the other and you wave a wing of a dead swan, or even a white rag, swans will come towards you. When they got close enough – bang – we'd shoot them with a single or double barrel shotgun.[247]

Lures were also used to mimic jumping fish. It was claimed that the Lower Murray hunter 'will decoy pelicans within his reach by imitating the jumping of fish by throwing mussel shells or splashing the water with his fingers'.[248] In 1939, Tindale received a similar account from Kuungkari man Jim Sweeny from the Thomson River in south-west Queensland:

> In the rivers and lagoons, pelicans were captured by men who concealed themselves under water with their heads protruding under some grass or small bushes. Small fish or mussel shells were thrown out to imitate jumping fish. Pelicans would be attracted and seized by the neck as they put their heads down to take the bait.[249]

In central New South Wales, a method to catch emus for the Yuwaalaraay-speaking (Euahlayi) people used an audio lure in conjunction with an emu feather decoy, noose and a leaf-covered pit. Langloh Parker explained that:

> Boobeen is a primitive cornet, a hollowed piece of Bibbil [*bibil*, bimble box tree] wood, one end partially filled up with [*Callitris*] pine gum, and ornamented outside with carvings. To blow through it is an art, and the result rather like a big horn. The noise is said to be very like an emu's cry, and this emu bugle will certainly, they say, draw towards it a gundooee, or solitary emu.
>
> The blacks used on the sandhills to make a deep hole to hide themselves in, usually only one though. From this hole they would run out a drain for about thirty yards [27 m]. The man with the Boobeen would have a little break of bushes round him; scattered over the leaves he'd have emu feathers, and then he would have a strong string, on the end of which he would have a small branch with this he would place about midway emu feathers on it; down the drain.
>
> When the emu answers the Boobeen's call, the bugler gets lower and slower with his call. The emu sees the feathered thing in the drain, comes inquisitively up

and sniffs at it. The man in the hole pulls in the string slowly; the emu follows, on, on, until heedlessly he steps on a Murrahgul, or string trap, and is caught.[250]

In harsh exposed desert areas, wide-ranging game species such as emus were particularly prone to the use of lures. In the Cooper Creek region of eastern Central Australia, anthropologist George (Poddy) Aiston remarked that:

> Anything moving on the desolate gibbers or sandy wastes attracts attention, and man, as much as any, comes in for the curiosity of the denizens of the wild. The dingo will follow a buggy or a motor at a distance for miles, the brolga, or 'native companion,' is inquisitive though he runs no risks, but the most foolish is the emu. Near the camp at Mungeranie I noticed little flags stuck on trees. These, I was told, attract the attention of the bird, which circles round them and may be dispatched with a *kirra* [bent wooden throwing-club].[251]

Tindale recorded that to attract emus into the game nets, the Kuungkari people in the Thomson River area in Queensland used a 'form of drone pipe made from a 4 ft [1.2 m] length of hollow log carefully burned out so that it became thin and c. 6 inches [15 cm] in diameter ... This had a mouth piece of spinifex gum'.[252] When blown, the instrument imitated the throaty call of the emu, which brought in birds to meet the challenge from a supposed rival. A record from Central Australia described the use of audio lures, whereby the 'Wild turkey [bustards] are often enticed with a lure made of a hollow root of the swamp marshmallow [flood mallow]. The hunter hides himself in the bushes and sounds the lure until the turkey comes within striking distance'.[253]

Sometimes a hunter would lure gamebirds by using his own voice to imitate their calls. The imitation of bird calls was frequently used in combination with other hunting strategies. In the Darwin hinterland region, Basedow noted that magpie geese 'can also be lured, by imitating their call, so close to a native seated motionless in high grass that they can be actually grasped by hand'.[254] Here, Basedow observed that:

> The note of the whistling duck (*Dendrocygna eytoni*) is also accurately reproduced, by which flocks of them are attracted and killed with a throwing-stick while hovering around the spot which conceals the native. Cockatoos, plovers, and many other birds are secured in a similar manner.[255]

Aboriginal hunters took advantage of the natural inquisitiveness of the Australian bustard, which was a favoured source of meat. In the Lower Murray, these large birds were lured away from their shade cover with a butterfly decoy, composed of white feather plumes – either swans down or pelican breast feathers – tied to a stick with sinew.[256] During the summer, bustards were more easily caught as they were preoccupied with catching butterflies and other insects. A similar method for catching bustards, involving a large moth as a lure, was recorded from upstream along the Murray River.[257] In the western districts of Victoria, feathers of a small bird or dead butterfly were attached to the bustard lure.[258] In parts of Victoria, a small bird was caught then tied up as a decoy for curious bustards.[259] The setting of a fire followed by the use of a moth lure was a bustard hunting technique employed in the south-east of South Australia.[260] Along the Murray River, hunters in a hide covered with weed lured these inquisitive birds in by starting a small fire near their nests, when 'the turkey [bustard] would be generally watching to and then come back and try to drive the fire away with its wings [and] when the turkeys wings were singed the blackfellow would have no trouble to catch the turkey and bring home the eggs as well'.[261] European colonists were also fond of the bustard as a gamebird, leading to the species becoming locally extinct over much of the continent, particularly in the south.[262]

Other large bird species could also be attracted with live bait decoys. In south-west Victoria, Dawson recorded that hunters used a mobile screen covered with shepherd's purse foliage when creeping up on gamebirds like the emu, bustard and 'gigantic crane' (brolga), 'exposed to view as a lure a blue-headed wren [superb fairy-wren], which is tied alive to the point of a long wand and made to flutter'.[263] It is recorded in south-eastern Australia that Aboriginal hunters dressed in an emu skin or a grasstree crown could approach inquisitive emus, then kill them with spears they had dragged across the ground by their toes.[264] Basedow remarked that:

> The natives also take advantage of the inquisitive nature of the [emu] bird by enticing it into a cul-de-sac or other trap by waving a conspicuous object, as for instance a corrobboree plume, from behind a boulder or bush. When the bird is near enough, it is either rushed with waddies or speared by a number of chosen, astute men.[265]

In the Great Sandy Desert of Western Australia, the Walmajarri people used a lure known as the *witawita*, which was a long thin wand with feather plumes fastened with hair string to the end. Here, according to Walmajarri man Jimmy Pike, it was used when hunting bustards or kangaroos. When the hunter:

> … found fresh tracks he would hold up the wand, feathers aloft, and flutter it to imitate a bird of prey. Small kangaroos or turkeys [bustards], seeing this apparent threat from the air, would attempt to hide in the spinifex, where the hunter using the deception could then easily find them.[266]

The Koombokkaburra (Kumbukabura) people living between the Belyando and Cape Rivers inland from Townsville in northern Queensland also used lures to capture game. Colonist James MacGlashan recorded that:

> Emu are speared in dry weather when water remains in but few holes. Having found by the tracks those commonly frequented, the Black, provided with a spear, ascends some tree near at hand, from which he suspends a bunch of emu

feathers with a string. When the birds come to water, he imitates their cry, and they, with the curiosity so characteristic of them, proceed to examine the bunch of feathers, when the Black hid amongst the boughs overhead spears one of them.[267]

In the Daly River area of the Northern Territory, hunters hid in a bush or behind a screen of branches, uttered guttural calls and shook a bunch of emu plumes to attract the inquisitive emus towards them.[268] In the early 20th century, Basedow described the hunting of magpie geese in the Darwin hinterland of the Northern Territory. He observed that:

> The clever imitation of the cries and calls of these birds, a '*nga ngang, ngang-nagang-ngang*' induces large numbers of them to be attracted, at dusk, close to the native, who sits in the branches of a tree, and kills the birds with a stick.[269]

As stated earlier, within the region stretching from Victoria River south to Lake Amadeus in the Northern Territory and across to the east Kimberley, Aboriginal people made rock-walled hides roofed by green bushes to capture hawks that were attracted to the spot by burning clumps of spinifex.[270] There are walled rock shelters recorded in the Pilbara region, which appear to have had some Aboriginal use in hunting.[271] Hawks were encouraged to land at the hide where they could be grabbed, having been enticed there by either waving a meat-baited stick in the air or tethering a small live bird, such as a dove, as a decoy.[272] This is essentially the same technique that people in other parts of the world have historically used for capturing birds of prey for taming in falconry.[273] In the 1960s, Australian bird watercolourist Robin Hill utilised the same decoy method in Victoria, involving a domestic pigeon, to capture a hawk for study.[274] Aboriginal people in Tasmania had a similar technique for catching ravens along the coast, using fish as a lure and hiding themselves in grass huts from which to grab their quarry.[275]

Hunters preyed on the fact that many of the wetlands existed in areas which lacked natural perches for birds. It was observed by Adelaide-based scholar Thomas Worsnop that 'In rivers where there are neither reeds nor trees the natives fix stakes in the water for resting-places for shags, cormorants, and other birds, and when they perch on these the natives would swim quietly up to them (their heads covered with grass), and seize them'.[276] A framework of stakes supporting perches was constructed in swamps in order to attract large flocks of roosting pelicans. These could be caught by hunters stealthily swimming up to and under them, then clubbing two or three birds before the rest took flight.[277] Hunters threw clubs at shags in flight, as well as targeting those sitting on a log or resting on the edge of the water.[278]

Charms and rituals

Not all devices utilised by Aboriginal peoples are, from a Western perspective, based upon scientific principles that explain their operation. In Aboriginal tradition, the essence of the ancestor's power is imparted into ritual objects, often small and unusual-looking stones such as quartz crystal, which knowledgeable people can then access for various purposes.[279] In the Western Desert region, it was an Aboriginal belief that australites, referred to as 'emu-stones' and 'emu-eyes', had some ritual power over emus when hunting. In a study of the Aboriginal uses of australites, mineralogist George Baker noted that:

> Mr. H.R. Balfour of Toorak, Victoria, who made enquiries among the natives of the Woomera region of Central Australia about the reason for their use of the term emu-stones, informs me that these aborigines wrap up australites in balls of emu feathers which are then thrown in the direction of flocks of emus. The particular natural inquisitiveness with which the emu is especially endowed, results in a close approach to these objects for near inspection and extraction of the contained australites. While absorbed in their investigations, the emus are speared by the aborigines.[280]

In the Cooper Creek area of Central Australia, a Diyari man known as Old Piltibunna described the use of 'obsidian bombs', or australites, as charms for blinding game during emu drives.[281] Here, it was recorded by anthropologists George Horne and George Aiston that:

> Obsidian bombs were called *warroo getti milki* (emu eyes), and were supposed to be eyes that the emu had lost when walking about looking for food. These when found were smeared with fat and red ochre and were stored in a net bag full of emu feathers, kept together by being wrapped around with hair rope.[282]

By another account, Aboriginal hunters threw australites that were still contained within a 'nest', which probably made their recovery easier. Naturalist Charles L Barrett explained that:

> ... they believe that the small rounded objects, which they carry in a nest of feathers, have the power of making emus blind. It is no trouble at all to get emus if only you possess a nest of emu eyes. The blacks throw among emus that have come to drink one of the feather nests containing two or three australites. The birds are supposed to become blinded and run into the water. Aborigines have strong faith in the power of emu eyes and are reluctant to part with them.[283]

Rituals were performed to bring birds into Country. There is a record of Diyari people using a ceremony to make wildfowl lay eggs, such as when large numbers of migratory birds come towards Lake Eyre in the north-east of South Australia during floods.[284] Stencils of birds found in the Djulirri rock shelter in western Arnhem Land may have been made to ritually improve a bird hunter's success.[285] Due to their shape and size, these paintings are thought to represent the singing honeyeater. Anthropologists have described 'increase' ceremonies as a means that Aboriginal peoples employ to ensure that their resources are replenished.[286] Anthropologist Mervyn Meggitt explained, in relation to the Warlpiri people of northern

Central Australia, that with these ceremonies the 'participants are simply concerned to maintain the supplies of natural species at their usual level, to support the normal order of nature'.[287]

Snares and traps

Precise knowledge of the feeding habits of game species was particularly important for hunting techniques such as snaring, as sometimes the hunter was either not around or close by to observe whether the snare was in the correct position.[288] In the Lower Murray, snares, *nongi*, were formerly used for catching a wide range of birds.[289] Snares were used in conjunction with a framework of sticks, often placed next to a hide made from branches and reeds. One method used by hunters to attract birds to land on the framework and become entangled was to produce noises.[290] Another practice was to situate the snare between two sticks driven into mud on either side of a narrow swimming channel.[291] Black swans in particular make these channels or paths among reeds. Wrens could also be caught in the grass, using 'traps similar to those used for mice, except that the *wirili* [possibly the white-browed scrub-wren] trap had a narrower pathway leading to a noose through which the bird put its head and was thus securely caught'.[292]

In the Lower Lakes area, a *ranari* or snaring-rod was a pole up 4.5 m long with a sinew noose. It was used to catch birds such as the magpie goose, purple swamphen, musk duck, black swan and Cape Barren goose, usually when the game was on the water.[293] Here, a method of catching swans involved the hunter, armed with a swan wing decoy and snaring-rod, hiding among bulrushes (water-flags) and reeds by the margin of a lagoon.[294] The splashing of the bird wing, imitating a bird in distress, attracted distant swans. Upon the approach of a swan, the loop of the snaring-rod was quickly extended over the bird's head, the rod pulled in, and the bird caught and killed. Swans are flightless while moulting from mid July to January, and so are easily caught at this stage.[295] At other times, these and other large bird species could be taken by spear.[296] When hunters used the snaring-rod method, Tindale said:

> They cover their head with a bundle of grass and close their eyes almost completely so that the bird would not see the glint of their eyes through the grass. They walk along in the water, slowly approaching the bird. They always place the noose over the bird's head from the back.[297]

In the Lower Murray, a shorter form of the snaring-rod was also used for bird catching.[298] It consisted of cord made from *yalkuri* (spiny flat sedge) fibres that was attached to a length of flexible *puyelangki* (possibly rough cryptandra) root, made flexible with fish oil or pelican fat, that was hafted to a handle of wild hop wood.[299] The snaring-rod was ~45 cm long, and 'Once the lasso was over the bird's head, any movement by the creatures would tighten the cord, as would any movement by the hunter'. A strategy was to place upright sticks a short distance into the lake in areas where shags and cormorants were known to frequent; hunters would swim out with weed-covered heads and snare the birds with rods.[300] It was observed that

'In this sport they frequently receive severe bites from the shags upon their naked limbs'.[301] Snaring-rods were also used in western Victoria, where colonist James Dawson recorded that:

> Ducks and the smaller waterfowl are captured among the reeds and sedges with a noose on the point of a long wand. The hunter approaches them under concealment of a bunch of leaves, and slips the noose over their heads, and draws them towards him quietly, so as not to disturb the others.[302]

In mallee country away from the Murray River, snaring-rods were used to catch nectar-feeding birds such as honeyeaters and lorikeets, with hunters constructing a hide next to a flowering bush and the first captive tied to the frame as a live decoy.[303] With the use of hides, snares and live decoys as lures, a large number of gamebirds could be quickly caught. In Victoria, colonist R Brough Smyth recorded that:

> Small birds of various kinds, which feed on the blossoms of the honeysuckle (*Banksia*), are caught by the natives living in the Mallee scrub in the following manner. A hole is dug in the ground sufficiently large to admit of a man's sitting in it comfortably, and over it is built a mia-mia [shelter] of green boughs and twigs. In front a number of small sticks are stuck in the ground slantingly and crossing each other. The native, having provided himself with a thin stick, furnished with a running noose of fine cord at the end, takes his seat in the hole, and imitates the chirping of the birds. After some trouble, he secures one, and he uses this as a decoy, fastening it by a cord to one of the long slanting sticks. It attracts numbers by its cries, and the native cautiously ensnares one after the other with the loop, until he takes perhaps three hundred or more. Having passed the loop over the head of the bird, he twists the stick and adroitly draws it into the hole. A patient hunter is always well rewarded when pursuing this method of capture.[304]

Deputy surveyor general Thomas Burr, who accompanied Governor George Grey and artist George French Angas to the south-east of South Australia in 1844, described coming across at Maria Creek (his Ross Creek) an Aboriginal 'trellis', which was constructed to bring birds within range of a snaring-rod. He said that:

> The trellis is formed by seven slender sticks, two of which are fixed to the ground about 5 or 6 feet [1.5 or 1.8 m] apart, and rise about 4 feet [1.2 m]; the tops of these are connected by a third, whilst the remaining four are placed diagonally across. At about 4 feet from the trellis a hollow is formed, which is screened by small branches of trees that rise about 2 feet [60 cm] from the ground, and a small hole is left at the back through which a native creeps, and thus concealed, places the first and second finger of his left hand across his lips, which are slightly opened, and by drawing in his breath, he makes a chirp that calls the birds, which, thus enticed, perch on the trellis work.[305]

In eastern Central Australia, traps, termed *mokwari*, were used to capture flock pigeons. These structures were made in the form of straight narrow drives 6–9 m long in the scrub alongside waterways.[306] Aboriginal hunters used resinous materials to help capture small birds that had been lured to specific places.[307] For instance, it is recorded that Pitjantjatjara people in the Western Desert gathered runners from the tar-vine creeper, which is 'a sticky plant, and is used as a tangle-foot to trap small birds. It is spread around waterholes'.[308] Similarly, in north-east Queensland, Aboriginal hunters spread sticky latex made from the sap of the Watkins strangler fig and hairy fig onto branches to trap small birds.[309] Resin from the aptly named umbrella catch-bird-tree (cabbage-wood) was similarly used in this rainforest area.[310]

Emu hunters in arid south-west Queensland dug deep pit traps alongside food plants, such as the emu apple-tree, that were favoured by their game.[311] In the rainforest region of Cairns in northern Queensland, Yidiny people made traps for the orange-footed scrubfowl (scrub-turkey) in the form of a tunnel constructed from sticks, situated under Queensland walnut trees.[312]

Netting of birds in flight

For Aboriginal hunters, the use of bird netting offered the chance to capture a large number of birds on a single occasion.[313] In the Lower Lakes, nets, *wakiya*, were particularly useful when catching flocks of waterfowl.[314] Here, groups of men using nets, presumably by dragging them through the water, caught Eurasian coots and dusky moorhens during the night, and nets were set up to entangle hoary-headed grebes.[315] Extensive net constructions on large wooden poles were placed near sand spits where the Eurasian coots and dusky moorhens roosted.[316] In the Lower Murray and elsewhere in the Murray–Darling Basin, the nets were strung between two trees in the flight path of ducks.[317] Hunters made shrill whistles that mimicked those of birds of prey, and threw bark boomerangs which imitated hawks or falcons, in order to flush ducks and make them fly low into the netting. Many waterfowl could be caught in this manner. According to Tindale, a duck flyway, *parthi*, was a 'hunting place or trap, said of a place where ducks are netted using boomerangs and imitation calls of duck hawk [swamp harrier] *wampanji*'.[318]

Along the Coorong, *ngarankura* were 'nets for ducks, used at night time when raiding duck roosting places'.[319] Duck roosting places were called *teingawandi*, with 'a well known one on Round Island Tarlajing', which is south of Hack Point in the Coorong.[320] Hunters caught large numbers of hardhead ducks in open swamp areas by connecting several nets to form a large catchment approaching 30 m across.[321] Each hunter would claim the birds that were caught in his section of the net. Hunters would wade out during the night and lure birds into the nets by whistling. Bonney recalled Alf Watson's description of duck hunting in the southern Coorong/upper south-east of South Australia area using nets strung low to the ground and downwind from the waterhole. He said that ducks always fly into the wind when approaching to land. The thinner the twine the better, so that the ducks will not see it, and the mesh should be just smaller than a man's fist. The hunters stayed hidden near the base of the net so that they could grab the entangled ducks, twist their heads to break the necks, and put them into a bag.

Inland explorer Major Sir Thomas Mitchell provided a detailed account from 1836 of the Aboriginal netting of waterfowl along the Murray River in northern Victoria:

> The natives had left in one place, a net suspended across the river, between two lofty trees, evidently for the purpose of catching ducks and other water-fowl. The meshes were about two inches [5 cm] wide, and the net hung down to within five feet [1.5 m] of the surface of the stream. In order to obtain water-fowl with this net, some of the natives proceed up, and others down the river, to scare the birds from other places; and, when any flight comes into the net, it is suddenly lowered into the water; thus entangling the birds beneath, until the natives go into the water and secure them. Among the few specimens of art manufactured by the primitive inhabitants of these wilds, none come so near our own as the net, which, even in quality, as well as in the mode of knotting, can could scarcely be distinguished from those made in Europe.[322]

Mimicking a sky-diving bird of prey in order to keep the ducks flying low and into nets was another technique used in Victoria. Smyth recorded that:

> A common method of catching ducks is by fixing a net, about sixty yards [55 m] in length across a watercourse, a river, a swamp, or a lagoon – the lower part being three or four feet [90 cm or 1.2 m] above the water. The ends of the net are either fixed to trees or held by natives stationed in trees. One man proceeds up the river or lagoon, and cautiously moves so as to cause the ducks to swim towards the net. When they are near enough, he frightens them, and they rise on the wing, and at the same time another native, near the net, throws up a piece of bark, shaped like a hawk, and utters the cry of that bird. The flock of ducks at that moment dip, and many are caught in the net. Four men are usually employed when this sport is pursued.[323]

Birds, other than waterfowl, were also caught using netting. Along the Murray River, it was recorded by explorer/artist Angas in 1844 that:

> Poles with nets are also put up in the passages leading to the water, and when the [common] bronze-wing and crested pigeons come at dusk to drink, the nets are let go as they fly past, and sand is thrown at the birds to prevent their escape, or to make them alter their course into the net.[324]

Poisons

In seasonally arid parts of Australia, the use of water-based poisons derived from toxic plants offered the opportunity for catching large numbers of gamebirds. In northern Western Australia, the Kukatja people rubbed together the stems of *yungku-yungku* (leafless mirbelia), which is a medium size bush with yellow pea flowers, to put into drinking water in order to stupefy emus.[325] Wild indigo is another poisonous member of the pea family (Fabaceae) that was used to drug emus to make them easy to catch. It is also effective as a fish poison, but apparently mammals and amphibians are largely unaffected.[326] Basedow wrote that:

> The northern tribes of Western Australia have discovered a simple means of capturing the big struthious bird [emu] in that they poison a water known to be frequented by the game. When the bird has quenched its thirst, it is stupefied to such a degree that it is an easy matter for the natives, lying in ambush, to overtake it and crack it on the head. The poison used is supplied by the leaf of *Tephrosia purpurea* [wild indigo], which the natives call 'moru'; the active principle is a saponine.[327]

Across the Western Desert and Central Australian regions, Aboriginal hunters submerged foliage from the pituri shrub in waterholes to poison emus.[328] These plants are rich in alkaloids, which is a common characteristic of its family, the Solanaceae, that also contains the commercial tobacco species and the infamous hallucinogenic garden plant, angel's trumpet.[329] In the eastern central deserts, pituri was widely traded as a narcotic and kept in specially woven bags.[330] Records of pituri use as an emu poison cover much of its natural range across the arid zone: from central Western Australia, Mount Liebig and the Finke River in the Northern Territory, to Ooldea in western South Australia.[331] In Central Australia, biologist/anthropologist W Baldwin Spencer claimed that:

> The chief use of the Pituri plant in this neighbourhood (apart from its value as an article of [narcotic] barter) seems to be that of making a decoction for the purpose of stupefying and then catching the emu. The leaves are pounded in water and the decoction is placed in a wooden vessel where the emu is likely to come across it, or else a small pool or a fenced portion of a larger one is used for the purpose. After drinking it up the animal becomes so stupefied that it falls an easy victim to the blackfellow's spear.[332]

The main part of pituri used as a poison was its foliage, comprising narrow leaves and twigs. Central Australian missionary Rev Friedrich W Albrecht detailed the care required when using pituri as emu poison:

> The twigs are cut or broken into short lengths, then with a little water the juice is squeezed into one of the open waters, and all the leaves and signs of this removed, as emus know this poison plant and will go away if they see signs of it. However, if everything is clear, the animals will come at night and drink, when they are poisoned and die. Natives may eat the meat, provided they remove the eyes and intestines. Old men report that they have caught as many as seven emus in one night in this way.[333]

It is likely that this technique was used only when desert dwellers knew there were enough alternative sources of drinking water that they could afford to poison one of their waterholes. Western Desert people would cover the other nearby water sources to ensure that game could access only the poisoned hole.[334] Water laced with pituri is reportedly not harmful to marsupials and other birds, such as pigeons and parrots, although it will affect fish and various introduced placental species, such as camels.[335] A colonist reported that along the Finke River in Central Australia 'a few bullocks were poisoned by drinking on one occasion from a waterhole prepared in this way for emus by the aboriginals'.[336] It is oral history from Central Australia that 'One old [Aboriginal] man killed over 100 head of cattle in a water hole using this [pituri] plant as a payback to a pastoralist who killed his best hunting dog'.[337]

There were probably many other toxic plant species used to kill or stupefy gamebirds. For instance, the Warlpiri people of the Tanami Desert in northern Central Australia scattered the dried and crushed leaves of the striped mint-bush over the surface of waterholes to stupefy gamebirds, and left branches near the water's edge to warn other human visitors of the poison's presence.[338] It was reported in a newspaper that poison corkwood, a plant from the same genus as the pituri, was the source of a medical anaesthetic, and that in outback Queensland and New South Wales the 'Early settlers in their day noted that when emus drank from water poisoned by the leaves of the tree they became stupified [*sic*], while white men who tried the queer abo. mess for a chew found it produced a state resembling intoxication'.[339]

*****

Indigenous foragers in Australia possessed information about the distribution and behaviours of birds that was useful to ornithologists. This is demonstrated in the investigations of the past distribution of the night parrot. The species was formerly widespread across arid Australia, and in the early 20th century it was recorded in the Arrernte languages of the Macdonnell Ranges in Central Australia as *tnaljurbara* and *tnukutulbara*, and in the neighbouring Loritja language as *murkunba* and *analtjirbiri*.[340] In semi-arid north-west Victoria, a local ornithologist used evidence provided by Wergaia man Jowley (Peter McGinnis), who was born in the late 19th century.[341] A historian remarked that Jowley's 'observations of Night Parrot calls, habitat use, breeding and flight are some of the earliest published natural history of this most mysterious of Australian birds'.[342] During the Calvert Scientific Exploring Expedition to the Great Sandy Desert of 1896, the night parrot was seen but a specimen could not be obtained due to the 'erratic flight of the bird through the scrub', leading naturalist Keartland to write 'I afterwards ascertained from the natives that these Parrakeets lay four eggs in a loosely-made cup nest under the shade of the spinifex'.[343] The night parrot was for a long time assumed to be extinct,[344] but it has recently been rediscovered in areas such as the Indigenous Protected Area in northern Western Australia.[345] Given the present known range of the night parrot, Aboriginal rangers will be actively involved in the conservation of this species into the future.

The environmental knowledge that Aboriginal foragers held concerning birds, all of which were potentially game, was immense. It encompassed the seasonality of each species and the behavioural characteristics that they needed to take advantage of when hunting. As discussed in Chapters 8 and 9, most parts of the killed bird's carcass would have been utilised, if not for food then as raw materials for making a wide range of artefacts and ceremonial decorations. Many of the foraging techniques described in this chapter were used concurrently, which greatly increased the hunters' success. In remote parts of Australia, some foraging practices have continued in modified form. However, even here much environmental knowledge has been lost since the early 20th century, when Aboriginal peoples were forced to abandon their lifestyle of seasonal movements across Country for a more sedentary existence. In many parts of Australia, small-scale harvesting of wild sources of food, medicine and artefact-making materials has the potential to add to the local Aboriginal economy.[346]

Endnotes

1 Summarised by Clarke (2003a), Hallam (1975), Rose (1987) and Sutton & Walshe (2021).

2 Thomson (1939), Davis (1989), Clarke (2003a:Ch.7, 2009c, 2018g), Bird Rose (2005) and Turpin *et al.* (2013).

3 Tindale (1978:160–162) and Clarke (2003a:57, 2007a:12–13).

4 Tindale (1931–62) and Gale (2009:42–43,87).

5 Tindale (1931–62) and Gale (2009:74).

6 Wilhelmi (1861:172), Tindale (1978:160–162), Goddard & Kalotas (1988:14–15) and Clarke (2003a:57).

7 McConnel (1931:21–22). For Gugu Gulunggur, McConnel used the name Koko-yalunyu.

8 Sutton (1995:154).

9 Pascoe (2014).

10 Sutton & Walshe (2021). Cahir *et al.* (2018) described the Aboriginal environmental knowledge of south-eastern Australia.

11 Meyer (1846:191), Thomson (1949:21), Lawrence (1968:215), Yengoyan (1968:187), Berndt & Berndt (1970:33–40,109–110), Berndt (1981), Rose (1987:183–184) and Clarke (1994:178–179).

12 Clarke (2008a:25–41). Yengoyan (1968:186–187) described the vast range of Australian food sources, while Lee (1968:31–35) did the same in relation to the !Kung San of the Kalahari Desert in southern Africa.

13 Lawrence (1968:219–223), Meehan (1982:Ch.3) and Rose (1987:49–51).

14 Lawrence (1968:101–102,Table 4), Berndt *et al.* (1993:78–79) and Clarke (1994:161–163, 2017:28–30).

15 Dawson (1881:92), Noetling (1910:289), Horne & Aiston (1924:59–60), Mountford (1948:173–175) and McCarthy (1965:16).

16 Smyth (1878:Vol.1:195,208), Noetling (1910:288–289), Cleland (1940:6), Hope & Coutts (1971:109) and Meagher (1974:20–21,29–32,52).

17 Hyam (1943:173).

18 Basedow (1925:125).

19 For instance, see Smyth (1878:Vol.1:xxxvi), Kenyon (1912:99), Spencer (1918:118), Bellchambers (1931:90), Mulvaney (1959:8), Hope & Coutts (1971:111), Thomson (1975:42–43), Hope *et al.* (1977:375–378), Hemming *et al.* (2000:30) and MA Smith (2013:326).

20 Cooper & Condon (1947).

21 Stirling (1911:12). The Australian bustard is also called the plains turkey and wild swamp turkey, particularly in the earlier literature.

[22] West & Sim (1995:17–18).

[23] Attenbrow (2010:80).

[24] Attenbrow (2010:74–76). Note that the Pacific baza is also known as the crested hawk.

[25] Roberts *et al.* (2001:1891), Low (2014:143) and Shute *et al.* (2017:57).

[26] Flannery (1994:119–121,202), Rich (1979, 1985) and Murray & Vickers-Rich (2004).

[27] Shute *et al.* (2017:58).

[28] Woods (1882:387).

[29] Stirling & Zeitz (1896:172,177) and Rich (1979:1–2,59).

[30] For example, see Mulvaney (1960:62,65), Cosgrove *et al.* (1998:246), Jones (1998:108) and MA Smith (2013:99,326).

[31] Horton (1978).

[32] Jones & Allen (1978) and Sherwood *et al.* (2016).

[33] Cosgrove *et al.* (1998:246) and MA Smith (2013:175–176).

[34] Mulvaney & Kamminga (1999:22).

[35] Basedow (1914), Mountford (1929), Mulvaney & Kamminga (1999:369–373) and Morwood (2002:57–59).

[36] Chaloupka (1993:184–187).

[37] Murray & Vickers-Rich (2004), Gunn *et al.* (2011) and Miller *et al.* (2016).

[38] Tindale (1951). See also Hall *et al.* (1951).

[39] Mountford & Edwards (1962, 1963).

[40] Bednarik (2013).

[41] Basedow (1925:308).

[42] Isaacs (1980:17, see also 14).

[43] For the antiquity of people in Australia, refer to Tobler *et al.* (2017) and Purnomo *et al.* (2021).

[44] Cocker & Tipling (2013:27).

[45] Wehi *et al.* (2018).

[46] Wehi *et al.* (2018:466).

[47] Anderson (1989:Ch.7, see also 149–150,189).

[48] Anderson (1989:191).

[49] Roberts *et al.* (2001), Miller *et al.* (2016) and Hocknull *et al.* (2020).

[50] Clarke (2018c:51).

[51] Tunbridge (1991:41).

[52] Bennett (1860:235).

[53] Langloh Parker (1905:43). See also Elkin (1977:89). Ash *et al.* (2003:141) identified as 'weedah' as *wiidhaa* the spotted bowerbird.

[54] Thomson (1935:76).

[55] Green (2010:20).

[56] Clarke (2014a:91).

[57] Thomson (1939), Davis (1989), Clarke (2003a:Ch.7, 2009c, 2015b, 2018g:Ch.15) and Woodward *et al.* (2012).

[58] Newsome (1980).

[59] Barber (2005:55).

[60] Kaberry (1935) and Morphy (1984) have provided detailed accounts of the organisation of Aboriginal funerals.

[61] Tindale (1981:1878–1880) and Clarke (2003a:Ch.8).

[62] Tindale (1981:1875–1878) and Clarke (2003a:Ch.9).

[63] Warner (1937:369–376), Thomson (1949:Ch.2) and Clarke (2003a:Ch.10).

[64] Jones (1969) and Clarke (2003a:65–67).

[65] Hardwick (2019b:64).

[66] Bird *et al.* (2004, 2009:13–15) and Low (2014:228).

[67] Woinarski (1999:61–62) and Woinarski & Legge (2013).

[68] Hassell (1936:698).

[69] Hallam (1975:16).

[70] Gill *et al.* (1999), Gott (2005) and Low (2014:215–226).

[71] White (1994:Ch.19).

[72] Clarke (2016b, 2018b).

[73] Dawson (1881:52).

[74] Clarke (2018e).

[75] Eyre (1845:Vol.2:294).

[76] Bates (1918a). The preferred spelling in the Wirangu language for 'wild turkey' or bustard is *wardiring* (Miller *et al.* 2010:81).

[77] Lawrence (1968:Table 4).

[78] Berndt *et al.* (1993:560).

[79] Basedow (1925:137), Clarke (2017) and Hardwick (2019a:35).

[80] Sackett (1979), Altman (1987:Ch.12) and Clarke (2003b, 2018h).

[81] See also Wiynjorrotj *et al.* (2005:123), Karadada *et al.* (2011:72), Doonday *et al.* (2013:130) and Hardwick (2019b:22–23).

[82] Puruntatameri *et al.* (2001:104).

[83] Raymond *et al.* (1999:103), Wiynjorrotj *et al.* (2005:122) and Karadada *et al.* (2011:71).

[84] See also Puruntatameri *et al.* (2001:98).

[85] Hiatt (1967:126).

[86] Gould (1969b:261).

[87] Harney (1959:133–135), Raymond *et al.* (1999:113) and Puruntatameri *et al.* (2001:93–94). The 2006 award-winning film *Ten Canoes*, directed by Rolf de Heer and Peter Djigirr and starring Crusoe Kurddal, featured a main narrative based on a magpie goose egg-collecting expedition.

[88] Wiynjorrotj *et al.* (2005:123) and Karadada *et al.* (2011:71).

[89] Puruntatameri *et al.* (2001:103,106).

[90] Hardwick (2019b:68).

[91] Memmott (2010:14).

[92] Dawson (1881:19).

[93] Brand Miller *et al.* (1993:200–201,206–207).

[94] Angas (1847b:83) and Clarke (2017:30–34).

[95] Tindale (1981:1879).

[96] JM (1856:7).

[97] Berndt *et al.* (1993:559).

[98] Bellchambers (1931:28).

[99] Beruldsen (1980:186).

[100] Tindale (1964, 1987a:11). The malleefowl is also known as the 'wild pheasant'.

[101] Bellchambers (1931:28).

[102] Clarke (2017:34). Eyre (1845:Vol.2:274–275) stated that the malleefowl egg shells are 'extremely thin and fragile'.

[103] Beruldsen (1980:161).

[104] Tindale (1987b).

[105] Fison & Howitt (1880:226).

[106] Dawson (1881:93).

[107] Clarke (2015b:225).

[108] Clarke (2017:30).

[109] Clarke (2003b:97–98, 2018h:12). See paintings by Ian Abdulla (1993).

[110] Abdulla (1993).

[111] Karadada *et al.* (2011:71,73,83–84).

[112] Karadada *et al.* (2011:71).

[113] Harney (1959:133–135).

[114] Hardwick (2019b:52–53).

[115] Hardwick (2019b:52–53, 2019d:22).

[116] Thomson (1935, 1983a:95–100, 1996:50–55). See also Olsen & Russell (2019:154–157).

[117] Thomson (1939:215–217).

[118] P Sutton (pers. comm.).

[119] Hale & Tindale (1934:108).

[120] Puruntatameri *et al.* (2001:94) and Raymond *et al.* (1999:113). For instance, in 1979 a dangerous 5.1 m crocodile, known locally as 'Sweetheart', was caught along the Finniss River in magpie goose Country (Museum & Art Gallery of the Northern Territory website, https://www.magnt.net.au/sweetheart-crocodile).

[121] Thomson (1935:30–31).

[122] Thomson (1935:30–31).

[123] Tunbridge & Coulthard (1985:67), Clarke (2007a:27,34–37, 2008a:63–64,66,74–75,98, 2012:43–44) and Hardwick (2019b:59).

[124] Tunbridge & Coulthard (1985:71).

[125] Weldon (1936:51).

[126] Basedow (1925:141).

[127] Eyre (1845:Vol.2:274).

[128] North & Keartland (1898:130).

[129] North & Keartland (1898:169). The authors referred to this species as the 'blood-stained cockatoo'.

[130] Hardwick (2019b:61, 2019e:54).

[131] Eyre (1845:Vol.2:274).

[132] Singer *et al.* (2021:24). For descriptions of Aboriginal sign language, refer to Roth (1897:Ch.4), Mountford (1949), Kendon (1988), Bauer (2014) and Green *et al.* (2019:Ch.10).

[133] Clarke (2003b:96, 2018h:11).

[134] Basedow (1925:389).

[135] Wiynjorrotj *et al.* (2005:122).

[136] Davis (1989:8).

[137] Thomson (1983a:95–100, 1996:50–51), with accounts from 1937.

[138] Hardwick (2019b:52–53, 2019d:22).

[139] Chaloupka (1993:11,185–189).

[140] Akerman *et al.* (2014:184).

[141] Hemming *et al.* (2000:13) and Clarke (2017:34–37).

[142] Berndt *et al.* (1993:560). Clarke (2019a:139) provided the identity for this species.

[143] Berndt *et al.* (1993:94).

[144] Hardwick (2019b:63).

[145] Berndt *et al.* (1993:557). The great egret also known as the 'white crane'.

[146] Berndt *et al.* (1993:557).

[147] Smyth (1878:Vol.1:193).

[148] Jones (1996a:98–99). Davidson (1936) provided an overview of the distribution of various types of throwing-stick and boomerang.

[149] Ramsay Smith (1909:12, 1930:221–222), McCarthy (1961:346), Jones (1996a:48–51) and Hemming *et al.* (2000:13,19).

[150] Basedow (1925:137).

[151] NB Tindale (cited in Gale 2009:33).

[152] Berndt *et al.* (1993:558–560). The great skua was referred to as the 'white-bellied seagull', the Caspian tern as the 'large red-legged white seagull' and the sharp-tailed sandpiper as the 'tern' (Clarke, 2019a:125,132,136–137,139).

[153] Tindale (1930–52:23,257).

[154] Berndt *et al.* (1993:557). The white-faced heron was described as the 'blue crane'.

[155] Berndt *et al.* (1993:561).

[156] Taplin (1859–79:31 December 1859).

[157] Berndt *et al.* (1993:558). The hardhead is also locally known as a 'widgeon'.

[158] Smyth (1878:Vol.1:194). Refer to Smyth (1878:Vol.1:195) for a list of waterfowl and other birds eaten at Lake Tyers. Also see Smyth (1878:Vol.1:197) for list of birds eaten by Victorian Aboriginal peoples.

[159] Smyth (1878:Vol.1:193).

[160] Parker & Reid (1983:138) and Blakers *et al.* (1984:73).

[161] Clarke (2017:36, 2018:6).

[162] Berndt *et al.* (1993:561).

[163] Slater (1978:18).

[164] Hardwick (2019b:22,58).

[165] Grey (1841:Vol.2:281–282).

[166] Cawthorne (1868:1,5).

[167] Cawthorne (1868:5).

[168] Smyth (1878:Vol.1:195). The *braak* (*breyak*) has been identified as the sulphur-crested cockatoo in the Gippsland languages (Mansergh & Hercus 1981:18).

[169] Ramson (1988:414).

[170] DR Horton (in Horton 1994:Vol.1:180) and Low (2014:283).

[171] West & Sim (1995:18–19).

[172] Clarke (1996a:64).

[173] Anonymous (1893:4, 1895:2), Barrett (1909:7), Smith (1965), West & Sim (1995:19–20) and Low (2014:283–287).

[174] Bonwick (1884:192).

[175] Smyth (1878:Vol.1:192).

[176] Cleland (1966:143), Thomson (1975:25,70,87–88, 1983b:66) and Raymond *et al.* (1999:106).

[177] Harney (1957:175).

[178] Jones (1996a:84–87).

[179] North & Keartland (1898:169). North described this species as the 'Cockatoo Parrot'.

[180] Thomson (1983b:73, see also 66).

[181] Hardwick (2019b:51).

[182] Hardwick (2019b:57).

[183] Hardwick (2019b:58).

[184] Thomson (1983a:101). Originally written in 1937.

[185] Ramsay Smith (1930:227) and Clarke (2009a:152).

[186] Smyth (1878:Vol.1:192).

[187] Clarke (1996a:59,64). See also Slater (1978:14).

[188] Mjöberg (1918:320), Wiynjorrotj *et al.* (2005:122) and Clarke (2012:42). For use of mud to disguise scent, see Doonday *et al.* (2013:129) and Clarke (2017:38–39).

[189] McCarthy (1965:17).

[190] Berndt *et al.* (1993:92–93,Fig.13).

[191] Mjöberg (1915:205).

[192] EKV (1884). As proof of the risk posed by emus, this species is listed as being one of the '6 of the world's most dangerous birds' (Britannica website, https://www.britannica.com/list/6-of-the-worlds-most-dangerous-birds#:~:text=Reports%20of%20emu%20attacks%20resulting,100%20occurring%20in%202009%20alone).

[193] Tindale (1941b:239).

[194] Smyth (1878:Vol.1:194, Vol.2:298), Schell (1914), Mjöberg (1918:320), Raymond *et al.* (1999:113) and Clarke (2017:34–35).

[195] Angas (1847b:69,90–91), Chenery (cited in Smyth 1878:Vol.1:194), Sullivan (cited in Smyth 1878:Vol.1:194), Ramsay Smith (1909:10, 1930:220–221), Atkins (1911), Berndt *et al.* (1993:94), Clarke (2009a:152) and Doonday *et al.* (2013:136).

[196] Anonymous (1913a).

[197] Clarke (1994:162, 2003b:87).

[198] Clarke (2003b:87, 2017:28).

[199] Wilson (1998:91).

[200] Basedow (1925:138–139).

[201] Hardwick (2019b:23).

[202] Clarke (2009a:151, 2017:37).

[203] Gould (1969a:42) and Patak & Baldwin (1998).

[204] Collins (1798–1802:462).

[205] Roth (1897:96).

[206] Giles (1875:43,71).

[207] Anell (1960) and Satterthwait (1986).

[208] Angas (1847b:99–100), Meyer (1846:194), Nott (1856–c.1935:25) and Beveridge (1883:44–45).

[209] Tunbridge & Coulthard (1985:31,33).

[210] Tunbridge & Coulthard (1985:31,70) and Tunbridge (1991:63–64).

[211] Langloh Parker (1905:116–117).

[212] Langloh Parker (1905:118).

[213] McCarthy (1965:17).

[214] Tindale (1937a:113).

[215] Tindale (1936b:23).

[216] Lycett (1820–22:Plate 17) and Attenbrow (2010:88,92–93).

[217] Hahn (1838–1839:133).

[218] Tindale (1981:1878–1879).

[219] Bellchambers (1931:90).

[220] Clarke (2017:37).

[221] Anonymous (1911a).

[222] Doonday *et al.* (2013:129) and Clarke (2017:38–39).

[223] Basedow (1925:138, see also 1907:21).

[224] Plomley (1966:751–752) and Lewis (1988:76).

[225] Angas (1847b:63).

[226] Tindale (1981:1880).

[227] Angas (1847b:139).

[228] Tunbridge & Coulthard (1985:65).

[229] Basedow (1907:21–22, 1925:138).

[230] Bird Rose *et al.* (2002, pp.88,94–95).

[231] Hardwick (2019b:52, see also 62).

[232] Basedow (1925:138).

[233] Browne (1895:12).

[234] Lewis (1988:74).

[235] Raymond *et al.* (1999:110–111).

[236] Hardwick (2019b:65).

[237] Clarke (2017:37–38).

[238] Collins (1798–1802:455). Also cited by Smyth (1878:Vol.1:197) with some alterations.

[239] Krefft (1862–65:369), Smyth (1878:Vol.1:193), Sullivan (cited in Smyth 1878:Vol.1:194), Beveridge (1883:45–46), Richards (1903:166) and Langloh Parker (1905:118).

[240] For fishing technology, refer to Clarke (2002:151–153).

[241] Basketry duck decoy, South Australian Museum specimen A23410, made by a Tangani (Tanganekald) man Milerum (Clarence Long) in 1936, collected NB Tindale.

[242] Atkins (1911).

[243] Ramsay Smith (1930:221–222) and Clarke (2009a:152).

[244] Clarke (2018h:10).

[245] Amateur Naturalist (1887).

[246] Bellchambers (1931:19).

[247] Abdulla (1994).

[248] Ramsay Smith (1909:10).

[249] Tindale (1938–60:21). Tindale wrote Kuungkari as 'Ku:ngkari'.

[250] Langloh Parker (1905:116).

[251] Horne & Aiston (1924:59–60).

[252] Tindale (1938–60:17,19).

[253] Horne & Aiston (1924:30).

[254] Basedow (1907:22).

[255] Basedow (1907:22).

[256] Berndt *et al.* (1993:93–94,556,Fig.13).

[257] Anonymous (1889).

[258] Smyth (1878:Vol.1:192).

[259] Mathews (1904:256–257).

[260] Davidson (1898:331).

[261] Nott (1856–c.1935:25).

[262] Slater (1978:34).

[263] Dawson (1881:91).

[264] Basedow (1925:139–140).

[265] Basedow (1925:139).

[266] Lowe (2002:25).

[267] MacGlashan (1887:22).

[268] Basedow (1935:62).

[269] Basedow (1907:22).

[270] Mathews (1901:76–77) and Mulvaney (1993).

[271] Bindon & Lofgren (1982).

[272] Basedow (1925:137, 1935:72–74) and Lewis (1988).

[273] Seddon & Launay (2008:199) and Cocker & Tipling (2013:166–167).

[274] Hill (1968:5–9).

[275] GA Robinson, 12 July 1833 (cited in Plomley 1966:751–752).

[276] Worsnop (1897:116).

[277] Berndt *et al.* (1993:94,Fig.14).

[278] Berndt *et al.* (1993:556).

[279] Howchin (1934:78) and Clarke (2019b:159–167).

[280] Baker (1959:190).

[281] Horne & Aiston (1924:135).

[282] Horne & Aiston (1924:135).

[283] Barrett (1938:42).

[284] Gason (1879:278) and Johnston (1943:270).

[285] Taçon *et al.* (2010).

[286] Spencer & Gillen (1927:Vol.1:145–146,148).

[287] Meggitt (1962:221).

[288] Clarke (2009a:152, 2017:38–39).

[289] Gale (2009:90).

[290] Angas (1847b:148) and Worsnop (1897:114–116,Fig.58).

[291] Ramsay Smith (1930:224–225).

[292] Berndt *et al.* (1993:559, see also 94). The possible identification of *wirili* is given by Clarke (2019a:142).

[293] Meyer (1846:193), Angas (1847b:90–91), Taplin (1859–79:6 January 1874), Atkins (1911), Tindale (1934–37:284), Berndt *et al.* (1993:556,561) and Hemming *et al.* (2000:13). Note that *ranari* has also been written as *runeri*.

[294] Ramsay Smith (1930:223–224) and Berndt *et al.* (1993:94,560).

[295] Penney (1842–43:70).

[296] Eyre (1845:Vol.2:283).

[297] Tindale (1934–37:284).

[298] Berndt *et al.* (1993:88–90,344,Fig.10C).

[299] Clarke (2017:39).

[300] Angas (1847b:91).

[301] Angas (1847b:91).

[302] Dawson (1881:93).

[303] Worsnop (1879:116).

[304] Smyth (1878:Vol.1:196–197).

[305] Burr (1845:168).

[306] Roth (1897:98) and Duncan-Kemp (1933:154).

[307] Cleland (1957:161), Goddard & Kalotas (1988:148) and Clarke (2012:58).

[308] Tindale (1941a:9). See also Cleland (1957:161) and Goddard & Kalotas (1988:148).

[309] Dixon (1991:198).

[310] Dixon (1991:200). The umbrella catch-bird-tree is also known as 'cabbage-wood'.

[311] Roth (1901c:26). Thomas (1906:100) provided a generalised account of 'emu pits'.

[312] Dixon (1991:178).

[313] Clarke (2017:39–40).

[314] Berndt *et al.* (1993:94,Fig.14), Tindale (cited in Gale 2009:159) and Taplin (1889).

[315] Berndt *et al.* (1993:557). The hoary-headed grebes are referred to as 'divers'.

[316] Berndt *et al.* (1993:94,96,Fig.12).

[317] Eyre (1845:Vol.2:286–287), Angas (1847b:100), Stanbridge (1861:293), Krefft (1862–65:368–369), Smyth (1878:Vol.1:193), Anonymous (1889), Worsnop (1897:92–93), Schell (1914), Ramsay Smith (1930:223), Tindale (1930–52:23), Mann (cited in Padman 1987:8) and Hemming *et al.* (2000:13). Note that pituri is sometimes written as 'pitjuri'.

[318] Tindale (1931–62).

[319] Tindale (c.1931–91a).

[320] Tindale (c.1931–91a).

[321] Berndt *et al.* (1993:94).

[322] Mitchell (1839:153).

[323] Smyth (1878:Vol.1:193).

[324] Angas (1847b:10).

[325] Peile (1997:80).

[326] Heuzé *et al.* (2018).

[327] Basedow (1925:139).

[328] Basedow (1925:157), Spencer & Gillen (1927:Vol.1:16), Cleland (1966:120), Warlpiri Lexicography Group (1986:66–67), Latz (1995:163) and Bindon (1996:108).

329 For data on Australian species of *Duboisia* and *Nicotiana* species, refer to Johnston & Cleland (1933–34), Peterson (1977), Watson (1983), Saitoh *et al.* (1985) and Ratsch *et al.* (2010). Everist (1981:622–685) discussed the poisonous species of Solanaceae in Australia, including *Datura* species.

330 Roth (1901c:30–31), Watson (1983), Clarke (2007a:108–109) and Silcock *et al.* (2012).

331 Hamlyn-Harris & Smith (1916:1), Spencer (1922:69), Basedow (1925:139,157), Cleland & Johnston (1933:123, 1937–38:211), Cleland & Tindale (1954:85), Cleland (1966:120), Tindale (1972:254), Webb (1973:294), Roheim (1974:43), Reid (1977:161), Goddard & Kalotas (1988:22), Goddard (1992:178) and Green (2010:164). Aboriginal artefact collection, South Australian Museum. Records of *Duboisia myoporoides* used as a poison in Central Australia probably relate to *D. hopwoodii*.

332 Spencer (1896:Vol.1:82). See also Cleland & Johnston (1933:116).

333 Albrecht (1959). Albrecht gave the Aboriginal name for the poison as *monanga*, which appears to be equivalent to the *monunga* specimen of *Duboisia hopwoodii* collected by JB Cleland at Mount Liebig in 1932 (Aboriginal artefact collection, South Australian Museum).

334 K Akerman (pers. comm.).

335 Tindale & Hackett (1933:103) and Basedow (1935:64–67). Everist (1981:637) stated that there are also field cases of pituri poisoning involving cattle, goats, sheep, horses and camels.

336 Newland (1890:23).

337 Smith (1991:22).

338 Henshall *et al.* (1980:20) and Bindon (1996:209).

339 ECS (1935:2).

340 Strehlow (1909:341,348).

341 Menkhorst & Ryan (2015:112).

342 Menkhorst & Ryan (2015:107).

343 North & Keartland (1898:171).

344 Slater (1978:60–61), Boles & Longmore (2014), Pyke & Ehrlich (2014) and Olsen (2018).

345 Refer to websites: http://www.nespthreatenedspecies.edu.au/news/talking-night-parrots-on-paruku-country; https://www.klc.org.au/paruku-rangers-discover-night-parrot.

346 Altman (1987, 2001), Wilson *et al.* (1992, 2004), Altman *et al.* (1995, 1997), Clarke (2003b, 2018h) and Jones & Clarke (2018).

7
Birds working with people

According to Aboriginal tradition, birds were prominent among the ancestors who shaped the world. Aboriginal peoples believe that after the creation period was over, birds continue to help them in diverse ways. Birds are watched intently for signs to help make deductions about such things as local flooding, changes in tides and weather, the onset of new seasons and the presence of game animals and other people. They also assist with hunting, finding water and the spreading of fire. This chapter provides an account of the myriad ways, both tangible and intangible, that birds impact the lives of Aboriginal peoples, excluding those chiefly concerned with food/medicine sources and artefact-making materials, which will be discussed separately in Chapters 8 and 9.

Controlling sea incursions and floods

In earlier colonial times, Christian missionaries who were engaged in cultural syncretism saw the Aboriginal beliefs in floods as Indigenous versions of 'the Flood of Noah's day' and a 'plan of salvation'.[1] As discussed in Chapter 2, scholars have suggested that the close and devastating interactions between early British colonists and Aboriginal peoples across south-eastern Australia created a greater emphasis on 'All-father' ancestors, such as Ngurunderi and Baiame, in this region. More recently, some scientists have similarly engaged in historical syncretism in the dubious exercise of equating the Aboriginal myths of apparent cataclysms with the purported ancient memories of past climate-changing events, such as formation of the current coastline at the end of the Pleistocene, dating back many thousands of years.[2] In the case of the latter, as discussed in Chapter 2 it is my opinion that such highly selective extractions of 'data' from Aboriginal mythology are highly speculative and therefore any findings derived from them are fanciful. The body of environmental knowledge that Aboriginal peoples maintained is dynamic, with relevant information being constantly augmented by insights gained from the life experiences of Aboriginal peoples. Within their own lifetime they had been subject to events like severe local flooding and extreme weather, and therefore incorporated or maintained them as themes in their mythologies.

In the creation mythologies, birds are among those ancestors who caused the rising of the sea. For instance, in the south-east of South Australia there is a myth whereby the brolga ancestor stabbed the ground at the coast near Kingston with its beak, causing the seawater to rise and drown the emu family.[3] Further south in the Mount Gambier area, it is a Bunganditj tradition that Croom the musk duck ancestor stopped the burning of Country caused by the first release of fire when he clapped and shook his wings, bringing in water that settled as swamps and lakes.[4] In south-east New South Wales and neighbouring eastern Victoria, Kaboka was an angry bird ancestor who created a flood to drown his companions, by creating a fire and dancing round it.[5] The Tambo River in Victoria floods each year because of Kaboka, who is described as a 'thrush'. This description illustrates the problem with many European accounts of myths, particularly those in the early records, as this bird could presumably refer to a grey shrikethrush, a Bassian thrush or another species altogether – we just do not know the species involved. In the above examples, from an Aboriginal perspective the avian descendants of those ancestors who controlled water have retained a close association with wetlands, while the descendants who suffered as a consequence have remained as dryland species.

Bird ancestors are widely said by Aboriginal peoples to have been involved in preventing the incursions of the sea and clearing Country of floods. During a fieldtrip across western South Australia in 2002, a Southern Pitjantjatjara man informed me that during the creation Nyiinyii the zebra finch men were responsible for creating the Nullarbor cliffs of the Great Australian Bight, when they made a wall from spears made from *ngalta* (desert kurrajong) roots to hold back the encroaching sea and regain their drinking water.[6] This is clearly a narrative from an inland Aboriginal group, as desert kurrajong trees do not occur as far south as the coast in this region.[7] In the Port Keats area south of Darwin in the Northern Territory, the Murinbata people have a myth concerning bird ancestors who took refuge during a flood on the summit of Table Hill. The water level subsided only when Karan the bush stonecurlew cut off the top joint of one of the young bird men's fingers so that the blood ran freely into the flood waters.[8] In the central Kimberley, it was the southern boobook ancestor who turned back the sea. Missionary James RB Love recorded that:

> On the brow of a hill near the Kunmunya Mission Station is a group of long stones erected in a circle, within sight of the salt water. The story of these stones is that once the tide came in exceptionally far, threatening to engulf all the land. Seeing the danger, the Boobook Owl (*Ninox boobook* [*Ninox novaeseelandiae boobook*]) flew to this hill crest and perched on a rock. Looking down on the encroaching tide, he uttered his awesome cry, 'Ngok! Ngok! Ngok! Ngok!' Overawed by his big eyes and fearsome voice, the tide turned and ran back, and so the Boobook Owl saved the land. The stones then arose to mark the site of his deliverance.[9]

In Aboriginal tradition, once the creation had ceased, birds continued to be involved with the change of tides. In the central Kimberley region, the spring tides are very high, and here Love observed that 'The Rufous Whistler (*Pachycephala rufiventris*) may be often heard near the mangroves, sending forth his lovely clear whistle that ends in a sharp upward lift. The Worora say that he is continually calling in the tide'.[10] In the Lower Murray region, it is the Australian raven ancestor in the Skyworld who controls the tides. Here, Ronald M Berndt and Catherine H Berndt recorded that:

> Crow [Marangani, Australian raven], metamorphosed as a star, is especially mentioned, since he is propitious for fishing and is responsible for the ebb and flow of the tides … Crow appears in June–July (winter) when fishing is sporadic, since he spends little time in the sky and most in the huts of various women: he disappears in October, coinciding more or less with Waiyungari [Mars].[11]

Forecasting weather and seasonal change

The life of a hunter-gatherer throughout the year was highly varied. Aboriginal peoples, as they moved across Country in foraging bands, took careful note of any changes they observed in the environment, and used this knowledge to time their movements and fine-tune their subsistence strategies.[12] For instance, when predicting incoming weather, Aboriginal people in south-western Victoria looked out for such phenomena as animal behaviour, wind direction, the appearance of the sun and moon, colour of the horizon at sunset and sunrise, and the time of day when rainbows occurred.[13] Often, it was not one but several factors that were involved. In 1876, a pastoralist gave an account of Aboriginal weather forecasting in the mallee country straddling the South Australia–Victoria border. He said:

> 'Rain come bailie [good] this time!' said the blackfellow, 'because pelican fly away from billabong, and wallows all quambie [stop] along o' ground. Big one rain come bailie, me know, because curlew [bush stonecurlew] yabber [crying out] along o' day time.' The sky at that time was like a great canopy of lead, and the heat was thick, intense, and oppressive; and there was considerable agitation amongst the birds, as if they expected a catastrophe.[14]

In arid Australia, when the Walmajarri people of the Great Sandy Desert in Western Australia see the fork-tailed swift or 'rain-maker bird' high in the sky they know the rains are coming, and if they hear the rainbow bee-eater sing 'tirn…tirn' they know that it will be a heavy fall.[15] When the Walmajarri see lots of Australian pelicans flying over their Country, they know it's a good time to fish.[16] The channel-billed cuckoo is also known as the 'storm bird' and 'rain bird', and when it is heard calling this means that rain will fall shortly.[17] In Central Australia, Arrernte-speakers refer to this bird as *lkwarrer-arrpwernene*, which literally translates as the 'yeller of bush bananas'.[18] The channel-billed cuckoo is an important ancestor for many Aboriginal groups, with a myth track that runs from the Country of the Mudburra people at Elsey Station in the Top End of the Northern Territory south towards Pitjantjatjara Country in northern South Australia.[19] The channel-billed cuckoo is also part of the creation mythology of the Worora people in the central Kimberley, concerning Kaluru the Wandjina spirit who created a flood to punish people for their ill treatment of the 'Winking Owl' (barking owl). According to Love, since then:

> The weird cry of the storm-bird (channel-bill cuckoo) is interpreted as her cry of mourning for the people who were lost in the flood. The storm-bird cries in the wet season, when the storms are severe.[20]

The oriental pratincole has been described in northern Western Australia as the 'little storm bird', because its sudden appearance is an indication of approaching rain.[21] Naturalist George A Keartland was part of the 1896 Calvert Scientific Exploring Expedition in northern Western Australia when he obtained specimens of the oriental pratincole in the southern Kimberley. With some amusement, he noted:

The natives were very indignant at my shooting these birds, and a deputation from the blacks' camp explained for my edification that if I killed any more a big rain would come and never stop until it had washed everything away. Although I was responsible for the death of about a score of birds the deluge had not occurred in the district at the time of writing. Perhaps because I left.[22]

Apart from birds giving signs of approaching weather, they also forecast the impending change in season. In the Lower Murray, for example, the onset of a new season was indicated by such phenomena as changes in wind direction and water flow, as well as by the flowering of certain plants.[23] Each season here was also associated with specific animal behaviour, such as the arrival of migratory birds. For instance, Aboriginal hunters knew that the blue-winged shoveler and freckled duck come down the Murray River to feed in the swamps of the Lower Murray during summer and autumn.[24] According to the Berndts, banded lapwings and dusky moorhens would be caught by boys wielding clubs as they converge on the river and lakes, when waterholes and creeks in the surrounding Country had dried out.[25] Brown bitterns are present throughout the year but were more easily caught during the summer, as they are not heard calling during the winter when they stay among the reeds.[26] Brolgas arrived in the Lower Murray from the south-east of South Australia in spring to dig out beetles from the flats, while Australian bustards were attracted to inland areas which had been burnt by bush fires.[27] Knowledge of the seasonal movements and preferred habitats of such birds was important for hunter-gatherers. The Berndts described the *kenigeri-on* as a 'large grey bird that whistled', possibly a grey shrikethrush, that came to the Lower Murray at the beginning of summer:

> The old people regarded them as the harbinger of summer. Two birds always made their appearance first, before the mass migration. They were found in the scrub and bush country.[28]

The close seasonal association of birds is also important for other parts of Aboriginal Australia. On the Adelaide Plains, Wilto the wedge-tailed eagle ancestor in the Skyworld was believed to have dominated spring time, which was fittingly known as Willutti, and similarly Wolta the Australian bustard from the heavens governed the summer season, Woltatti.[29] In south-western Victoria, it was reported in a regional newspaper in 1885 that 'There is a solitary pelican wandering about Hotspur River north of Heywood at present, and the aborigines at Condah [Lake Condah Mission] say that it is a sure sign of unusual drought in the Mallee'.[30] Here also, colonist James Dawson remarked that local Aboriginal people knew that:

> ... when mosquitoes and gnats are very troublesome, rain is expected; when the cicada sings at night, there will be a hot wind next day. The arrival of the swift, which is a migratory bird, indicates bad weather. The whistle of the black jay [grey currawong], the chirp of the little green frog, the creak of the cricket, and the cry of the magpie lark indicate bad weather; wet weather is more likely to come

after full moon. It is a sign of heat and fine weather when the eagle [probably wedge-tailed eagle] amuses itself by towering to an immense height, turning its head suddenly down, and descending vertically, with great force and with closed wings, till near the earth, then opening them and sweeping upwards with half-closed wings to the same height. This movement it repeats again and again, for a long time, without exertion and with apparent pleasure. The aborigines call this movement 'warroweean,' and always expect warm weather to follow it.[31]

Colonist and writer Katherine Langloh Parker described the perceptions of the Yuwaalaraay-speaking (Euahlayi) people of central New South Wales concerning the origins of the weather, some of which reflected the antagonism of the wedge-tailed eagle and crow/raven ancestors, as described in Chapter 2. She claimed that here:

> The big mountainous clouds when they come from the south-west are said to be Mullyan, the eagle-hawk [wedge-tailed eagle], who makes the south-west wind claimed by Maira, paddy [native] melon totem, one whose multiple totems Mullyan is.

> The crow keeps the cold west wind in a hollow log, as she was too fond of blowing up hurricanes; she escapes sometimes, but the crow hunts her back. But they say the log is rotting and she will get away yet, when there will be great wreckage and quite a change in climates.[32]

In Central Australia, Aboriginal people imitated the call of the masked lapwing to ritually bring in the rains,[33] and in central New South Wales the arrival of 'swifts' was the 'harbingers of rain'.[34] In the north central Kimberley, the Wunambal Gaambera name for both the fork-tailed swift and the frigate-bird is *wunjuwunju* and it refers to the monsoonal rain, *wunju*, with which these birds are closely connected.[35] The call of the blue-winged kookaburra in the south-west Kimberley announces the arrival of the wet season, and it is closely associated with rain.[36] Here, it was recorded from Butcher Joe Nangan that:

> As a human in the Bukarikara [creation] the kookaburra was a healer and rainmaker. If a kookaburra is killed or harmed, storms will result. If its eggs or chicks are harmed, it will 'belt you with rain'. It will use its magic *binjabinja* to alert the rainbow serpent at Lanin, which will then cause the rain to fall. The rain will disperse any game in the area, making it hard to hunt.[37]

Each season, as recognised by Aboriginal peoples, is intimately connected to a set of edible foods. For instance, in the rainforest region inland from Cairns in northern Queensland, the Yidiny-speaking people said that the brown cuckoo-dove will 'call out when *jambun* grubs are ready'; these are generally found in candlenut and milky pine trees.[38] During the 'build-up' or pre-wet season of the Top End, Aboriginal people note that the 'singing bushlark [Australian skylark] sings a song to tell that this is the time when things should grow'.[39] In 1960, Norman B Tindale was doing fieldwork on Mornington Island of the Wellesley Islands in the Gulf of Carpentaria in Queensland when he observed a Lardil performance of the frigate bird dance. He noted that 'This bird is one which indicates the beginning of the monsoonal storms of the N.W. [wind] season'.[40] Tindale also observed here that 'This bird, appearing in numbers at the end of the dry season indicates storms with rain and is therefore a harbinger of rain and good times to come'.[41] Biologist/anthropologist Donald Thomson remarked that 'Frigate-birds were often noted, particularly in windy weather, and they are believed by the sea-faring aborigines of the east coast [of Cape York Peninsula] to be the bringers of wind'.[42] Concerning the frigate bird, the Mawng people of coastal western Arnhem Land say 'When they see a big wind, they fly and stop the wind'.[43] The presence of such large birds with red throats is easily observed, even during storms, as I found in February 2020 when from an Islander's boat in Torres Strait I saw a pair of frigate-birds flying high above the seas off Muralag (Prince of Wales Island) during a monsoonal downpour.

In the tropics, when Aboriginal people hear the distinctive call of the Asian koel, commonly known in English as the 'stormbird' or 'rainbird', it indicates to them that the wet is rapidly approaching.[44] For the Jawoyn and Wardaman people living in the Katherine region of the Northern Territory, the koel 'sings up' the black plum fruit just before the first rains of the wet, so that it is large and ripe for eating.[45] When the common emerald dove's distinctive breeding and territorial call is first heard during the early to mid wet season on Melville and Bathurst Islands, it is taken by the Tiwi people as a sign that the kurlama yams are fully developed.[46] The pied imperial pigeon disappears from many parts of its range across the tropics in the late wet, but returns in the late dry to eat the fruit of the Carpentaria palm.[47] In the coastal area of western Arnhem Land, the Mawng people explain that the magpie-lark leaves their Country before the wet season rains in November and returns when it is time to gather the water chestnut tubers in June and July, which is also when it is time to shoot magpie geese.[48]

Changing bird behaviour signalled the onset of the next season. For example, in the coastal south-west Kimberley the arrival of the Asian koel from south-east Asia heralds the beginning of the wet (September–October) and is a sign for fishers to use plant poisons to kill fish in the reef pools.[49] At Borroloola in the southern Gulf of Carpentaria, the call of the crow is said by Yanyuwa people to bring out the cold season.[50] In the Yidiny language, spoken in the Cairns region of northern Queensland, *jalnggan* is a dark-coloured species of swamp bird, unidentified by linguists, that is 'said to give warning of an approaching cyclone by flying uncharacteristically low over the ground'.[51] Here also, *jawajawa*, jointly the Australian magpie and the pied currawong, is said to be a 'rain-maker'.[52] There are potentially many signs to indicate a change of season. Anthropologist John Bradley stated that for the Yanyuwa people at Borroloola the:

> *Julaki* – birds, bats and flying foxes are seen, in Yanyuwa eyes, to be the markers of seasonal change and variation. Their movements tell the Yanyuwa which seasons are approaching, which plants are flowering and seeding, and the location of lagoons in the dry season that still may have water.[53]

Aboriginal perceptions of the seasons in Australia are generally more complex than those of the European colonists, in both the number of recognised seasons and their associations with Country.[54] For instance, in 1949 Thomson noted that Aboriginal people in Arnhem Land divide the year into six or more seasons and:

> State without hesitation the appropriate occupational sites at each of them, and their activities and food supplies which depend upon the seasonal conditions, with an insight and a precision that no white man who is not a trained ecologist and well versed in botany and zoology, could approach.[55]

Due to the recognised close relationship between the flora and birds, Aboriginal peoples can look at plants for an indication of what is happening in the bird world. For instance, during my 1980s fieldwork in the south-east of South Australia, Ron Bonney stated that when the yellow flowers of a large daisy bush (possibly *Senecio* species) appeared in mid July, it was a sign that black swan eggs were available in swamps, such as in the lakes near the coast at Teeluc, just north of Kingston.[56] The linking of two environmental events, both subject to the same annual changes in temperature and rainfall, took into account any variation caused by an early or late commencement of a season. Aboriginal foragers also recognised that gamebirds were better eating in seasons when they had abundant food supplies, such as insects. Colonist George F Fletcher recorded in the south-west of Western Australia the Nyungar word *dangyl* for lerp sugar produced by a scale insect found on eucalypt foliage. He stated that it was a favoured food for Aboriginal peoples and that 'Birds feed upon it and are in excellent condition during the season when it abounds'.[57] The birds eating the lerps were probably pardalotes.[58]

In north-east Arnhem Land, the flowering of the country mallow shrub is a sign that terns (probably greater crested terns) are laying eggs, which for the Yolngu people are favoured foods.[59] The yellow colour of the flower on this bush is said to be the same as the yolk of the tern egg. In the swampy parts of northern Australia, where there is a marked distinction between the wet and dry periods, the appearance of tall grass in the landscape during the early rainy season is quite sudden and linked to other biological events, such as birds laying their eggs. In Yolngu, a synonym of *djarrak* for 'gulls and terns' is *ritharr*, which also means 'spear grass', because 'Terns lay their eggs when Spear Grass flowers'.[60] Similarly, linguist Barry Alpher, who worked with the Yir-Yoront people of western Cape York Peninsula, explained that:

> For Aborigines, there is a close relationship between biological events that they use to inform themselves about the readiness of food resources. For example, 'When spear grass (*Sorghum plumosum/S. laxiflorum*) has ripened, it's time the Magpie Geese laid their eggs'.[61]

Many fishing birds are migratory, and this is reflected in their associated myth tracks. For instance, Saara (also known as Shiveri) the silver gull (seagull) and Nyunggu the pied imperial pigeon (Torres Strait pigeon) as ancestors were said to have travelled along the western coast of

Cape York Peninsula to the Torres Strait islands during the creation. In terms of sites they are particularly associated with the area round Weipa and Mapoon on the coast of north-west Cape York Peninsula.[62] The myth tracks of Saara and Nyunggu replicate the natural movements of the associated bird species from Cape York Peninsula through Torres Strait to Papua.[63] These northern Queensland links between birds and seasons are explained by myth. Thomson remarked that:

> The fact that the Pheasant Coucal calls regularly with the first rains of the north-west monsoon [from late October] explains the association of this bird with water in the mythology of certain of the native tribes (e,g. Koko Ya'o [Kuuku Ya'u]) of this area. It is known as *Pull'o* an onomatopoeic word, in imitation of the call notes of the bird, which the natives render as 'pull, pull, pull, pull, pull, pull,' a sound suggestive of rapidly bubbling water.[64]

Aboriginal environmental knowledge concerning the link between birds and seasons is encoded in songs, which are handed down through the generations. As described by anthropologist Deborah Bird Rose, 'Some are beautiful little pieces which convey one piece of information but which evoke a whole season and a set of places, people and activities'.[65] Bird Rose gave an example from the southern Gulf of Carpentaria, where the approaching wet season is marked by the arrival of the migratory pied imperial pigeon. She said that in Yanyuwa man Jack Baju's song, the bird is first heard before it is sighted:

> Where is she?
>
> *The Torres Strait Island Pigeon* [pied imperial pigeon],
>
> She was away,
>
> now she has returned,
>
> And is calling,
>
> From island to island.[66]

Bush intelligence

The understanding of bird behaviour can provide hunters with cues for the presence of game animals.[67] For instance, in the north central Kimberley, the Wunambal Gaambera people claim that the black-tailed treecreeper climbs up tree trunks, where it will look for kangaroos and call out if it sees one.[68] In the Walmajarri language of the Great Sandy Desert, the yellow-throated miner is called *piyirn-piyirn*, meaning 'people people', because of its habit of uttering a loud repetitive two-note call when humans approach game animals, thereby drawing attention to their location.[69] It is recorded from Walmajarri people that *wiikwiik* the pallid cuckoo 'often makes a loud call when it sees a *jarrampayi*, large [sand] goanna, or something unusual, in the bush, it can tell you that food or game is close by'.[70] Similarly, according to the Walmajarri the bird they recognise as *murungkurr*, which is both the white-plumed honeyeater and white-winged triller, 'calls out 'wirtij ... wirtji' when you are in the bush to indicate you are near food like a *murntuny*, the black-headed python or *jarrampayi* [sand goanna]'.[71] The green figbird in northern Australia has an alarm call that indicates that raptors are in the area.[72] When the Yidiny-speaking people from the rainforest region inland from Cairns in northern Queensland hear a *girrgirr* (Lewin's honeyeater) call, they think that 'perhaps a snake is crawling along and has been seen (by the bird)'.[73]

Certain birds can warn hunters of impending dangers from human sources. Some of these 'messenger' birds are small, but their behaviour attracts attention from people. The Wiilman people at Bremer Bay in the south-west of Western Australia had a tradition concerning the behaviour of the male splendid fairy-wren, 'When a little blue bird is seen hopping about by himself it is always a sign that strangers are coming, but if he has all his little wives with him everything is all right and safe'.[74] In western Victoria, Dawson reported that 'The shepherd's companion [willie-wagtail] belongs to both men and women, and is never killed, because it attacks snakes, and gives warning of their approach'.[75] As discussed in Chapter 3, there are many other accounts from across Australia of the willie-wagtail, as a spirit bird, delivering unwanted messengers and even being considered as the source of misfortune and despair. According to the Yidiny-speaking people in the Cairns rainforest region of northern Queensland, *dabuy* is 'a brown bird which calls out "*dudidudi*" if anyone is coming'.[76]

In northern coastal New South Wales, linguists recorded a myth in the Gadang language concerning the 'Mail Bird', a little grey species in the forests described as 'a kind of wagtail', *guri djugi*.[77] From this account it can be perhaps identified as the grey fantail. The linguists were told that this bird once came to a man sitting on a log and said 'Guri djugi, guri djugi'. But the man took no notice of him, so the 'Mail Bird' said '*Ts-ts-ts-ts-ts ... guri djugi, guri djugi*'. The man then turned around and replied to the small bird, 'A man is coming from the west, a man is coming soon? I understand. I understand what you mean'. It was believed that the 'Mail Bird' could talk in the Gadang language and that it would appear with messages if anything unusual was about to happen and deliver a warning of impending dangers. According to anthropologist Ursula H McConnel, the Wik Mungkan people of western Cape York Peninsula in Queensland have a myth which explains how the magpie-lark (peewit) 'sits aloft in a tree where he can see around and warn husbands when their wives are unfaithful – the "bush policeman"!'[78]

Cockatoos were widely known to convey messages to Aboriginal peoples about incoming visitors. In western Victoria, 'The cries of the banksian [red-tailed black cockatoo] and white [sulphur-crested] cockatoos announce the approach of friends'.[79] In the north central Kimberley it is said that the red-tailed black cockatoo 'acts as a look out for Wunambal Gaambera people when they are out bush, if something or someone is sneaking up they will sing out and let you know', and 'If you are lost in the bush they will call out until someone finds you'.[80] In the Litchfield area south of Darwin in the Northern Territory, the Mak Mak Maranunggu people know when a 'Featherfoot' or sorcerer is approaching because of the screech of the 'white cockatoo', which may be the little corella.[81] The Jawoyn people living round Katherine in the Northern Territory have a related observation concerning the sulphur-crested cockatoo, which 'often signals to people in the bush that someone is approaching by flying in circles and calling out'.[82] Similarly, in coastal western Arnhem Land, the Mawng people consider the sulphur-crested cockatoo to be 'a very clever bird. When he sees someone coming, he warns people'.[83] As described in Chapter 2, Aboriginal peoples have considered that their individual totemic bird species looks after them when on their own Country and gives them signs. This avian warning behaviour was also recognised by European Australians, who coined the term 'cockatoo' for someone who acted as a lookout during illegal gambling meetings in order to warn of approaching police.[84]

Hunting partners

Particular bird species did more than provide signs to people – they were actively involved in helping the hunt. Aboriginal peoples had close foraging relationships with various bird species, some of which were considered to have shared ancestors with people. Lower Murray people had a special mutually beneficial relationship with *wampanyi* the swamp harrier, which was described as 'a species of hawk which operated with Aborigines in driving ducks down to their spears at Purung'.[85] Here, two of the names that appear to have been given to the peregrine falcon ('duck hawk') were *puri* and *puwuri*; these were said to have 'also applied to duck netting place' for the placement of *wurki* or 'duck net'.[86] Other Lower Murray birds were also seen as useful, for hunters 'will find and capture snakes by watching the movements of their companions the [grey] butcher birds'.[87] In the coastal western Arnhem Land region, the Mawng people say that *jitpuruluk*, which is used for both the grey butcherbird and pied butcherbird, 'makes a call and tells us that someone has caught *inyarlgan* [turtle or dugong]'.[88] Tindale recorded on the Murray River that it was said that 'Ordinary men track each other by tracks and by observing the signs of birds disturbed by people', while 'powered' men used 'magic' means.[89] Ramindjeri fishers in Encounter Bay were said to watch the movement of the 'blue crane' (white-faced heron), as this species will wait on rocky perches for shoals of fish.[90]

In the Great Sandy Desert of Western Australia, the wedge-tailed eagle is renowned for helping Aboriginal hunters. An anthropologist interviewed Walmajarri man Jimmy Pike, who said that:

> When they spot someone hunting, they fly down and land in a nearby tree to watch. If the human being flushes out a wallaby or small kangaroo, the eagle swoops down and seizes it in her claws. The hunter calls out and the eagle drops the animal. Leaving it for the hunter. This may happen three of four times, but the next time the eagle dives and catches a prey animal. She carries it up and flies away with it. In this way the eagle helps people to hunt, and in return receives an animal for herself.[91]

Nuisance birds

Not all bird species are so helpful to Aboriginal hunters. In the Lower Murray region during the late 19th century, a renowned bird hunter was a Ramindjeri man known as Pantonie. It was reported that on one occasion when hunting ducks a 'plover' (probably the masked lapwing) had cried out and alerted the game, which then escaped.[92] In anger, Pantonie caught one of the birds and stripped off most of its feathers before letting it go. This bird species appears to have annoyed Aboriginal bird hunters across Australia. From the opposite end of the continent, Thomson remarked in relation to the masked lapwing that on Cape York Peninsula in northern Queensland the bird:

> … has the habit of the allied Spur-wing Plover (*L. novaehollandiae*) of rising into the air when alarmed, and wheeling around uttering strident cries, and on this account is not regarded with favour by native hunters, for it often gives the alarm to game such as the Jabiru [black-necked stork], Native Companion [brolga], or ducks that they may be stalking.[93]

Ramindjeri people at Encounter Bay in South Australia treated *wate-eri-on* the grey currawong as a spirit, 'If a hunter killed one, the other *wate-eri-on* would have their revenge and make it very difficult for him to catch other birds'.[94] It was said that 'This bird warned a kangaroo when it was being tracked down by a hunter by calling out *kiling-kildi*, which was also the name of this bird on the Lower Murray'.[95] In other parts of Australia, the willie-wagtail was regarded as a nuisance species for hunters.[96] When describing the response of the noisy miner to Aboriginal hunting, environmental writer Tim Low said 'No wonder noisy miners complained when Aborigines appeared, voicing warnings understood by the other birds'.[97] These birds were a nuisance to hunters, rather than a direct threat.

Some Australian birds may, in certain situations, have presented more of a direct threat to Aboriginal peoples. In Chapter 4, malevolent bird spirits were described and an account given of alleged wedge-tailed eagle attacks upon young people. As discussed in Chapter 6, emus are recognised as being dangerous to humans, mainly when being pursued.[98] Today, the southern cassowary is considered as threatening to people, although most of the documented attacks are from birds whose behaviour has been modified by human feeding.[99] It therefore appears unlikely that either of the large flightless birds would have been a serious threat to a band of armed foragers.

Finding water

In the desert, sightings of 'diamond birds' – a group that consists of species such as the zebra finch, diamond firetail, spotted pardalote and striated pardalote – were taken as a welcome sign that there was a source of drinking water somewhere nearby.[100] The zebra finch in particular is a sedentary species, with a range probably limited to about 20 km, making any observation of it highly significant for those seeking water in the desert.[101] An Arrernte person in the Macdonnell Ranges of Central Australia said that *nyingke* the zebra finch:

> … can show you where there is water, if you get lost and don't have any. If you hear those zebra finches singing just go and see because you might find water, or even wet mud there. The zebra finch is a bird that tells you where there is water.[102]

During a fieldtrip to the Anangu Pitjantjatjara Yankunytjatjara Lands in the north-west of South Australia in 1994, my Aboriginal guides showed me an active nest of zebra finches in the brush roof on an Aboriginal shelter near a rock hole. I was struck by the apparent tameness of the small birds as they flew back and forth past my face. The fearless nature of the 'diamond birds' is legendary. The outback colonist Alexander T Magarey noted that:

> So docile are these little fellows (diamond birds) that they have been seen sitting on the bare skin of an aborigine while he was bending to his work of digging out with his wooden scoop a hole in the sand of a creek-bed for water supply. And the observer states that they seem to be in no way disconcerted as he throws the sand over his shoulder (where they are perched) as he worked. Nor does the native resent at all the presence or familiarity of his trusting little feathered companions. He wants water, and they are waiting to drink with him.[103]

Observations of birds were a matter of Aboriginal life or death in arid Australia, leading Magarey to remark that 'As thoroughly reliable guides to water in very dry regions the birds have no rivals. Some varieties of these should be named "the bushmen's water-finders."'[104] The significance of 'diamond birds' was part of the bush lore that European explorers learnt from their Aboriginal guides and drew upon when crossing harsh regions. On 6 March 1861, the trans-Australian explorer John McDouall Stuart was in Central Australia and desperately short of water. He wrote in his journal:

> Followed the creek up into the gorge, and found it very dry. Our former tracks are still visible in the bed of the creek. No rains seem to have fallen here since last March. I had almost given up all hopes of finding any water, when, at seven miles [11 km], we met with a few rushes, which revived our sinking hopes; and, at eight miles [13 km], our eyes and ears were delighted with the sight and sound of numerous diamond birds, a sure sign of the proximity of water.[105]

The dusk sighting of desert birds such as the budgerigar, little corella, galah, crested pigeon, common bronzewing and flock bronzewing is taken as a good sign of the presence of nearby water, as the birds are returning from daytime foraging to drink.[106] In contrast, the observation of emus and laughing kookaburras in arid areas is not considered a reliable indicator of the proximity of freshwater, as both species habitually forage further away from water and are able to survive on the moisture absorbed from their food alone.[107] In the central Kimberley, Love recorded that the Worora people have a belief that 'The Bronzewing Pigeon is the special friend of the Red Kangaroo, and when the kangaroo is distressed through thirst the natives believe that the pigeon takes water between his shoulders and flies with it to his relief. He can be often seen flying fast on his errand of mercy'.[108] Magarey noted that pigeons and bronzewings are heavier once they have drunk and consequently they find flying more difficult. This meant that:

> An ordinary intelligent and alert observer can form a fairly accurate judgment from the style of the flight whether the pigeon is going to water or coming away from it, and in either case direct his course accordingly should he be in need of water. The bird does not drink during the heat of the day.[109]

Foragers could also determine much through the examination of animal tracks. For instance, the trails of animal prints could indicate the presence of surface water. Magarey claimed:

> … if a bird, or dingo, kangaroo, or camel is thirsty, and travelling eagerly towards water to drink, the stride is long and direct. If after it has had a drink at a water the animal is satiated or browsing, the trail is wandering hither and thither, and the shortened stride shows leisure and presence of water at the time the trail was made in their vicinity.[110]

In arid regions, birds as spirit beings could help Aboriginal rainmakers with rituals to bring rain clouds. Kim Akerman wrote that:

> By hanging engraved pearl shell ornaments (*riji*) from a rack of bush timber, a *jalnganguru* ['clever man' doctor] can seek the aid of the *rai* [spirits] of the trees, rocks, flowers and water in making rain. The [red-backed] wren (*jit-jit*) is also a great *jalnganguru*. He helps rainmakers when they carry out their ceremonies in public during the dry season of September–October (*laja-laja*).[111]

Birds as 'firestick farmers'

Aboriginal peoples in Australia routinely burnt parts of their foraging territory as a tool for food production.[112] This was termed 'firestick farming' by archaeologist Rhys Jones.[113] Over time, the change in fire regime in the Australian environment, whether by natural or human causes, would have profoundly affected local bird populations. An increase in burning would favour those species that forage round the edge of fires and others more closely associated with the fire-loving open eucalypt forests and savannah woodlands.[114] It is apparent that established fires in some areas are also spread by birds. In the tropical Australian savannahs several raptors, in particular the brown falcon, black kite, square-tailed kite and whistling kite, are collectively referred to as 'firehawks' as it is claimed that they intentionally spread fire by picking up burning sticks with their talons or beaks and dropping them up to 1 km from the fire front.[115] According to the Mak Mak Maranunggu people of the floodplains south of the Finniss River in the Northern Territory, 'fire-makers' like the brown goshawk and brown falcon will 'swoop down, grab a fire-stick, carry it away, and drop it ahead of the fire, thus encouraging it to keep moving'.[116] These birds live on the small animals fleeing from the fires and those killed by it.

In the Yanyuwa language spoken in the south-western Gulf of Carpentaria, the term *buyukawu* is used for several falcon and kite species and means 'of the fire'.[117] In Aboriginal thinking, this category relates to both mythological events and actual habit. Bradley explained that:

> Species such as the brown hawk [brown falcon] and black kite are often seen in the vicinity of burning fires, searching for insects, reptiles and small animals that are trying to escape the fire. The peregrine falcon, and the other species of falcon listed are, in Yanyuwa Law, mythologically associated with the making of fire using firesticks, and are sometimes seen close to burning country. The black falcon is associated with the making of fire in the [nearby] Garrwa peoples' Law. The ceremonial body designs for these species are two long strips painted on the torso of the dancer representing fire sticks.[118]

Knowledge of the fire-spreading behaviour of birds of prey is widespread among northern Aboriginal communities. It is part of the corpus of Aboriginal environmental knowledge and is supported by their creation traditions.[119] As discussed in Chapter 3, across Aboriginal Australia there are many recorded beliefs of spirit ancestors who released fire into the landscape for the later benefit of the people who became their descendants in human form.[120] Some of these fire creation traditions involve raptors. For instance, in Western Desert mythology Kipara the Australian bustard had a firestick stolen by two 'hawk men', known as the Multja-didju, who then placed the property of fire into dry branches so that it would be accessible to all.[121] The Warumungu people who live in the region surrounding Tennant Creek in the Northern Territory have a tradition whereby two 'hawks', Warapulapula (probably the whistling kite) and Kirkalanji (probably the brown falcon), made the first fire by rubbing two sticks together so they could 'walk about in the smoke', but the fire escaped and caught the moon man and the bandicoot woman, killing Kirkalanji.[122]

Among my own anthropologist colleagues who are experienced fieldworkers, some claim to have observed raptors intentionally spread fire in the manner described above. There are others who believe that the birds probably pick up burning branches only by mistake, thinking they are lizards, then drop them. Apart from the more recently recorded observations by non-Aboriginal people,[123] there is evidence dated to 1963 in the Roper River area of south-east Arnhem Land from Alawa man Waipuldanya (Phillip Roberts), who asserted that 'firehawks' deliberately spread fire with smouldering sticks for the purpose of hunting.[124] Similarly, when recording Wik Ngathan bird names on the north-west tip of Cape York Peninsula in 1978, linguist/anthropologist Peter Sutton recorded that *thiimoenthenh* the peregrine falcon 'carries burning grass & drop it to start new fire – smart bird'.[125] The weight of evidence favours the finding that raptors regularly spread fire,[126] and this is another example of the avian use of foreign objects as tools.[127] More recently, with the increasing role of Indigenous peoples in land management, it has been claimed that 'Aboriginal rangers and others who deal with bushfires take into account the risks posed by raptors that cause controlled burns to jump across firebreaks'.[128]

Birds who help to collect and find food

Here, the focus is on foods that are found through bird behaviour, not the bird itself being eaten as a meat source. The consumption of plants by birds has led to the production of some partially processed food sources which Aboriginal peoples could exploit. A small discarded feather, picked up from the local area when required, could be used as a visual guide during the search for bee hives. These possibilities are discussed more fully below.

Collecting seeds

Western Desert people extracted preserved desert kurrajong seeds from the crow dung found at waterholes, then processed them to eat since they had lost most of their irritating coat of hair.[129] Tindale recorded that to Western Desert people in the north-west of South Australia, 'The [desert kurrajong] seeds, deposited with the dung of crows on granite rock surfaces bearing water holes, constitute an acceptable source of supply of grain which is milled on the rocks and made into native bread'.[130] Seeds that were voided in regurgitated pellets also appear to have been used. In an account of the Pitjantjatjara people, Tindale claimed that in this region 'the hard ripe kurrajong seeds, deposited in the dungs and coughings of ravens around rockholes, may be washed clean of their coatings of dung and so are freed for human consumption, either after being roasted in hot ashes or pounded into a meal between stones and baked as a cake'.[131] Concerning the use of kurrajong seeds in the desert, botanist/ doctor John Burton Cleland stated that 'As the crows have to fly back to the ranges for water, the seeds are there voided by them'.[132] He remarked in biblical terms that 'One is reminded strikingly of Elijah being fed by the ravens'.[133]

Other seeds may also have been used in a similar manner. It was suggested by the naturalist on the 1891–92 Elder Expedition, Richard Helms, that the large number of quandong stones that were voided by emus and found alongside waterholes in Central Australia would have been an important Aboriginal food source 'during times of severity'.[134] On the floodplains south of the Finniss River in the Northern Territory, Mak Mak Maranunggu foragers gathered *pelangu* nuts of the *mirrwana* tree (possibly the emu-apple) from the manure of emus which had eaten them.[135] The fruit surrounding the nut is digested by the emu but the kernels inside the seed are not; they pass intact through the gut. Concerning these nuts, Mak Mak Maranunggu woman Nancy Daiyi said 'Emus eat it too. Sometimes we used to pick those nuts up out of the emu shit. We get that nut again, we bust it up, we open it up and we eat it, that "miyi" (food from a plant)'.[136] Her daughter Kathy Deveraux added that 'They were sharing with emus. They scrounge it out of the emu shit and cook it up for themselves' and 'My grandfather, Old Wigma, he cooked it on a leaf and gave it to me'.[137] They also gathered cycad nuts from the same avian source.[138] For the Mak Mak Maranunggu people, the use of common foods by them and birds increased the sense of relatedness they had with their bird ancestors.

Meals from inside bird bodies

The meals of gamebirds, recovered from within the carcass, were also a source of food for Aboriginal peoples. The Tiwi people on Melville and Bathurst Islands in the Northern

Territory often find the crops of magpie geese to be full of *kirlinja* (spike-rush or water chestnut) corms, which were removed, cooked and eaten.[139] They were favoured foods and suitable for consumption by the young. In the Wadeye area south-west of Darwin, the spike-rush corms removed from the gullets of magpie geese are given to children, who wash them before eating.[140]

Children were also involved in the procurement of eucalypt nectar through the agency of nestlings. In western Victoria, Dawson recorded that:

> Another sweet liquid is obtained by mischievous boys from young parrakeets [lorikeets] after they are fed by the old birds with honey dew, gathered from the blossom of the trees. When a nest is discovered in the hole of a gum tree, it is constantly visited, and the young birds pulled out, and held by their feet till they disgorge their food into the mouth of their unwelcome visitant.[141]

Bee markers

White feather down, which is easily seen by foragers, had a practical use in the gathering of honey from the hives of native bees in rock cavities and hollow trees. It was recorded by early colonist/explorer Edward John Eyre that in the Murray River area of South Australia:

> The method of discovering the hive is ingenious. Having caught one of the honey bees [native bees], which in size exceeds very little the common house fly, the native sticks a piece of feather or white down to it with gum, and then letting it go, sets off after it as fast as he can: keeping his eye steadily fixed upon the insect, he rushes along like a mad-man, tumbling over trees and bushes that lie in his way, but rarely losing sight of his object, until conducted to its well-filled store, he is amply paid for all his trouble. The honey is not so firm as that of the English [honey] bee, but is of very fine flavour and quality.[142]

There are similar accounts of the use of feathers, or feather-like plant materials, as bee markers from across Aboriginal Australia.[143] In 1860, naturalist George Bennett described the 'tracking' of bees in New South Wales. He wrote:

> They use a small portion of the lightest down – that of the eagle in preference; the down is twisted into two small points, like two minute feathers, and when they find a bee whose thighs are laden with pollen, they carefully attach the feathers on each side. To allure a bee for this purpose, they cut a piece off the bark of a [*Callitris*] pine-tree, which appears, from some cause or other, to be highly attractive to this insect. When the feathers are inserted, they start the bee off homewards, and, accompanied by one or more natives, who never lose sight of the insect, they track it to the nest, which is invariably discovered by this mode of proceeding.[144]

As with many Aboriginal practices, the origin was said to be due to the actions of the ancestors during the creation. A supreme male ancestor is recorded to have used a feather to find a bee hive in central New South Wales. Here, concerning Baiame the creator, the late 19th century anthropologist Robert H Mathews claimed that:

> It is related that Baiame started for Cobar after a wild bee, on the feet of which he had put bird's down. He followed the insect all the way to a large rock at Wittaguna, in a cleft of which was the honey comb, which Baiame succeeded in securing.[145]

Amusements

Birds, as a source of materials and as themes, were incorporated into forms of amusement in Aboriginal camp life. A popular activity in the Lower Murray was the contested ball game known as *puldjungi*, and the men of the winning side 'were given a *priyinggi* (headband) made of white swansdown and skin, with white feathers fixed upright around it'.[146] The *kukelani* game, which was named after a bunch of emu feathers, was another activity here when people gathered for initiations; it was played as a team sport to see which clan was strongest.[147] A related game, involving 10–20 men, was described upstream in the Murray River area.[148] In the North Wellesley Islands of the Gulf of Carpentaria, a similar game was played with a ball made from netting or string bag stuffed with grass, between two teams that each represented one of the patri-moieties.[149] Jacob Karruck Harris, from the Wutaltinyerar clan of the Lower Murray, described Aboriginal games:

> During the day the men played at a game we call Cook-ka-limee [*kukelani*]. A man has a bunch of feathers in his hand and with this he challenge another by saying 'waung-kun-peejull', so and so and mentions the name of the person's country or tribe; the meaning is the emu has landed on your country. The man has to pursue the other that has the bunch of feathers and try to get it away from him, no matter how tired he felt he has to try and get it from him, a party of ten or twelve a side plays it.[150]

Chiefly for the benefit of an audience of children, younger Aboriginal women played string games on their hands, like those that Western Europeans know as 'cratch-cradles' or 'cat-cradles'.[151] In the Lower Murray the games were called *tjamburung*, with 'young swans' being one of the recognised string configurations.[152] In northern Queensland, birds were commonly featured in Aboriginal string games, including forms representing 'roosting cockatoos', 'ducks in flight' and 'bird nests'.[153] Birds are also major subjects in the string games played in the Wadeye area south-west of Darwin.[154]

Games were not just for amusements, but served a serious teaching function for children. Ronald M Berndt remarked that among Aboriginal peoples the 'child's school is its playground, and his playground is everywhere'.[155] Researcher Claudia Haagen described Aboriginal play

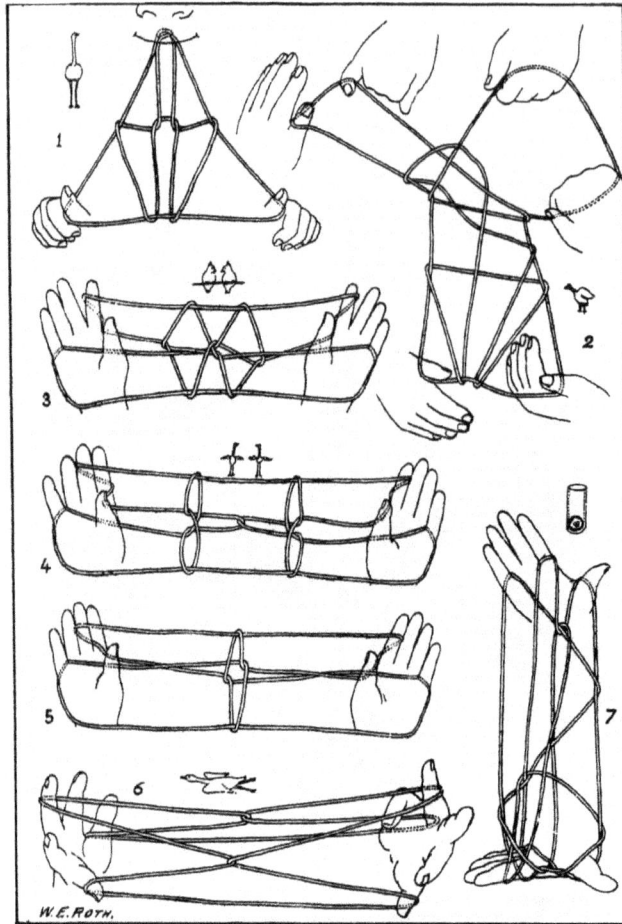

Plate V.—Animals: birds.
 1. Cassowary. Atherton.
 2. Eagle-hawk. Atherton. For Fish-hawk, see Pl. XI. 4:
    Hawk's Foot, see Pl. XII. 7.
 3. Two cockatoos roosting side by side. (l.) Tully River.
 4. Two white cranes. (l.) Tully R.
 5. Giant crane. (l.) Tully R.
 6. Duck in flight. Pr. Charlotte B., (mid.) Palmer R.
 7. Bird's nest, in the bottom of a hollow stump. Pr.
    Charl. Bay.

String games featuring birds. Examples from northern Queensland. Roth WE (1902) Games, sports and amusements. *North Queensland Ethnography Bulletin 4.*

as 'a preparation for the toil of subsistence'.[156] Based on their fieldwork with the Gunwinggu people of western Arnhem Land, the Berndts remarked that:

> Children learn early to read the signs in their natural environment. The songs contain only a fraction of this knowledge. By the time they can run about they should be able to identify ordinary tracks in and around the camps, starting with people, dogs, and small things like ants, beetles and lizards.[157]

Adults taught children how to track animals by imitating their prints in the sand, sometimes using their hand or a twig as the only tool.[158] Experienced hunters taught young people how to recognise game behaviour, such as travelling direction and the amount of time that had elapsed since the beast had passed through.[159] Cultural information, such as from myths, was also communicated via the medium of sand. In the deserts of Central Australia, a drawing that a woman makes on the ground for an audience of children is what anthropologist Nancy Munn described as a 'sand story', a genre of storytelling which is accompanied with gesture signs and songs.[160] Biologist/educationalist Rob Morrison remarked that the teaching skills required for the sand drawings stem from the ability of Aboriginal peoples to record and reproduce the tracks of animals from their Country, which 'are often painted in realistic or stylised form'.[161] Similarly, Basedow explained that:

> Commendable pains are taken by the adults in imitating the tracks of all the animals of chase, and the children are invited to compete in reproducing them. For instance, an 'emu track' is obtained by pressing the inner surfaces of the index finger and thumb, held at an angle of about forty-five degrees, into a smooth patch of sand; then, without lifting the index finger, the thumb is moved to the opposite side and there pressed into the sand, at about the same angle as before. Often the impression of the 'pad' of the bird's foot is indicated by dabbing the round point of the thumb into the sand immediately behind the intersection of the three 'toes.'[162]

The majority of games that Aboriginal children played were primarily based upon the foraging activities of the adults in their community, and were therefore not just for amusement but were also educational.[163] In this context, play has a crucial role in Aboriginal society. Basedow remarked that:

> I have seen the little fellows stalk a flock of foraging cockatoo and, when within range, fling several of the toy weapons into the birds as they are rising; invariably one or two birds are brought to fall … The trimmed stalks or bullrushes and reeds make excellent toy spears, which are thrown with the heavier end pointing forwards and the thinner end poised against the index finger of the right hand. With these 'weapons' the lads have both mock fights and mock hunts. In the latter case, one or two of their number act the part of either a hopping kangaroo or a strutting emu and, by clever movements of the body, endeavour to evade the weapons of the hunting gang.[164]

Ronald M Berndt described children at Ooldea in western South Australia making an 'emu' to ride, similar to a European hobby horse, by locating a sturdy sapling and trimming off the branches.[165] In this example, features included a pad of leaves as a seat strategically placed on the mid-section of the bent-over tree, foliage left on as a 'tail' and the discarded branches left on the ground nearby to cushion the child's fall. According to Basedow, a favourite pastime of the Kokatha people at Mount Eba in northern South Australia was to play 'emu'. This involved a man dressing up as the bird by using feathers, sticks, grass and brushwood.[166] He would then walk through the camp while mimicking the behaviour of emus, to the great enjoyment of the children. Similarly, anthropologist Frederick McCarthy noted of Aboriginal Australia that:

> In a popular game among young and old a man or boy playing the part of an emu was stalked by the other males in the group. In gesture language, the emu was denoted by the shape of its track, the motion of its head, or by its wobbly motion when running.[167]

Among the Guugu-Yimithirr people of Cooktown and Cape Bedford of Cape York Peninsula in Queensland, it was observed that spearing the birds roosting on a branch was the theme of the 'Catching Cockatoos' game. This is played by a group of children who make a stack of 'birds' from one of their hands, while others try to 'kill' the top 'bird' on the end of the 'branch' using the fork of their fingers from the free hand as the 'spear', and then 'eat' it.[168] Another recorded amusement in this part of Queensland was the 'Duck game', that involved the use of a stick to represent the switch with a noose for catching ducks. Children caught in this manner lay as if dead and then ran back towards their home.[169]

*****

In Aboriginal Australia, people and birds occupy the same space. This began with the major role of bird people who were part of the pantheon of ancestors who shaped and gave meaning to the land during the creation, and continued afterwards when people and birds as the descendants of those ancestors became co-inhabitants of Country. The possession of detailed environmental knowledge enabled the reading of Country, with foragers understanding the signs from animals, particularly birds, in relation to such things as changes in weather and seasons. For Aboriginal foragers, their partnership with certain bird species involves finding water, game and honey, and alerting people to the arrival of others. Like people, birds spread and used fire. Much of this is explained by the events of the creation. Avian bodies also yielded foods, such as cleaned seeds, corms and nectar. The cultural importance of birds is reflected in Aboriginal games and amusements.

Endnotes

[1] Peterson (1969:18).
[2] Isaacs (1980:26), Nunn & Reid (2016) and Nunn (2020).
[3] Wilson (1937). See Chapter 3 for a fuller account of the emu and brolga mythology.
[4] Smith (1880:22).
[5] Howitt (1904:495) and Massola (1968:88).
[6] Cane (2002:88–92). See also Isaacs (1980:87–89).
[7] For the distribution of desert kurrajongs (*Brachychiton gregorii*), see the Atlas of Living Australia website (https://bie.ala.org.au/species/https://id.biodiversity.org.au/node/apni/2889211).
[8] R Robinson 1968 (cited in Isaacs 1980:26).

[9] Love (1946:258).

[10] Love (1946:258).

[11] Berndt *et al.* (1993:367).

[12] Chase & Sutton (1981), Davis (1989), Jones (1985:195–199), Latz (1995:Ch.2) and Clarke (2003a:Ch.7, 2009c, 2018g).

[13] Dawson (1881:98) and Clarke (2009c).

[14] Grassie (1876:4).

[15] Doonday *et al.* (2013:145,148).

[16] Doonday *et al.* (2013:131).

[17] Green *et al.* (2009:184), Turpin *et al.* (2013:10,24) and Si (2019:29).

[18] Turpin *et al.* (2013:21).

[19] Green *et al.* (2009:184).

[20] Love (1946:261).

[21] North & Keartland (1898:160).

[22] North & Keartland (1898:161).

[23] Clarke (2003a:131–135, 2014a:24–25,32).

[24] Parker & Reid (1983:138–139) and Berndt *et al.* (1993:558). The freckled duck is also known as the 'whistle duck'.

[25] Berndt *et al.* (1993:560). The banded lapwing was also known as the 'banded plover' and the 'water hens' are almost certainly dusky moorhens (Clarke 2019a:124,133,135,138,141).

[26] Berndt *et al.* (1993:556).

[27] Berndt *et al.* (1993:556–557).

[28] Berndt *et al.* (1993:561). Clarke (2019a:133) identified the *kenigeri-on* as possibly the grey shrikethrush.

[29] Teichelmann & Schürmann (1840:55,57–58) and Clarke (1997:137).

[30] Anonymous (1885a:4).

[31] Dawson (1881:98).

[32] Langloh Parker (1905:94, see also 110). Ash *et al.* (2003:106) identified *maliyan* (Mullyan) as the wedge-tailed eagle. It is unclear whether a certain species of crow or raven was being referred to, or whether all of them were included.

[33] Kimber (1997:8–9). The masked lapwing is also known as the 'spur-winged plover'.

[34] Mathews (1904:346).

[35] Karadada *et al.* (2011:83,85).

[36] Akerman (2020:134–135).

[37] Akerman (2020:134).

[38] Dixon (1991:159). This author referred to the brown cuckoo-dove with the common name of 'brown pigeon'. For a description of *jambun* grubs, refer to Dixon (1991:170).

[39] Davis (1989:6).

[40] Tindale (1960:323).

[41] Tindale (1960:59,61).

[42] Thomson (1935:30).

[43] Singer *et al.* (2021:63).

[44] Puruntatameri *et al.* (2002:95) and Karadada *et al.* (2011:77).

[45] Raymond *et al.* (1999:104) and Wiynjorrotj *et al.* (2005:129).

[46] Puruntatameri *et al.* (2002:98).

[47] Puruntatameri *et al.* (2002:98) and Wiynjorrotj *et al.* (2005:138). The pied imperial pigeon is also commonly known as the 'Torres Strait pigeon'.

[48] Singer *et al.* (2021:214).

[49] BJ Nangan (cited in Akerman 2020:150–151).

[50] Bradley with Yanyuwa Families (2010:36).

[51] Dixon (1991:161).

[52] Dixon (1991:161).

[53] Bradley *et al.* (2006:81).

[54] Clarke (2003a:Chs8–10) and Prober *et al.* (2011).

[55] Thomson (1949:16).

[56] Clarke (2015b:225).

[57] Moore (1884:Vol.2:18). For lerps, refer to Tindale (1966:180), Clarke (2018e:75–76) and Faast *et al.* (2020).

[58] Low (2014:36–37).

[59] Yunupingu *et al.* (1995:13).

[60] Zorc (1996:97, see also 240). The spear grass is *Heteropogon triticeus*.

[61] B Alpher 1987 (cited in Tidemann & Whiteside 2010:160).

[62] McConnel (1957:20–27), Sutton (1988a:20,233) and Swain (1993:76,87–89,93–99,104). The identity of 'seagull' as the silver gull is apparent in a sketch of associated sites (McConnel 1936:218).

[63] McConnel (1936:219).

[64] Thomson (1935:10).

[65] Bird Rose (1996:59).

[66] Bird Rose (1996:59).

[67] Clarke (2017:40–41).

[68] Karadada *et al.* (2011:85).

[69] Lowe (2002:50).

[70] Doonday *et al.* (2013:143).

[71] Doonday *et al.* (2013:148).

[72] Raymond *et al.* (1999:103).

[73] Dixon (1991:161).

[74] Hassell (1934b:322). The *ter ter* wren was identified by Abbott (2009:259).

[75] Dawson (1881:52–53).

[76] Dixon (1991:161).

[77] Holmer & Holmer (1969:38). The Gadang language is also known as 'Kattang'.

[78] McConnel (1957:74).

[79] Dawson (1881:53).

[80] Karadada *et al.* (2011:74).

[81] Thurman (2014:33). From my own fieldwork in this area, the 'white cockatoo' mentioned by the author is most likely to be the little corella, which in flocks is particularly apparent in the trees along the creeks and rivers.

[82] Wiynjorrotj *et al.* (2005:135).

[83] Singer *et al.* (2021:151).

[84] Ramson (1988:153).

[85] Tindale (1931–62). The name for swamp harrier, *wampanyi*, has also been written as *wampanji* (Clarke, 2019a:142).

[86] Tindale (1931–62).

[87] Ramsay Smith (1909:10).

[88] Singer *et al.* (2021:60, see also 52).

[89] Tindale (1953:3).

[90] Berndt *et al.* (1993:561). According to this source, the 'blue crane' is *tjirbuki*, like an 'ibis', but the described feeding habits identify it as the white-faced heron – also popularly called a 'blue crane' (Clarke, 2019a:134).

[91] Lowe (2002:50).

[92] Anonymous (1918:4). For the identity of Pantoni (Pantonie), see Berndt *et al.* (1993:523).

[93] Thomson (1935:31). Note that the spur-winged plover is now classified with the masked lapwing.

94 Berndt *et al.* (1993:124). The grey currawong is described as the 'black magpie'.
95 Berndt *et al.* (1993:234).
96 Raymond *et al.* (1999:105) and Tindale (2005:372).
97 Low (2014:289).
98 Tindale (1941b:239).
99 Kofron (1999).
100 Magarey (1895:74). The striated pardalote is also known as the 'striped pardalote'. Ramson (1988:198) defined 'diamond bird' as 'PARDALOTE, esp. the spotted pardalote'.
101 Thomson (1975:151–152) and Bayly (1999:18).
102 Henderson & Dobson (1994:516).
103 Magarey (1895:74–75).
104 Magarey (1895:74).
105 Stuart (1865:253).
106 Magarey (1895:74–76) and Bayly (1999:18). The crested pigeon is also commonly known as the 'topknot pigeon'.
107 Magarey (1895:76).
108 Love (1946:258).
109 Magarey (1895:75).
110 Magarey (1899:121).
111 BJ Nangan (cited in Akerman 2020:154,sketch on 155).
112 Gould (1971), Hallam (1975), Kimber (1983) and Clarke (2003a).
113 Jones (1969, see also 1980).
114 Woinarski (1999:62).
115 Raymond *et al.* (1999:112), Wiynjorrotj *et al.* (2005:142), Karadada *et al.* (2011:89), Gosford (2016) and Bonta *et al.* (2017).
116 Bird Rose *et al.* (2002:28, see also 18–19,29). This source used the common name 'chicken hawk' for the brown goshawk.
117 Bradley *et al.* (2006:84).
118 Bradley *et al.* (2006:84).
119 Bonta *et al.* (2017:704–708).
120 Isaacs (1980:102–108) and Clarke (1999b).
121 Mountford (1976a:506,511) and Isaacs (1980:106–107).
122 Spencer & Gillen (1904:619–620). In the related Warlpiri language, *kirrkirlanji* (cognate of *kirkalanji*) is the brown falcon (Turpin 2012:27). Similarly, in the Kukatja language, *pulapula* or *pula-pula* refers to the whistling kite (Peile 1997:289, Cataldi 2004:272).
123 Bonta *et al.* (2017:706–708).
124 Lockwood (1963:92).
125 Sutton (1978b:9).
126 Bonta *et al.* (2017:712–714).
127 For overviews of avian tool use, refer to Lefebvre *et al.* (2002) and Striedter (2013).
128 Bonta *et al.* (2017:700).
129 Latz (1995:133) and Clarke (2007a:91).
130 Tindale (1941a:11).
131 Tindale (1972:234).
132 Cleland (1957:153, 1966:146,148).
133 Cleland (1940:6).
134 Helms (1896:291).
135 Bird Rose *et al.* (2002:34). Smith *et al.* (1993:37) recorded emus eating the fruit when ripe.
136 Bird Rose *et al.* (2002:34).
137 Bird Rose *et al.* (2002:34).

[138] Bird Rose *et al.* (2002:94).

[139] Puruntatameri *et al.* (2001:45).

[140] Hardwick (2019c:46).

[141] Dawson (1880:21).

[142] Eyre (1845:Vol.2:273–274).

[143] Clarke (2012:41).

[144] Bennett (1860:193).

[145] Mathews (1904:343).

[146] Berndt *et al.* (1993:168).

[147] Berndt *et al.* (1993:168–169,373–375).

[148] Eyre (1845:Vol.2:228).

[149] Memmott (2010:62,Fig.82).

[150] Harris (1894–95:21 May 1895).

[151] Roth (1902:10–11,Plates 3–12), Berndt (1940d:292), Davidson (1941) and Clarke (2003a:43).

[152] Harvey (1939). The term *tjamburung* can also be written as *jamburung*.

[153] Roth (1902:10–11,Plate 5).

[154] Hardwick (2019d:43).

[155] Berndt (1940d:293).

[156] Haagen (1994:2).

[157] Berndt & Berndt (1970:33).

[158] Roth (1902:493), Love (1936:153), Berndt (1940d:293), Morrison (1981:Ch.6) and Haagen (1994:10–12).

[159] Haagen (1994:11).

[160] Munn (1973:Ch.3).

[161] Morrison (1981:161).

[162] Basedow (1925:71).

[163] Berndt (1940d:290–292).

[164] Basedow (1925:81).

[165] Berndt (1940d:293).

[166] Basedow (1925:81–82). The Kokatha people are also described in some texts as the 'Kukata'.

[167] McCarthy (1965:17).

[168] Roth (1902:14,Plates 22–24). In early texts, the Guugu-Yimithirr were known as the 'Koko-yimidir'.

[169] Roth (1902:14).

Australian raven. In many parts of Aboriginal Australia, this species totemically represents one of the two moieties through which the known world is divided. Philip A Clarke, Adelaide Plains, South Australia, 2019.

'Lyre bird dance' by William Barak, Coranderrk Mission, Victoria, 1890s. AA795/1/1, South Australian Museum Archives.

Australian pelicans. This large bird is among the most frequently recorded species in Aboriginal mythology and is a major totemic species. Philip A Clarke, Raukkan, Lower Murray, South Australia, 1990.

Cape Barren goose. In the Lower Murray, this bird, called *lawari*, is a totemic species and a highly favoured food. Philip A Clarke, Adelaide Hills, South Australia, 1990.

Crested pigeon. In the creation mythology of the Flinders Ranges in South Australia, the ancestor of this species stole large grindstones from the diamond dove and left them at sites where Aboriginal people later quarried them to make grindstones for preparing grass seed. Philip A Clarke, Adelaide Plains, South Australia, 2014.

Black swans. Birds on the water could be lured in by hunters splashing a swan wing. In the Lower Murray region, *kungari* the swan is a totemic species for several clans. Philip A Clarke, Mark Point, Coorong, South Australia, 1991.

Wedge-tailed eagle. In both myth and everyday life, the bird exists in a state of perpetual conflict with other birds, such as ravens and crows. Mark Crocombe, Thinhi Creek, near Fossil Head, Northern Territory, 2018.

White-backed magpie. Aboriginal creation mythology explains how these birds, originally white, gained their black plumage and red eyes. Philip A Clarke, Laratinga Wetlands, Mount Barker, South Australia, 2021.

Australian bustard footprint. This bird, often called a plains turkey, has a major role in myths concerning how the emu lost the ability of flight, and accounts of how fire came into the world. Philip A Clarke, Shaw River, west Pilbara, Western Australia, 2015.

Mornington Island dance group, performing in 2009 at a public festival held in New South Wales. The 'corroboree caps', which only the male performers wear, are topped with emu feathers. Mirndiyan Gununa Aboriginal Corporation, Mornington Island Art.

Emu. An ancestor in many Aboriginal myths, often in opposition to other large birds, such as the brolga or Australian bustard. Philip A Clarke, Adelaide Hills, South Australia, 1986.

Australasian grebe and chick. In a myth from the Flinders Ranges, the black duck man attacked the grebe man with boomerangs, causing the latter to become a diving bird. Philip A Clarke, Tomato Lake, Perth, Western Australia, 2020.

Rock painting of what is believed to be Baiame the creator, who sometimes took the form of an eagle. Note the large eyes and feather-like extensions to the long arms. Prior to the end of the creation period, many ancestral beings had human form. The rock shelter also contains welcome swallow mud nests. Philip A Clarke, Baiame Cave, near Bulga, Hunter Valley, New South Wales, 2021.

Common bronzewing. A creation ancestor who, when wounded, is said to have produced gold from its blood and quartz from its feathers, and to have made opal by throwing a firestick. Philip A Clarke, Adelaide Hills, South Australia, 1983.

Papajara myth site. The granite boulders represent the emu ancestor and its chicks, who were all killed when the brolga ancestor caused the sea to rise during the creation. Philip A Clarke, The Granites near Kingston, South Australia, 1990.

Johnson grasstree. In Aboriginal mythology, bird ancestors placed the property of fire into grasstrees, from where it could be released when using the dried flower stalks in fire-drills. Philip A Clarke, Isla Gorge, south of Theodore, Queensland, 2016.

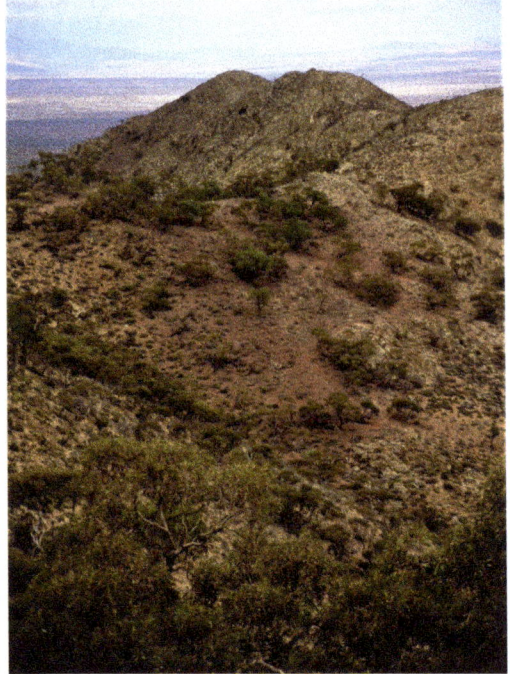

Pukardu ochre mine. The ochre, seen here 'bleeding' from the hill, represents the blood of an ancestral emu killed during the creation. Philip A Clarke, Flinders Ranges, South Australia, 1984.

Emu footprint. The emu print, in this instance made by a bird walking across the clay of a lakebed, is a common motif in the Panaramittee rock engraving tradition that dates back many thousands of years. Philip A Clarke, Pipeclay Lake near Salt Creek, South Australia, 2021.

Brolgas and magpie geese. Both birds are a major bird ancestors and highly desired game. Philip A Clarke, Kakadu wetlands, Northern Territory, 2009.

Black swan. In Aboriginal mythology, this bird variously received its red beak through the carrying of fire or as a wound caused by a wedge-tailed eagle. Philip A Clarke, Perth, Western Australia, 2020.

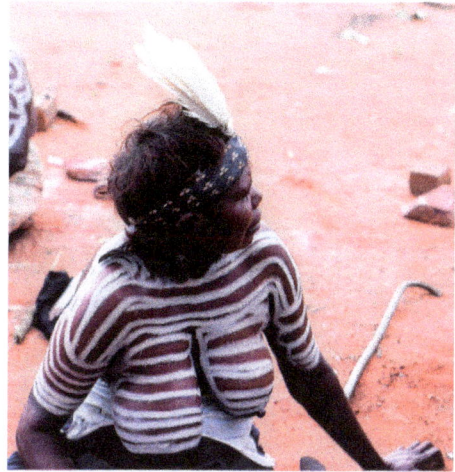

Warlpiri woman performer wearing a Major Mitchell cockatoo feather head ornament at the formal opening of an arts centre. Mary Laughren, Mount Allan, Northern Territory, 1988.

Warlpiri women dancers, wearing Major Mitchell cockatoo feather head ornaments during a Yawulyu performance at the Yuendumu Sports weekend. Main performers (left to right): Mona Napurrurla Poulson, Peggy Napurrurla, Judy Nangala Tex, Ruby Nangala Spencer (Lirramunju), Emily Nangala Watson, Rosie Milanjiya Nangala Fleming and Dolly Pirrkiriya Nampijinpa Daniels. Mary Laughren, Yuendumu, Northern Territory, 1970s.

Australite core. Known as an 'emu eye', it was found by the author on Middleback Station, northern Eyre Peninsula, in 1980. Philip A Clarke, Adelaide, South Australia, 2021.

Magpie-lark. In the creation mythology of some parts of Australia, this bird was originally white, but during fighting with other ancestors it became covered with burning ashes, giving it the black patches seen today. Philip A Clarke, Jubilee Lake, Perth, Western Australia, 2020.

Warlpiri and Anmatyerr men performing an open *purlapa* (corroborree) with emu feather headdresses. Mary Laughren, Mount Allan, Northern Territory, 1988.

Southern boobook. A widely feared spirit bird in Aboriginal Australia. In the creation mythology of south-west Victoria, Gartuk the southern boobook attacked the Brambambult brothers by releasing storms from his kangaroo skin bags. Philip A Clarke, Adelaide Plains, South Australia, 2013.

Australian magpie. Bird species such as this exhibit social behaviours, like group playing, akin to those of people, so they are sometimes seen as spirit beings. Philip A Clarke, Theodore, Queensland, 2016.

Australian crow. This species frequently picks up objects left by humans. An Aboriginal interpretation is that this is a behaviour of sorcerers. Philip A Clarke, Pentland, Queensland, 2020.

Willie-wagtail. Contemporary Indigenous folklore contains many accounts of this species. While generally seen as mischievous, some Aboriginal people believe that the bird will warn them when snakes are close by. Philip A Clarke, north Cairns, Queensland, 2014.

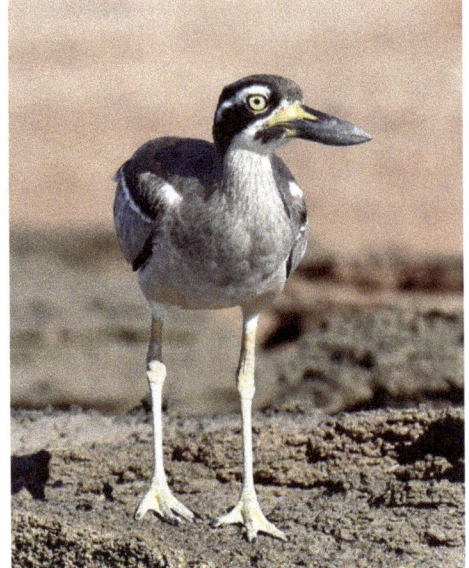

Beach stonecurlew. An omen of death across much of Aboriginal Australia, this bird is active at night and has a mournful call. Mark Crocombe, Cape Dombey, Thamarrurr, Northern Territory, 2019.

Australian crows. In various parts of Aboriginal Australia, crows and ravens were seen as forming groups with complex human-like interactions. Philip A Clarke, Ormiston Gorge, Macdonnell Ranges, Northern Territory, 2019.

White-bellied sea-eagle. Large eagles are seen as being capable of carrying the souls of the dead. Philip A Clarke, Yellow Water Lagoon, Kakadu, Northern Territory, 2009.

Large-tailed nightjar. Aboriginal people treat nightjars as spirit beings, and when the birds call out at night it is taken as a sign that somebody is sick or has died. Jeff Hardwick, Upper Finniss River, Northern Territory, 2019.

Wanjina spirits wearing head ornaments made from red-tailed black cockatoo feathers, painted in Otilyiyalyangngarri Cave. Kim Akerman, Mount Barnett, Kimberley, Western Australia, 1985.

Bush stonecurlew. An Aboriginal death spirit, it has a human-like face with particularly large, slightly front-facing eyes and a mournful call. Mark Crocombe, Nightcliff, Darwin, Northern Territory, 2014.

Royal spoonbill. In many Aboriginal languages, this species is classified with other waterfowl, such as the yellow-billed spoonbill and white ibis. Philip A Clarke, Laratinga Wetlands, Mount Barker, South Australia, 2015.

Blue-billed duck. In the Lower Murray, the name of this species, *pulki-nyeri*, referred to *pulki* ('well' or 'trench') and -*inyeri* ('belonging to'). Philip A Clarke, Tomato Lake, Perth, Western Australia, 2020.

Kookaburra. Its Australian English name is believed to have been 'borrowed' from *kukuburra* of the Wiradhuri language spoken across inland New South Wales. The name is based on its call. Philip A Clarke, Murgon, near Cherbourg, Queensland, 2016.

Superb fairy-wren. In the Lower Murray, the name for this small bird, *wetjungali*, referred to lignum bushes (*wetji*), a link which is explained by an episode of the local mythology. As an important ancestor, these birds were generally not hunted. Philip A Clarke, Narrung, South Australia, 1990.

Galahs. The Australian English bird name, galah, was 'borrowed' from *gilaa* in Yuwaalaraay and the neighbouring languages of northern New South Wales. Philip A Clarke, Alice Springs, Northern Territory, 2008.

Rainbow bee-eater. In the Jawoyn language spoken in the Katherine area of the Top End, this species is called *wirrirtwirrit*, which is a rendering of its call. Philip A Clarke, Rapid Creek, Darwin, Northern Territory, 2020.

Black-winged stilt. In some of the Nyungah languages spoken in the south-west of Western Australia, this bird was called *djandjarok*, which is possibly based on its high-pitched call. Philip A Clarke, Jualbup Lake, Perth, Western Australia, 2020.

Australasian darter. In the Antikirrinya language of the Western Desert, a group of waterbirds with long necks, such as this species, the white-faced heron and others, are collectively called *ngurntiwarlarta*, from *ngurnti*, 'neck', and *warlarta*, 'long'. Philip A Clarke, South Alligator River, Northern Territory, 2009.

White ibis. In many Aboriginal languages, this bird shares a name with other species of ibises and spoonbills. Philip A Clarke, Jualbup Lake, Perth, Western Australia, 2020.

Sacred kingfisher. In some desert languages, birds such as the sacred kingfisher, red-backed kingfisher and rainbow bee-eater are all called *luurn*, possibly because all of them nest in burrows dug into riverbanks. Philip A Clarke, Berry Springs, south of Darwin, Northern Territory, 2009.

Silver gulls. In some Aboriginal languages, there are different names for the immature and adult forms of some birds, particularly when they look markedly different. Philip A Clarke, Bowen, Queensland, 2020.

Masked lapwing. In the Walmajarri language of Western Australia, this species is known as *tintirrpari*, with its night time call being 'tintirr … tintirr … tintirr'. Philip A Clarke, East Point, Darwin, Northern Territory, 2020.

'Aborigines hunting waterbirds'. Note the use of spears and spearthrowers. Watercolour painting by Joseph Lycett, coastal New South Wales, 1817. National Library of Australia, NLA.obj-138499073.

Western bowerbird. Aboriginal hunters generally left bowerbirds unmolested, probably due to their human-like practice of hoarding special objects. The bower of this bird contains quandong fruit, animal bones and many pieces of plastic. Philip A Clarke, Alice Springs, Northern Territory, 2004.

Australian pelican tracks. For Aboriginal hunters, the land is full of signs and can be read like a text. Philip A Clarke, Woodrow Point, Lower Murray, South Australia, 1988.

Australasian darter. Hunters seasonally visited major waterbird colonies comprising these and other waterfowl to harvest eggs and nestlings. Birds drying their wings or moulting were more easily caught. Philip A Clarke, Adelaide Hills, South Australia, 1986.

Australian pelican. This large bird is one of the most frequently recorded species in Aboriginal mythology, and generally acknowledged for its fishing prowess. The feathers were used to make decorations. Philip A Clarke, Centenary Lakes, North Cairns, Queensland, 2019.

Waterfowl, with black-necked stork (central left), pied cormorants (centre), little black cormorants (central right) and pied herons (foreground). Wetlands are places rich in gamebirds, used for sources of food and artefact-making materials. Philip A Clarke, Fogg Dam, east of Darwin, Northern Territory, 2011.

Emu manure with white currant seeds. Hunters knew that emus seasonally migrated from inland regions to the coast when the wild currant was in fruit. Philip A Clarke, Teeluc, south-east of South Australia, 1987.

Pacific gull, immature. In southern Australia, this bird produces middens of shellfish remains that could be mistaken as Aboriginal middens. Philip A Clarke, Port Lincoln, South Australia, 2016.

Little egret. In western Arnhem Land, the rock paintings of birds such as the little egret are linked to the period from 1500 years ago to the present, when freshwater swamps replaced the existing saltwater environment. Philip A Clarke, Cahills Crossing, East Alligator River, Northern Territory, 2009.

Eurasian coot. These highly active birds were often caught by nets. Philip A Clarke, Herdsman Lake, Perth, Western Australia, 2020.

Black kite. This species is one of several raptors collectively referred to as 'firehawks' in the savannahs of tropical Australia, through their association with wildfires. Philip A Clarke, Camooweal, Queensland, 2015.

Sulphur-crested cockatoo. This bird is a major source of feathers for decorations. In the bush, these birds warn of the presence of other people. Philip A Clarke, Crafers, Mount Lofty Ranges, South Australia, 2021.

White-faced heron. Aboriginal fishers watched the movement of this bird, as it will wait on rocky perches for shoals of fish. Philip A Clarke, Glen Helen Gorge, Macdonnell Ranges, Northern Territory, 2013.

Pied currawong. In northern Queensland, this bird is known as a rainmaker and is believed to 'sing in' the wet weather. The name 'currawong' was probably derived from *garrawang* in the Yagara language of Brisbane. Philip A Clarke, Bowen, Queensland, 2020.

Whistling kites hunting on the edge of a grass fire. This species would have benefited from Aboriginal 'firestick farming' practices. Philip A Clarke, West Kakadu, Northern Territory, 2009.

Pied imperial pigeon. This species disappears from many parts of its range across the tropics in the late wet, but returns during the late dry to eat the fruit of the Carpentaria palm. Philip A Clarke, Cairns, Queensland, 2020.

Great egret. Large birds such as this were generally roasted in earth ovens. Philip A Clarke, Tomato Lake, Perth, Western Australia, 2020.

Magpie goose. A major game species and as an ancestor its call is associated with the didjeridu. Philip A Clarke, Darwin, Northern Territory, 2017.

Royal spoonbill, white ibis and great egret. Wetlands across Australia are rich places to hunt a variety of waterfowl. Philip A Clarke, Centenary Lakes, North Cairns, Queensland, 2019.

Purple swamphen and chick. Lower Murray people cooked birds such as this by placing the carcass on grass-lined heated stones, before being sealed in the earth oven for roasting. Philip A Clarke, Adelaide Plains, South Australia, 2010.

Pacific black duck. Some Aboriginal groups had a ban on hunting this species during the egg laying season, as the eggs were highly valued as food. Philip A Clarke, Jubilee Lake, Perth, Western Australia, 2020.

Emu. In most Aboriginal classification systems, the emu is not classed among the birds, probably because it is very large and flightless. Philip A Clarke, Alice Springs, Northern Territory, 2019.

Australian bustard. This species is highly prized as food. In some communities, males were not allowed to consume its flesh until they had been initiated. Jamie Robertson, between Mulan and Halls Creek, Western Australia, 2020.

Kimberley man cleaning out the body cavity of a gutted emu prior to cooking. Kim Akerman, Wombarella Creek, Kimberley, Western Australia, 1977.

Lardil man Lindsay Roughsey on Mornington Island in the 1970s. As a ceremonial leader, he is wearing white cockatoo plumes on the top of his head and an ornament of brolga feathers on the back of his head, and holding a spear with attached emu feathers. Mirndiyan Gununa Aboriginal Corporation, Mornington Island Art.

Ceremonial pole with tassels made from strings decorated with red-collared lorikeet and magpie goose feathers. Displayed at a public sea rights celebration in north-east Arnhem Land. Philip A Clarke, Yilpara, Blue Mud Bay, Arnhem Land, Northern Territory, 2009.

Dhukal Wirrpanda with a tassel made from red-collared lorikeet and magpie goose feathers. Performing at a public sea rights celebration in north-east Arnhem Land. Philip A Clarke, Yilpara, Blue Mud Bay, Arnhem Land, Northern Territory, 2009.

Dhukal Wirrpanda wearing a hair ornament made from feathers painted with ochre. Performing at a public sea rights celebration in north-east Arnhem Land. Philip A Clarke, Yilpara, Blue Mud Bay, Arnhem Land, Northern Territory, 2009.

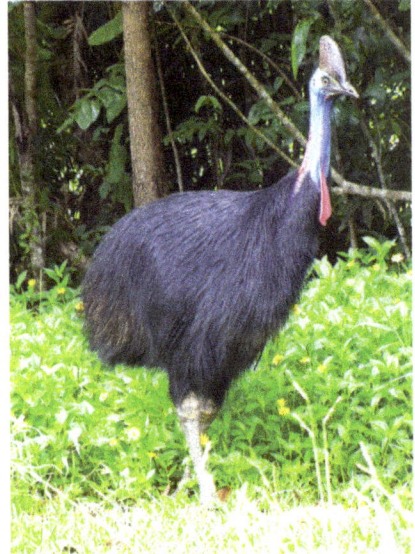

Southern cassowary. Large flightless birds are generally not classified with other birds in Aboriginal classification systems. The leg bones were used for making tools and weapons. Valerie Boll, Maadi, Queensland, 2018.

Tiwi mourning ring (*palmajina*), made from coiled pandanus fibre with hawk feather pendants. These large rings are worn during Pukumani mortuary ceremonies. Kim Akerman collection, Bathurst Island, Northern Territory, 1976.

Hair ornament, made from emu feathers. The plumes are fixed to a bone point, which is the part attached to the wearer's hair. The soft feathers from the emu's chest have been used, not the coarser tail plumes. Kim Akerman collection, Wiluna, Western Australia, 1972.

Hair ornament, made from hawk feathers and bone. Kim Akerman collection, Docker River, Northern Territory, 1972.

Ornamental nose-pegs, made from Australian bustard wing-bones, which are light in weight due to their hollowness. Kim Akerman collection, Warburton Ranges, Western Australia, 1972.

Kimberley man setting *jalimpara* (corella) plumes of split feathers, tied to a bone pin, in his hair. Kim Akerman, Barnett Gorge, Kimberley, Western Australia, 1977.

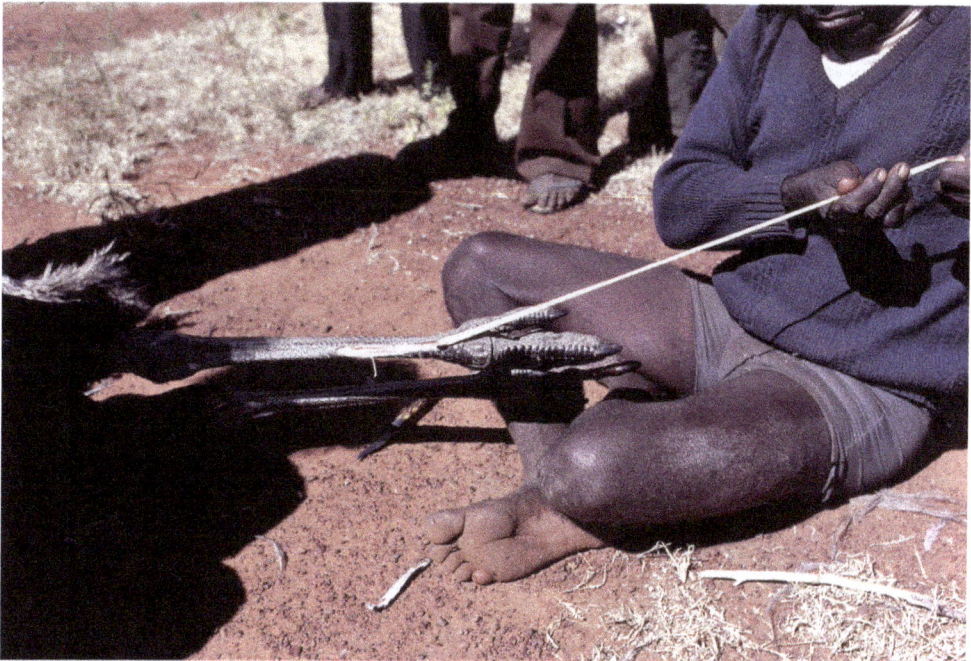

Emu sinew extraction. The sinew is pulled free from the muscle attachment and out of the leg. The attachment at the lower digit is then cut free. Emu (*kananganja*) sinews (*pulyku; jiliwa*) were used for lashing spear barbs, spear shafts and pegs on spearthrowers. Kim Akerman, Balgo, Western Australia, 1981.

Larry Lodi Juburula, wearing an emu feather hair ornament (*nyalpi*). Kim Akerman, Balgo, Western Australia, 1981.

Warlpiri man decorated with a headdress and feather down. Taken after a public performance at the Purlapa Wiri dance festival. Mary Laughren, Warrabri (Alekarenge), Northern Territory, 1978.

Emu feather pad, *wanya*, worn on a man's forehead, chiefly as protection from the sun. Made by Neville Poulson. Robert Graham collection, Yuendumu, Northern Territory, c.1993.

Forehead ornament, made from Major Mitchell cockatoo feathers. Note the red ochre on the plumes that has transferred from the body of the woman who wore it when dancing. Mary Laughren collection, Yuendumu, Northern Territory, 1970s.

Hair ornament/hand-held ceremonial paraphernalia, made from emu tail plumes and hawk feather vanes which have been torn along the rachis. Made by Robert Rallah. Robert Graham collection, Ringer Soak, Western Australia, 1985.

Headdress centre/hand-held ceremonial paraphernalia, comprising coarse emu tail-feathers bound with vegetable fibre string and decorated with pipeclay. Made by Neville Poulson. Robert Graham collection, Yuendumu, Northern Territory, c.1993.

Feather flowers, comprising dyed pelican feathers. Made in 1987 by Ngarrindjeri craftsperson Glenda Rigney at Meningie in the Lower Murray. Philip A Clarke collection, Adelaide, South Australia, 2017.

Adnyamathanha man Lynch Ryan demonstrating emu egg engraving using a penknife. Philip A Clarke, Nepabunna School, Flinders Ranges, South Australia, 1983.

Brolga dance performer. The brolga is a major ancestor for many Aboriginal groups and is often celebrated in public dances. Philip A Clarke, Numbulwar School, Gulf of Carpentaria, Northern Territory, 2010.

Frank Wolmby and other outstation residents at Watha-nhiin Outstation in 1979, farewelling federal government minister Fred Chaney and his advisor Jeremy Long. Jeremy is carrying the prestigious gift of magpie goose eggs the people had gathered in the wetlands (visible behind) during their overnight stay. Peter Sutton, Watha-nhiin Outstation from Aurukun, northern Queensland, 1979.

8

Food and medicine from birds

Studying perceptions of what can be consumed by people provides deep anthropological insights into culture. The old English proverb 'One man's meat is another man's poison' asserts that the definition of good food is not shared across the world's cultures.[1] Anthropologists Peter Farb and George Armelagos stated that 'by knowing how people eat, anthropologists can know much about them and their society'.[2] Anthropologist Mary Douglas described food as a system of communication, meaning that what and how we eat is heavily influenced by our cultural perceptions of food in general.[3]

In Aboriginal Australia, distinct regions utilised different combinations of food procurement and production technologies .[4] For example, museum-based scholar Norman B Tindale described the Panara culture as an arid zone belt across Australia where there was much reliance upon the grinding of hard grass seeds, like Australian millet, and the equally tough sporocarps of the nardoo fern.[5] In the worst of the deserts, Aboriginal foragers relied heavily on insects, reptiles and small burrowing mammals for their daily protein, with birds and larger mammals being more irregular and seasonal sources.[6] In contrast, people living in the better-watered areas of temperate and tropical Australia were able to seasonally gather tubers, fruits and shellfish, and could more readily procure a wide range of game such as waterfowl, mammals, turtles and fish.[7] Such differences are reflected in the Aboriginal material culture across the continent, which in the past has been a theme of several state museum exhibits.[8] The associated environmental knowledge held by groups across Australia was equally varied.

In times of need, Aboriginal peoples would have consumed most sources of protein, and therefore all bird species could in this sense be seen as a potential food source.[9] Nonetheless, in normal periods particular birds were seen as highly desirable foods, and as a result much effort was expended in procuring them. For instance, colonial naturalist George Bennett observed that in New South Wales 'the aborigines prefer the flesh [of the emu] with the skin upon it, regarding it, as the Esquimaux [Inuit of the Arctic] do the flesh of whales and seals, as a highly luscious treat'.[10] In 1959, Rev Friedrich W Albrecht from the Hermannsburg Aboriginal Mission (Ntaria) in the western Macdonnell Ranges of Central Australia commented on the meat favoured by Arrernte people, 'Of course there is no bird to the native like the emu and the turkey [Australian bustard]'.[11] Similarly, in the deserts of central Western Australia and the Kimberley, emus and bustards are among the most highly sought-after avian species for food.[12] Doctor/anthropologist Herbert Basedow, observed in 1925 that 'The flesh of an emu is valued, if for no other reason than for the size of the carcass and the large amount of grease which lies beneath the skin'.[13] In the emu, the femur/pelvic section of the carcass yields the most meat, and the high quantity and quality of fats associated with this part renders it unnecessary to fracture any bones for marrow extraction.[14]

Not all birds were so eagerly sought after by all Aboriginal groups as primary sources of meat under normal seasonal conditions. In the Wadeye area south-west of Darwin, the meat from the nankeen heron is said to be good eating while meat from the straw-necked ibis is considered 'fishy' and the brolga a bit 'stringy';[15] however, on Cape York Peninsula the brolga at least was said to be 'much relished by natives … in texture and flavour it resembles beef'.[16] In Wadeye, egrets in general were not hunted for food as they are considered too 'skinny', although their eggs were consumed.[17] Another species avoided by some Aboriginal groups is the Australian pelican, which many people consider is too fishy to eat.[18] In the Lower Murray, during my own fieldwork, both the pelican and the shag were said to be 'too fishy' for eating.[19] There are no universal rules on what is suitable as food –the Jawoyn people from the Katherine area of the Northern Territory consider pelican to have 'very tasty meat',[20] as do the Walmajarri of the Great Sandy Desert in Western Australia, who eat it after roasting.[21] As will be explained in this chapter, in earlier times there were strict rules for the division of food within each Aboriginal community, with the consumption of various sources being either prohibited or limited to certain sections or cuts.

Like food, medicine is an essential requirement for sustaining people. Definitions of what constitutes both medicines and treatments are shaped by cultural perceptions of the potential causes for a person's loss of wellbeing.[22] Generally, people across the world recognise three main causes of disease: natural, human and supernatural.[23] In many non-Western societies, the origin of serious illness is perceived as a lethal combination of human and supernatural agencies.[24] In the case of the latter, sickness is seen as the result of things such as sorcery, a serious breach of cultural sanctions and behavioural rules, spirit attack, influence of disease-objects, and even the loss of the soul. Aboriginal peoples generally attributed the sudden and inexplicable onset of serious illness to supernatural causes.[25] They routinely used a wide variety of medicinal materials to treat common ailments, such as abrasions and stomach-ache,[26] with specially trained healers focused on curing people using psychic means.[27] Plant-based medicines were widely used in Aboriginal Australia;[28] however, there are also other sources of medicine, including materials derived from birds.

Food preparation

Analysis of the techniques that people use to prepare food provides deep insights into cultural beliefs and traditions.[29] In Aboriginal Australia, it is common for meat foods, such as emu, kangaroo and fish, to have individually named cuts associated with practices determining who, within the kinship system, should most appropriately receive them.[30] Anthropologist Daisy M Bates remarked that 'All meat is distributed according to well established rules, a community of food existing throughout the West [of Australia]. The portions assigned to each person vary but slightly in the different tribes'.[31] Bates said that in the south-west of Western Australia, 'When an emu was divided by the Bridgetown hunter his mother's brother got the back, stomach and fat. Brothers-in-law got a side of the bird, grandparents got the upper parts, the hunter keeping the thigh bone and other smaller bones'.[32] In Tasmania, women regarded the roasted intestines of the black swan as a delicacy for them.[33] Anthropologist

Alfred W Howitt described the butchering traditions of the Ganai of Gippsland in eastern Victoria, and stated that in the case of the emu carcass:

> The intestines, liver, and gizzard are eaten by the hunter. The legs go to the wife's father as *Neborak* [father-in-law's cut] and the body is the share of his parents … If several swans are killed by a hunter, he keeps one or more, according to the wants of his family. The remainder go to his wife's parents, or, if many have been procured, most of them, and the lesser number go to his parents.[34]

In even greater detail, Howitt described the division of an emu carcass among the Ngarigo people of south-east New South Wales:

> The backbone to the hunter; left leg, left shoulder, and left flank to his father. The neck and head, right flank and right ribs to his mother. To his elder brother, the left rib; younger brother, part of the backbone; elder sister, part of the right thigh; younger sister, the right shin. The left thigh and left shin went to the young men's camp. The father and mother shared their part with their parents.[35]

Anthropologist Charles P Mountford remarked that on one occasion when he was out hunting with Pitjantjatjara men in the Western Desert, he was successful when shooting an emu, but back in camp he found that 'As I was considered to be a single man (my wife not being with me) my share of the emu was half the skin – which to European teeth and jaws was as inedible as sponge rubber'.[36] In this way, the random nature of success was spread within the group, although not necessarily evenly.

In the Lower Murray, there was much ritual involved in the butchering of certain bird species. It was recorded by Howitt that:

> … when an emu is killed, it is first plucked, then partly roasted, and the skin taken off. The oldest men of the clan, accompanied by the young men and boys, then carry it to a retired spot away from camp, all women and children being warned not to come near them. One of the old men undertakes the dissection of the bird, and squats near it, with the rest standing round. He first cuts a slice off the front of one of the legs, and another piece off the back of the leg or thigh; the carcass is turned over, and similar pieces cut off the other leg. The piece off the front of the legs is called *Ngemperumi*; that off the back of the leg or thigh, *Pundarauk*. The bird is then opened and a morsel of fat taken from the inside and laid with the sacred or *Narumbe* portions already cut off on some grass. The general cutting up of the whole body is then commenced, and whenever the operator is about to break a bone, he calls the attention of the bystanders, who, when a bone snaps, leap and shout and run about, returning in a few minutes only to go through the same performance when another bone is broken. When the carcass has been cut up into convenient pieces for distribution, it is carried

by all to the camp, and may then be eaten by men, women, and children, but the men must first blacken their faces and sides with charcoal.[37]

The acts of killing and eating an emu are highly symbolic. In central New South Wales, colonist and writer Katherine Langloh Parker recorded that among the Yuwaalaraay (Euahlayi) people:

> When a boy, after his first Boorah [Buura, Bora = initiation], killed his first emu, whether it was his Dhe, or totem, or not, his father made him lie on the bird before it was cooked. Afterwards a wirreenun (wizard) and the father rubbed the fat on the boy's joints, and put a piece of the flesh in his mouth. The boy chewed it, making a noise as he did so of fright and disgust; finally he dropped the meat from his mouth, making a blowing noise through his lips of 'Ooh! Ooh! Ooh!' After that he could eat the flesh.[38]

In the case of the Australian bustard, it was recorded in the Lower Murray that 'Before cooking it, all young boys were sent away; the bird's feathers were plucked, its body singed, opened with a flint or reed knife and cooked on the coals'.[39] It is likely that such practices here represent an acknowledgment of the high importance of these bird ancestors in the ceremonial life of the region.[40] The cultural significance of consuming ritually important meat imparts a value to the food that goes far beyond its nutritional properties.

Egg cooking

Eggs are a rich source of protein.[41] It was apparent to early British colonists that they were a favoured Aboriginal food source. In 1789, British marine officer Watkin Tench wrote that Aboriginal people in the Sydney area were 'ravenously fond of eggs, and eat them wherever they find them'.[42] It is likely that eggs were generally eaten only after cooking, to decrease the chance of making the consumer sick.[43] This was the view of colonial scholar Robert Brough Smyth, who said that 'Eggs are never eaten raw. They are always cooked in the ashes until hard, and they are eaten in all stages of incubation'.[44]

When engaged in fieldwork in the Lower Murray during the 1980s, I found that it was still a practice for elderly Ngarrindjeri people to eat the contents of black swan eggs just before the nestlings hatched.[45] According to historical records, this practice of eating the chicks extracted from eggs was widespread. For example, in the Murray–Murrumbidgee Rivers region, colonist Peter Beveridge remarked that Aboriginal foragers loaded their bark canoes with waterfowl eggs when in season, and 'It is of but small moment to them whether the eggs have birds in them or not, as they are consumed with a relish all the same'.[46] In the Wadeye area south-west of Darwin, it was observed that apart from freshly laid eggs of the magpie goose, the 'Partially to almost incubated eggs are also favoured as food'.[47] This was also the case for Aboriginal people who liked eating the advanced eggs of terns in the Kimberley,[48] and those from the magpie goose in Arnhem Land and the Australian bustard on Cape York Peninsula.[49] In the Goldfields of central Western Australia, on one occasion emu eggs with well-developed chicks inside them were presented for eating to visiting Europeans.[50] Similarly, in the Macdonnell Ranges of Central Australia, Albrecht said that:

> Emu eggs are highly valued, too, and are eaten if there is a chicken inside already half hatched. Fathers, if they find emu eggs while away from camp hunting for several days, will bake the eggs in hot ashes and bring them back to their children.[51]

In Tasmania, bird eggs, such as those from the 'mutton-bird' (short-tailed shearwater), were broiled in the hot embers of a cooking fire.[52] Basedow has provided a detailed account of Aboriginal techniques for cooking the eggs of the emu, black swan and magpie goose. He said:

> The eggs of the larger birds mentioned are laid upon, or into, hot sand and frequently turned to ensure them cooking on all sides. The desert tribes of the Kimberley district have a knack of snatching the egg, as it lies upon the hot ashes, spinning it in the air, catching it again, and replacing it on to the ashes. The process might be repeated two or three times. The idea is to stir up the contents of the egg, in order that they may cook uniformly, much after the style of an omelette or scrambled egg.[53]

According to Langloh Parker, emu eggs were a favourite food for the Yuwaalaraay people of central New South Wales. She reported that:

> Emu eggs the blacks roll in hot ashes, shake, roll again; shake once more, and then bury them in the ashes, where they are left for about an hour until they are baked hard, when they are eaten with much relish and apparently no hurt to digestion, though one egg is by no means considered enough for a meal in spite of its being equal to several eggs of our domestic hen.[54]

On Groote Eylandt in the north-west Gulf of Carpentaria, Tindale noted that the Anindilyakawa people ate the eggs of the orange-footed scrubfowl, and that by 'breaking a hole in one side ... [they] are cooked, hole uppermost, on the fire'.[55] According to anthropologist Alison Harvey, the Garawa people of the Wearyan and MacArthur Rivers region of the Northern Territory prepared eggs for longer-term use. She said:

> Eggs also are preserved, which during the appropriate season occur in thousands from turtles and sea and land birds. For this purpose the eggs are broken into a shallow bark dish, beaten up and then poured into a paper-bark container whose sides are bent upwards to meet at the top, forming a hollow cylinder. When this is placed for a short time on hot ashes the cooked egg becomes a hardened mass sealed in paper-bark, and is available for transport and trade.[56]

Meat cooking

Cooking techniques for birds varied primarily according to the size of the game. The carcasses of most birds in the Lower Murray were simply plucked, singed and opened out along the underside, then placed top side downwards with the skin on top of the coals.[57] Here, the act of plucking feathers was called *teriltun*.[58] Very small birds, such as mallee emu-wrens from the Murray River scrub, appear to have been eaten raw, feathers and all.[59] The preference for retaining the moisture in the bird's carcass had to be balanced with the preference to remove the feathers before eating. In the northern Western Desert, the Pintupi people throw budgerigars on the cooking fire to singe their feathers off and then eat the whole bird; only the outer horny covering of the beak and cere are discarded.[60] Similarly, in the Great Sandy Desert of Western Australia, the Walmajarri people lightly cook budgerigar nestlings before eating them.[61] This appears to be the practice for small birds elsewhere in Aboriginal Australia. In the North Wellesley Islands in the Gulf of Carpentaria, small birds were cooked in the hot ashes of a cooking fire without much preparation.[62]

There are accounts of the cooking of medium-sized birds that suggest there were only minimal preparations when the carcasses were to be cooked in earth ovens. It was claimed that in general 'The natives seldom stop to pluck the birds of their feathers before cooking, but placing damp grass on the hot stones of the oven, put the bird thereon, and laying on more wet grass and placing heated stones on it, cover up the whole with earth. In this way they are half stewed'.[63] In the case of the blue-winged shoveler, the *waringki* (tapeworms) had to be removed first.[64] During my 1980s fieldwork in the Lower Murray/south-east of South Australia, Lola Cameron-Bonney and Ron Bonney described for me one method of cooking duck, which was to remove the head and wings, cover the body with clay and then bury it in the ashes of the outdoor fire.[65] After the duck was cooked, the clay, along with the feathers embedded in it, was taken off 'like a banana' and the fleshy saline leaves of the *nganingi* (pigface) were squeezed over the meat for seasoning. Similarly, Langloh Parker recorded in central New South Wales that 'Ducks were plucked by our tribe [Euahlayi, Yuwaalaraay], but in some places they were encased thickly in mud, buried in the ashes to cook, and, when done, the plaster of mud would be knocked off, and with it would come all the feathers.[66]

Bigger birds, such as emu, Australian bustard, brolga and sometimes the magpie goose and large ducks, were generally roasted in earth ovens in the same manner as kangaroos and wallabies.[67] Protector of Aborigines/writer WE (Bill) Harney described how a European bushman, who had worked with the Wardman people in the Katherine area of the Top End, adopted an Aboriginal style of cooking large birds, such as magpie geese and ducks. This technique used lemon grass foliage as a flavouring, by soaking it in water then placing it with hot stones inside the carcass of a split bird, which was then tied up and put into an earth oven on a bed of steaming leaves for at least 45 minutes.[68] Slabs of butchered meat could simply be cooked on hot coals.[69] In the case of cooking the emu, museum-based anthropologist Frederick D McCarthy detailed the initial preparation of the carcass:

> Some of the feathers were plucked for ornaments, intestines and leg sinews withdrawn, the body singed over the fire, the legs cut off at the knee, and the

No. 8.– Group of Animals.

Drawing by " Yertabrida Solomon," an Aboriginal of the Coorong, in 1876. [From original in possession of Rev. Geo. Taplin.]

SURVEYOR GENERAL'S OFFICE, ADELAIDE. Fraser S.Crawford. Photo-lithographer

Emus, kangaroos, wallabies and an Australian bustard. A group of game animals drawn by an Aboriginal artist Yertabrida Solomon from the Coorong, Lower Murray of South Australia. The different animals were cooked by similar methods. Taplin G (1879) *Folklore, Manners, Customs, and Languages of the South Australian Aborigines.*

head skewered to the body. Ashes were then scraped out of the fire-pit, the carcass placed therein and covered with feathers, over which hot ashes and sand were strewn, and left to cook. The meat was shared either among the families of the hunter's local group or his relatives.[70]

In the Lower Murray, anthropologists Ronald M Berndt and Catherine H Berndt described emus being cooked by the *maramin* method:

… its legs first broken and doubled up and placed on the hot oven stones, alternatively, a bed of grass could be used. According to another method, grass could be strewn over the bird and more heated stones placed on it, then it would be covered up as for a kangaroo but with the head left protruding. When steam issued from its beak, the meat was regarded as being cooked.[71]

Anthropologists W Baldwin Spencer and Francis J Gillen provided a detailed account of emu cooking by Central Australian Aboriginal people, who used the feathers as both stuffing and covering material in the earth oven:

When cooking an emu the first thing that is done is to roughly pluck it; an incision is then made in the side and the intestines withdrawn, and the inside stuffed with feathers, the cut being closed by means of a wooden skewer. A pit is dug sufficiently large to hold the body and a fire lighted in it, over which the body is held and singed so as to get rid of the remaining feathers. The legs are cut off at the knee joint, and the head brought round under one leg, to which it is fastened with a wooden skewer. The ashes are now removed from the pit, and a layer of feathers put in; on these the bird is placed resting on its side; another layer of feathers is placed over the bird, and then the hot ashes are strewn over. When it is supposed to be cooked enough, it is taken out, placed on its breast, and an incision is made running round both sides so as to separate the back part from the under portion of the body. It is then turned on to its back, the legs taken off and the meat cut up.[72]

According to Langloh Parker, the Yuwaalaraay people of central New South Wales would stuff heated stones, eucalypt leaves and emu feathers into the body cavity of the emu carcass during cooking in an earth oven. She reported that:

Emu were plucked, the insides taken out, and the birds filled up with hot stones, box [*Eucalyptus* species] leaves, and some of their own feathers. A fire was made in a hole; when it was burnt down, leaves and emu feathers were put in it, on top of these the bird, on top of it leaves and feathers again, then a good layer of hot ashes, and over all some earth.[73]

On Cape York Peninsula, in the absence of suitable stones as hearths, slabs of heated 'ant bed' (earth from termite mounds) were used to cook large carcasses, such as emu, in earth ovens.[74] Mountford provided an account of emu cooking on the southern Eyre Peninsula of South Australia, which included using a stick to make an exit hole to monitor the steaming. He wrote:

In emu cooking a hole some fifteen to eighteen inches [38–46 cm] deep was dug. In this a large fire was lit and allowed to burn down. The coals and ashes were then raked out and the emu placed in the oven, feet upwards. The body of the bird was then covered with hot sand, coals and ashes, and allowed to remain in the oven for from three to four hours. As in the case of the fish, a green stick was pushed into the oven adjacent to the body.[75]

A description from the Wardaman people in the Katherine area of how to cook the Australian bustard outlined the use of plant materials, as both flavour enhancers and a protective layering, in the cooking process. It is stated that:

When the coals have died down the turkey [bustard] meat is placed in the ground oven and the hot rocks and coals spread under and around it, small rocks may also be placed in the stomach cavity. Young, fresh leaves of Dimarlan,

Eucalyptus camaldulensis, are also placed in the ground oven with the meat to provide flavouring. Paperbark [i.e. *Melaleuca argentea* and *M. leucadendra*] sheets are used as a covering to protect the meat from sand, dirt and burning. The bark also retains the moisture and flavour in the meat while imparting flavour itself. The ground oven is completely covered by soil and sand and the meat allowed to slowly cook for several hours.[76]

Some birds of medium size could also be cooked using the earth oven method. For instance, Lower Murray people used the *maramin* (earth oven) method when cooking a wide range of birds such as brolga, bustard, purple swamphen, cormorants, ducks, grebes, gulls, parrots and pelican; apart from the emu, the heads were included in the cooking.[77] Cormorants, which were seldom eaten by preference, were either cooked on the coals or steamed, with the latter method good for steaming out their fishy smell and taste.[78] According to colonial artist George French Angas, a Lower Murray delicacy was 'sausages made from the entrails of the pelican, stuffed with fat'.[79] The cooking of *kungari waltjeri* (swan intestines), stuffed with vegetables and fat, was still considered a delicacy by Ngarrindjeri people during my own 1980s fieldwork in the Lower Murray.[80] In the Katherine area of the Northern Territory, the Jawoyn people in earlier times used earth ovens to lightly roast the carcasses of waterfowl such as the plumed whistling duck, black duck, radjah shelduck, green pygmy goose, magpie goose, little black cormorant, little pied cormorant, great egret, little egret, white ibis, straw-necked ibis, yellow-billed spoonbill and royal spoonbill.[81]

In the Wadeye area south-west of Darwin, the magpie goose is prepared for cooking by having its breastbone pulled out to expose the heart and liver which are quickly cooked and consumed as snack food.[82] The gullet sometimes contains spike-rush corms, which are given to children, who wash them before eating.[83] The gizzard is then extracted for cooking, and the gravel is discarded.[84] The goose carcass is placed face down on the hot coals with the head and lower legs still attached, although these are not eaten. After grilling, the drumsticks are usually given to children. The taste of the magpie goose flesh is best just after the wet season, when the carcass contains much fat. By the end of the dry, the goose has less fat and is therefore not as tasty. Aboriginal cooks across Australia routinely used certain non-toxic and highly aromatic plant materials, such as river red gum and 'smoke tree' (silver-leaved box) leaves, as wrappers and sources of steam in earth ovens in order to enhance the flavour.[85] It is likely that the soil chemistry in the area where cooks chose to site their earth ovens was also a factor in determining the taste of the food.[86]

Food prohibitions

The eating of certain birds was avoided due to practices ranging from conservational to cultural in origin.[87] For instance, due to the recognised value of their eggs, some Lower Murray birds, such as Pacific black ducks, were not hunted during nesting time, and malleefowl hens were left entirely alone throughout the year.[88] This is consistent with other parts of Australia, such as at Port Lincoln on the Eyre Peninsula in South Australia where Aboriginal foragers took care not to destroy the nests of meat game species.[89] Similarly, on Groote Eylandt in the northern Gulf of Carpentaria, Tindale observed that in the case of the mound-building orange-footed scrubfowl the Anindilyakawa people 'do not usually molest the birds, but the nests are frequently rifled', with both eggs and hatchlings eaten.[90]

Many plants and animals were effectively protected by the existence of 'sacred' places where foraging was prohibited,[91] as well as by the cultural restrictions placed upon hunting in a clan estate after the death of a clan member.[92] A person's totemic affiliations also dictated whether they had to avoid hunting certain animals in normal times. For instance, in the Gulf Country in north-west Queensland, Walter Edmund Roth recorded that the social structure determined what an individual could eat, with the four marriage sections (which he called classes) being Koopooroo, Woongko, Koorkilla and Bunburi.[93] He wrote that:

> Among the Pitta-Pitta blacks and their messmates [related groups] throughout the Boulia District, the Koopooroo are not allowed to eat iguana [goanna], whistler-duck [plumed whistling-duck], [Pacific] black-duck, 'blue-fellow' crane [white-faced heron], yellow dingo, and small yellow fish 'with-one-bone-in-him': the Woongko have to avoid scrub-turkey [Australian brush-turkey], eagle-hawk [probably wedge-tailed eagle], bandicoot or 'bilbi [possibly lesser bilby],' brown snake, black dingo, and 'white-altogether' duck [possibly Burdekin duck]: the Koorkilla have to do without kangaroo, carpet-snake, [grey] teal, white-bellied brown-headed duck [possibly Australian wood duck], various kinds of 'diver' birds, 'trumpeter' fish [probably western trumpeter whiting], and a kind of black bream: the Bunburi dare not eat emu, yellow snake, galah parrot, and a certain species of hawk.[94]

Particular avian characteristics were considered transferable to the people who killed or consumed the birds, and for this reason some bird species were actively avoided by people deemed to be most at risk. In western Victoria, colonist James Dawson recorded that 'Children are severely punished if they kill and eat the magpie lark, for it makes their hair prematurely white'.[95] The avoidance of killing other birds was due to their close association with the creation ancestors. In the Lower Murray, the Berndts recorded that:

> The *piwi* or *piwingi* (the swamp or fish hawk [swamp harrier]) was not eaten, because of its cruelty to other birds and animals; the *puri* (duck hawk [possibly peregrine falcon]), which attacked like a thrown *puri* club, was not eaten for the same reason. The *wate-eri-on* (or *wati-eri*, the Ramindjeri term for this bird; the Yaraldi called it *killing-kildi*, and Karloan called it black magpie [grey currawong]) was taboo to the Ramindjeri Watierilindjera clan, as it was to all the Kukabrak ... The *wuldi* (eaglehawk [wedge-tailed eagle]), *kurki* (*keki*, [collared] sparrow hawk), *maranani* ([*marangani*] crow [Australian raven]), *kraldama* (night owl) and *watji-puldjeri* (lignum bush bird [superb fairy-wren]; the cock with blue feathers, the hen with green) were not eaten since 'they were all people once' and appeared in the local mythology.[96]

There were severe repercussions for consuming birds which were closely associated with ancestors and spirits. The Berndts went on to say that:

> The Lignum Bush Bird was 'a very cunning man' and in his bird form retained this trait, but that could be said of many other *ngatji* [*ngaitji* = clan emblem] … Also because of mythological associations, the *wetjungali* (whispering bird [superb fairy-wren]) and *retjurukeri* (willy wagtail) were rarely caught: in the first case, because of its connection with Ngurunderi [supreme male creator]; in the other because it was said to cause constipation … there was a general taboo on the killing of the white owl (*kroldambi*); anyone eating its flesh was said to become large-eyed. Old men would warn others against doing this, '*Kroldamb-wolamb pil*' (You will have eyes like owl).[97]

In Aboriginal Australia, each community had laws that specified what types of food could be eaten by an individual, which went beyond those that were to be avoided due to totemic associations.[98] Adherence to the rules, whether intended or not, resulted in the wider distribution of food sources across the whole community. Youths and young adults were generally subjected to the greatest number of food prohibitions. In the north-west Cape York Peninsula region of Queensland, anthropologist David McKnight noted that:

> Young children may eat practically anything, for the Wik-mungkan stress that they must be made strong and healthy. But as they grow older, and particularly after they have been initiated, they must follow certain food taboos. As the years go by there are fewer and fewer restrictions, so that in old age the cycle is completed and old people may eat what they please. In times of sickness most food taboos are temporarily set aside.[99]

In the Lower Murray, it was recorded that boys, before being initiated, were forbidden to eat many foods. These included birds such as hardhead, mountain duck, female musk duck, rainbow lorikeet, Australian bustard, brolga and freckled duck.[100] If boys consumed any of this *narambi* (sacred) food, which was forbidden, it was said to give them 'premature greyness of the hair and/or to make them ugly'.[101] If they ate bustard flesh, then *pulkuli* (sores or a rash) would spread from their mouth across the whole face, and consuming Australasian bittern meat would cause *pombongwolini* (large ulcers) to appear and the throat to become like the bittern's neck.[102] The bittern was never cooked in the main camp as the meat contains a lot of fat that may drip onto the grass, from where it might be sucked by children, who would then suffer from the 'supernaturally induced *narambi* disease'.[103] In the case of the willie-wagtail, which is a spirit bird, the Berndts recorded that in the Lower Murray it was:

> Caught only on rare occasion by boys, but sometimes by men. The wagtail was disliked by the Yaraldi. When someone tried to spear it but missed on each

throw, the bird wagged its tail at the hunter. At the same time, the hunter's anus was said to become blocked and he would be constipated for several days.[104]

In western Victoria, young Aboriginal women did not consume the flesh or eggs of the brolga ('gigantic crane'), and it was believed that if babies were placed in close proximity to this bird's body they would develop sores.[105] In central New South Wales, Langloh Parker recorded that 'Should a boy or girl eat plains turkey or bustard eggs while they were yet wunnarl, or taboo, he or she would lose his or her sight'.[106] It is recorded that in the coastal western Arnhem Land region, the Mawng people have a rule whereby the bustard meat is not eaten by women until their sons have finished ceremonies,[107] presumably meaning when they are initiated. The prohibitions were highly nuanced, with the time of the year sometimes being a deciding factor. In the region surrounding Darwin in the Northern Territory, Basedow recorded that:

> After the first appearance of wild geese [magpie geese] at the billabongs and lagoons, the women are not allowed to eat of their meat, believing that if they did so the geese would become lean and bony. Only after the geese have settled for some considerable time in a certain locality are the gins [Aboriginal women] allowed to partake of this food.[108]

Some foods were reserved for children and elderly members of the band. In the Lower Murray, meat from the masked lapwing was specifically for eating by children[109] and senior men were able to consume foods that were restricted to others. A colonist remarked that:

> In most of the lower Murray tribes the old men forbid the young men and women to eat certain parts of different game, viz., the breast and thighs of a duck … When out hunting or fishing, if there are no old men in the party to eat these tit bits, the young men are supposed to throw away the forbidden parts; or when practicable, to bring them into camp for the use of the old men.[110]

Lower Murray people avoided certain parts of the emu's flesh, as the 'sacred pieces *Ngemperumi* [off front of legs] and *Pundarauk* [off back of leg or thigh] can only be eaten by the very old men, and on no account even touched by women or young men'.[111] Fatty portions of meat, such as from the emu, would not be consumed by women.[112] This practice is similar to that recorded elsewhere in Australia, such as in western Victoria where women were not permitted to eat emu meat or eggs until they had grey hair.[113] Similarly, in the Kimberley region women of child-bearing age were not allowed to eat emu meat.[114] In spite of these restrictions, across Aboriginal Australia the emu as a food source was highly regarded for its thick layers of yellow fat, its innards and rump meat.[115] Most prohibited foods were animal-based, although the eating of some edible plants associated with birds was also forbidden. For instance, it is recorded from the Murray River area of South Australia that an unknown vegetable food, possibly a crucifer,[116] that the *war-itch* (emu) foraged on was subject to a prohibition – young women were not allowed to eat it.[117]

In the Lower Murray, once initiated, the young men themselves became *narambi* (sacred) and therefore had to eat a different set of foods for two years.[118] During this period they were considered to be *kaingani* (young man at puberty) and were allowed to eat only the most difficult-to-obtain game animals. Among the recorded forbidden food items at this stage were the Pacific black duck, teal (probably grey teal), magpie goose, hardhead, female musk duck, rainbow lorikeet, water hen (a Rallid species), Australian bustard, brolga, mountain duck and freckled duck.[119] The *narambi* youths were not allowed to eat anything touched by women, such as duck meat, or they would be considered as defiled.[120] Lower Murray women also avoided certain foods, and it was recorded that 'The *kalperi* (shoveller duck) was another such *narambi* food; if eaten by women, it would cause the lips of their vulvas to become so distended as to resemble the bird's bill'.[121] The women avoided eating any game that had been caught or handled by *narambi* novices.[122] Many of these customs were shared by Aboriginal groups based further upstream along the Murray River.[123] There are also similarities with the practices of Aboriginal peoples further east, such as the Kulin (probably the Taungurong group) of Goulburn River in central Victoria who 'believed that if the novice ate the spiny anteater [short-beaked echidna] or the [Pacific] black duck, he would be killed by the thunder'.[124]

Among the Gunggari, Gungadidji and Karangura people of south-west Queensland, young men were not allowed to eat emu eggs for fear of their hair turning prematurely grey,[125] and young people were forbidden to eat eagle flesh.[126] Similarly, in Central Australia the Arrernte people believed that eating the flesh of *irritcha* the wedge-tailed eagle would stop breast development in girls and young women, and make their bodies thin; however, boys could eat the legs of the bird and gain strength in their growing limbs.[127] On Bathurst Island in the Northern Territory, Tiwi people believed that a man had contracted leprosy through having consumed a prohibited food, which was meat from a kangaroo killed by a wedge-tailed eagle.[128] Among the Wardaman people in the Katherine region of the Northern Territory, older boys cannot eat emu meat unless they are first smeared with its fat, and young boys cannot consume it at all.[129] In the north-west of Cape York Peninsula, the Wik Mungkan people have rules that restrict children giving magpie goose eggs they found in nests to their parents, and prevent the gift of these eggs between marriage partners. These rules are apparently at least partly due to the symbolic association of the goose eggs with human testicles.[130]

As discussed in Chapter 4, bird bones from meals were carefully disposed of for fear of them being used for sorcery. There were also other considerations with the safe discarding of food remains. In 1960, Tindale recorded that for the Lardil people on Mornington Island of the Wellesley Islands in the Gulf of Carpentaria in Queensland, the scraps of 'land' food (originating from terrestrial sources) were never thrown into the sea, for fear of the resulting sickness it was believed would result.[131] The throwing of sea food (presumably fish, shellfish etc.) scraps onto land did not cause the same problem, as it was considered that the 'sea hawk' or 'white bellied sea eagle' as the 'leader' of the sea would presumably retrieve it.[132] Such cultural practices were in keeping with the perceived basic separation between land and sea Country.

Preventative and medicinal treatments

Aboriginal peoples used a vast pharmacopeia to help maintain their wellbeing.[133] While many have a demonstrable basis for their success in treatment, others were seen as effective chiefly through the access to ancestral powers that they provided. Some of the substances Aboriginal healers used were technically tonics rather than what Western practitioners would term medicines, but during my fieldwork in the Lower Murray and south-east of South Australia they were often still called 'blood medicine' in Aboriginal English.[134] Aboriginal peoples understood that blood was crucial for the proper functioning of the human body. In the deserts surrounding Balgo in northern Western Australia, anthropologist Father Anthony Rex Peile explained that:

> The Kukatja consider blood is derived principally from meat, but also from vegetable foods. These and meat become blood when they go through the body and not in any particular organ. 'Throughout the stomach, throughout the intestines, throughout the body everywhere. Blood develops from vegetables and from meat, and a person becomes strong without sickness or fainting.'[135]

Fat/oil

Animal fat or grease as rubs provided a covering for Aboriginal peoples, to protect from the cold and provide relief from insect bites.[136] Oils were routinely used medicinally for massaging areas of pain, as compresses for covering lacerations and as an inhalant to treat colds. Aboriginal peoples routinely used emu fat, which was rendered into oil through heat, to help protect their bodies from exposure.[137] Emu oil was highly prized. McCarthy stated that 'Lumps of fat from its body were valued as a salve to protect people's skin from the sun, cold and insects, and in some tribes the fat was reserved for use by the older men'.[138] Langloh Parker noted that among Aboriginal people in northern central New South Wales, 'Emu fat [was used] in cold weather to save their skins from chapping'.[139] She also wrote that Aboriginal people 'smear themselves over with the fat of fish or of almost any game they catch. It is supposed to keep their limbs supple, and give the admired ebony gloss to their skins'.[140] Basedow explained that:

> Although the aboriginal does not wear much clothing, he is very particular about regularly anointing his supple skin. This precaution no doubt gives him greater protection against the changes of weather than all the modern ideas of clothing could do. What he principally applies is fat of emu and goanna, and on the north coast that of some of the larger fish as well. The emu in particular, and especially during a good season, accumulates masses of fat under its skin, which are readily removed, when slain by the hunter. This grease the native rubs over the whole surface of his body to shield the skin from the painful sting of the broiling sun and of the arid wind.[141]

During cold nights, Aboriginal people in the Lower Murray rubbed bird oils on their bodies to keep warm.[142] Indigenous Tasmanians rubbed 'muttonbird' (short-tailed shearwater) oil onto themselves to treat rheumatism.[143] In some parts of Aboriginal Australia, the fat from birds such as emu chicks and boobook owls was used medicinally to treat aches and pains.[144] The Aboriginal panacea for all wounds was emu fat and ochre.[145] In areas south-west of Darwin in the Northern Territory, Basedow recorded emu fat being used, along with paperbark, clay and hot ashes, to staunch the bleeding of initiation wounds on young men.[146] Doctor/historian Roger Byard noted that across Australia:

> Aborigines used animal derivatives in their healing practices. Animal fat liniments for 'rheumatism' and musculoskeletal pain were in wide use, with the types of oil used depending on the availability of local animals. For example, in Tasmania the oil of the 'mutton bird' [short-tailed shearwater] was used, while on the mainland goanna oil, obtained from a lizard, was a remedy adopted by early settlers. Snake and emu fat were also used as liniments and wound dressings.[147]

It is likely that Europeans living on the frontier of settlement first learnt from local Aboriginal people about the effectiveness of emu oil to treat a broad range of ailments.[148] As a medicine, European settlers and bushmen used the clear bright yellow emu oil to treat a variety of aches and pains,[149] particularly rheumatism.[150] It was claimed by a former colonist that 'The fact is that the "rubbing in" is the beneficial agent, and the oil prevents abrasion of the skin'.[151] One bush doctor, known in the newspaper as 'The Quack', had a remedy for rheumatism which was 'a concoction made from emu eggs, gum leaves, kangaroo fat, goanna oil, and emu oil. This mixture could also be used as an anaesthetic. One sniff and the patient fell into a deep sleep'.[152] There were early suggestions in South Australia for the greater use of pelican products, such as the therapeutic use of its oil.[153]

In New South Wales and Queensland, colonists used emu oil as a liniment for sprains and bruises in horses and cattle, either mixed with turpentine or alone.[154] From the 19th century, the European use of emu oil as a leather-softening product and as medicine became popular in Australia after the earlier successes of dugong and goanna oils.[155] In 1932, naturalist Walter W Froggatt described the colonial history of emus, and wrote that:

> Emu oil was a universal remedy among the bushmen for rheumatism, and, mixed with turpentine, was used as an embrocation for sprains and bruises. It was also burnt in their lamps. [Naturalist George] Bennett says that the skin of a full-grown emu yielded six to seven quarts [5.7–6.6 L] of clear yellow oil. [The explorer] Leichhardt wrote: 'Several times when suffering from excessive fatigue I rubbed it into the skin all over the body, and its slightly exciting properties proved very beneficial'.[156]

Throughout the 20th century, emu oil was widely advertised in newspapers for its healing value in the treatment of arthritis.[157] Animal oils, such as from the emu, had many potential commercial uses in the age before a wide range of new products became available through the emerging pharmaceutical industry. In New South Wales, emu oil was much sought-after for burning in lamps instead of whale oil, and had the advantage of being odourless.[158] In the mid 20th century, a European man at Moree in north-east New South Wales apparently made a good living from selling a variety of emu products, which included 'an embrocation made chiefly from emu eggs and emu oil'.[159] It was stated that on the frontier of settlement the emus 'were esteemed for the oil they yielded, which was used as an embrocation. I was assured that emu oil was the best thing known for external application, as "it would penetrate anything."'[160] Pharmaceutical research since the late 20th century has proven that emu oil possesses antiviral and other medicinal properties, so interest in its human medicinal use remains strong.[161]

Feathers

Feathers were used medicinally as absorbent bandages. Langloh Parker noted that in northern central New South Wales, Aboriginal people treated wounds by packing them with birds' down to stop bleeding.[162] In the Gulf Country of north-west Queensland, bleeding was staunched with emu feathers plugged into the wound.[163] The structure of feathers is akin to the gauze of bandages used by Europeans.

Bird manure

Historian and nurse Jennifer Hagger wrote that when an Aboriginal person was injured, 'the flow of blood was stopped by packing the wound with a mixture of clay, gum leaves, powdered bird excrement and grease'.[164] The Pitjantjatjara people in the Western Desert used as medicine the dried manure from an abandoned zebra finch nest, which is often found in prickly wattles such as dead-finish.[165] Mixed with water, the manure is made into a paste then applied to the head of a patient suffering from headache or fever. The Kukatja people in northern Western Australia prepare the zebra finch manure in a similar way, and rub the paste into sore eyes and onto skin lesions. To improve its efficacy, the Kukatja mix it with other plants. For headache and fever, the manure is ground up with sap from the ghost gum and the paste is put onto the head. For boils and external sores, it is mixed with wet macerated leaves from the snake-vine. It is oral history that in 1979 a European stockman at Yuendumu in the Tanami Desert was advised by his Aboriginal co-workers to use zebra finch manure to treat his cattle dog's mange, which would not clear up using conventional veterinary treatments – the treatment was successful.[166]

*****

The Aboriginal selection of avian materials as food and medicine conforms to the rules that apply to all sources of food and medicine, in the way local cultures define what is appropriate to be consumed or utilised externally as a medicine. In terms of cooking methods, there was much uniformity across Aboriginal Australia in the use of open fires and earth ovens to cook birds and their eggs. The use of oil, much of it from birds, as a preventative treatment for human bodies was also widespread. The historical record for the medicinal uses of bird materials is sparse, but what does exist suggests that many Aboriginal medicines had some physical and chemical basis for their efficacy, and their use was greatly enhanced by notions of ancestral healing powers. The relevant Aboriginal environmental knowledge therefore encompasses what Westerners would consider to be empirical facts as well as information solely pertaining to the Aboriginal spiritual realm. Some Indigenous bush medicines were adopted on the colonial frontier by hard-pressed early British settlers, and a few of these continue to be used to the present.

Endnotes

[1] Refer to the Proverb Hunter website (https://proverbhunter.com/one-mans-meat-is-another-mans-poison/).

[2] Farb & Armelagos (1980:14). See also Mintz & Du Bois (2002).

[3] Douglas (1982:Ch.4).

[4] Clarke (2003a:Pt3) and Jones & Clarke (2018:47–50).

[5] Tindale (1977). See also Clarke (2003a:Ch.7, 2013b:62–66) and MA Smith (2013:98–202).

[6] Gould (1969a:5–21, 1969b:258–265), Roheim (1974:Ch.2), Tindale (1981:Pt8) and Clarke (2003a:Ch.9).

[7] Lawrence (1968), Tindale (1981:Pt8), Clarke (1988:71–73, 2003a:Chs8,10, 2013a:98–99) and Gott (2008:216).

[8] Amateur Naturalist (1886, 1887). These differences in technology were highlighted in the Australian Aboriginal Cultures Gallery exhibition at the South Australian Museum (Clarke 2000).

[9] Von Sturmer (1978:226).

[10] Bennett (1834:297–298).

[11] Albrecht (1959).

[12] Cane (1987:423), Chapman et al. (1995:358) and Karadada et al. (2011:70,72).

[13] Basedow (1925:139).

[14] Garvey et al. (2011) and Hardwick (2019b:58).

[15] Hardwick (2019b:59,62).

[16] Thomson (1935:34).

[17] Hardwick (2019b:59–60).

[18] Dawson (1881:19) and Puruntatameri et al. (2001:105).

[19] Clarke (2003b:96).

[20] Wiynjorrotj et al. (2005:149).

[21] Doonday et al. (2013:131).

[22] Berndt (1982), Biernoff (1982), Meehan (1982b), Reid (1982b), Scarlett et al. (1982) and Clarke (2008b:4–12).

[23] Clements (1932), Foster (1976) and Joralemon (2015:Ch.1). In the case of Aboriginal Australia, refer to Eastwell (1973b:1013–1014), Cawte (1974:Ch.5, 1996:Ch.2) and Clarke (2007a:96–97).

[24] For example, see the study of witchcraft and sorcery by Evans-Pritchard (1937) among the Azande of the Sudan in northern Africa.

[25] Roth (1903), Reid (1978b, 1979, 1982a, 1983), Wiminydji & Peile (1978), Berndt (1982), Devanesen (1985, 2000) and Maher (1999).

[26] Webb (1969), Henshall et al. (1980) and Barr et al. (1988).

[27] Berndt (1947), Cleland (1953), Maddock & Cawte (1970), Eastwell (1973a, 1973b), Cawte (1974, 1996), Elkin (1977), Reid (1978a), Akerman (1979), Soong (1983), Tonkinson (1994) and Ngaanyatjarra Pitjantjatjara Yankunytjatjara Women's Council Aboriginal Corporation (2003).

[28] Bailey (1880), Webb (1960), Lassak & McCarthy (1983), Clarke (1987, 1989, 2007a:Ch.8), Barr et al. (1988) and Kyriazis (1995).

[29] Clarke (2017:41–42).

[30] Bates (1901–14:243), Howitt (1904:756–770), Rose (1987:184–189) and Clarke (2003a:59).

[31] Bates (1901–14:242).

[32] Bates (1901–14:243)

[33] Bonwick (1870:17).

[34] Howitt (1904:758).

[35] Howitt (1904:759).

[36] Mountford (1948:174–175).

[37] Howitt (1904:763). Howitt stated that he wrote his account from information he gained from George Taplin's son, Frederick.

[38] Langloh Parker (1905:41).

[39] Berndt *et al.* (1993:556).

[40] Berndt *et al.* (1993:211–213,Fig.31) and Clarke (2016a:281).

[41] Brand Miller *et al.* (1993:Table 3).

[42] Tench (1789:241).

[43] See Wiynjorrotj *et al.* (2005:125).

[44] Smyth (1878:1:208).

[45] Clarke (2003b:97, 2018h:12). For other accounts of swan egging, see Abdulla (1993) and Clarke (2014a:28,32).

[46] Beveridge (1883:36).

[47] Hardwick (2019b:52).

[48] Love (1936:183).

[49] Thomson (1935:24, 1983a:100). Thomson used the name 'brush-turkey' instead of Australian bustard.

[50] Anonymous (1930a:4).

[51] Albrecht (1959).

[52] Bonwick (1870:17).

[53] Basedow (1925:126).

[54] Langloh Parker (1905:125).

[55] Tindale (1925:80).

[56] Harvey (1945:191). The name of the Garawa people is also written as 'Karawa'.

[57] Berndt *et al.* (1993:104–105,Fig.21).

[58] Tindale (c.1931–91).

[59] Amateur Naturalist (1886).

[60] Thomson (1975:88).

[61] Doonday *et al.* (2013:142).

[62] Memmott (2010:35).

[63] Worsnop (1897:116).

[64] Berndt *et al.* (1993:558).

[65] Clarke (2017:41).

[66] Langloh Parker (1905:125).

[67] Meyer (1846:195), Angas (1847b:89–90), Smyth (1878:1:192–195), Spencer & Gillen (1927:1:19–20), Thomson (1939:220–221), Berndt *et al* (1993:555–557), Palmer (1998:35–36), Raymond *et al.* (1999:102), Wiynjorrotj *et al.* (2005:145), Memmott (2010:37), Karadada *et al.* (2011:70,72,80,82), Doonday *et al.* (2013:129–130) and Hardwick (2019b:28,55,62–63).

[68] Harney & Thompson (1960:32).

[69] Memmott (2010:34).

[70] McCarthy (1965:17).

[71] Berndt *et al.* (1993:104–105).

[72] Spencer & Gillen (1899:24).

[73] Langloh Parker (1905:124).

[74] Thomson (1939:221).

[75] Mountford (1939:199–200).

[76] Raymond *et al.* (1999:100).

[77] Berndt *et al.* (1993:555–561).

[78] Berndt *et al.* (1993:556).

[79] Angas (1877).

[80] Abdulla (1994) and Clarke (2003b:96, 2018h:10).

[81] Wiynjorrotj *et al.* (2005:124–125,145–147,149).

[82] Hardwick (2019b:55).

[83] Hardwick (2019c:46).

[84] Hardwick (2019b:55).

[85] Clarke (2012:80–83).

[86] J McEntee (pers. comm.).

[87] Clarke (2017:42–44).

[88] Tindale (1987a:11) and Berndt *et al.* (1993:124,558).

[89] Wilhelmi (1861:177).

[90] Tindale (1925:80).

[91] Newsome (1980) and Clarke (2003a:64).

[92] Baker (1999:49) and Clarke (2003a:64).

[93] Roth (1897:57).

[94] Roth (1897:57–58).

[95] Dawson (1881:52).

[96] Berndt *et al.* (1993:124). For species identifications, refer to Clarke (2019a:Table A2).

[97] Berndt *et al.* (1993:124).

[98] Bates (1906), Thomson (1936a:378–379), Berndt & Berndt (1970:35,37,52,66,93,96,115,165,180–181) and Clarke (2003a:48–50).

[99] McKnight (1973:196).

[100] Taplin (1859–79:13 November 1861). In this source, the rainbow lorikeet was described as the 'Blue Mountain parrot', the Australian bustard as the 'turkey', the brolga as the 'native companion' and the freckled duck as the 'pink eyed duck'.

[101] Howitt (1904:673–675) and Berndt *et al.* (1993:126). For more details on *narambi*, refer to Berndt (1974:26) and Berndt *et al.* (1993:Chs7,10).

[102] Berndt *et al.* (1993:127–128).

[103] Berndt *et al.* (1993:556).

[104] Berndt *et al.* (1993:560).

[105] Dawson (1881:53). In this source, the brolga was simply described as a 'gigantic crane'.

[106] Langloh Parker (1905:40).

[107] Singer *et al.* (2021:143).

[108] Basedow (1907:22).

[109] Berndt *et al.* (1993:560).

[110] Anonymous (1906).

[111] Howitt (1904:763).

[112] Berndt *et al.* (1993:128).

[113] Dawson (1881:53).

[114] Mjöberg (1915:206).

[115] Mjöberg (1915:112–113).

[116] Refer to Clarke (1986b:9, 2018d:64–65).

[117] Eyre (1845:2:295).

[118] Meyer (1846:187), Taplin (1859–79:5 January 1860) and Berndt *et al.* (1993:126–127).

[119] Taplin (1859–79:13 November 1861). In this source, the magpie goose is described as the 'Murray goose', the rainbow lorikeet as the 'Blue Mountain parrot', the Australian bustard as the 'turkey' and the brolga as the 'native companion'.

[120] Harris (1894–1895:1 June 1894).

[121] Berndt *et al.* (1993:124).

[122] Berndt *et al.* (1993:125–126,128,180).

[123] Eyre (1845:2:293–295).
[124] Howitt (1904:612).
[125] Heagney (1886:375) and Johnston (1943:271).
[126] Heagney (1886:375).
[127] Gillen (1896:180).
[128] Harney (1957:22).
[129] Raymond *et al.* (1999:103).
[130] McKnight (1973:196–197).
[131] Tindale (1960:63).
[132] Tindale (1960:63).
[133] Roth (1903), Webb (1960, 1969), Reid (1977), Henshall *et al.* (1980), Clarke (1987, 2007a:Ch.8, 2008b, 2015c), Barr *et al.* (1988), Byard (1988) and Kyriazis (1995).
[134] Clarke (1987:5,9–10, 2008b:7–8, 2014a:31,34–35,45).
[135] Peile (1997:74).
[136] Peile (1997:201).
[137] Langloh Parker (1905:54,133) and Basedow (1925:324).
[138] McCarthy (1965:17).
[139] Parker (1897:118). See also Hagger (1979:24) and Ash *et al.* (2003:63).
[140] Langloh Parker (1905:119).
[141] Basedow (1925:115).
[142] Berndt *et al.* (1993:17).
[143] Bonwick (1870:89).
[144] Peile (1997:200–201) and Hardwick (2019b:58).
[145] Hagger (1979:24).
[146] Basedow (1925:243).
[147] Byard (1988:794).
[148] Raven *et al.* (2021).
[149] Nature Lover (1918) and Anonymous (1940).
[150] Molineux (1905), Sorensen (1909:2), Anonymous (1915) and Barrett (1932).
[151] Molineux (1905).
[152] Eureka (1938). See also Thornton (1940).
[153] Molineux, (1905:6).
[154] Leichhardt (1847:298) and Taylor (1985).
[155] Folkmanova (2015:106).
[156] Froggatt (1932). For the original references cited, see Bennett (1834:1:297) and Leichhardt (1847:252).
[157] For examples of 20th century advertisements, refer to the advertisements of Tost & Rohu (1909), Marshallsea (1991), Mount Romance (1994) and Fruits of the Earth (1999). Raven *et al.* (2021) described the commercial exploitation of Aboriginal knowledge concerning therapeutic emu oil.
[158] Taylor (1985).
[159] AT (1950).
[160] Anonymous (1932).
[161] Whitehouse *et al.* (1998), Bennett *et al.* (2008), Abimosleh *et al.* (2012), Mashtoub *et al.* (2016) and Raven *et al.* (2021).
[162] Parker (1897:118). See also Hagger (1979:24).
[163] Roth (1897:175).
[164] Hagger (1979:24).
[165] Hilliard (1968:136) and Peile (1997:86–87).
[166] Peile (1997:86–87).

9

Material culture

In early Aboriginal Australia, each band of foragers was self-reliant by necessity, and while seasonally moving across their Country they utilised their detailed environmental knowledge in order to gain most of the material required for their daily subsistence from the local environment. In addition to birds being a major food source, the materials extracted from them were used to help make tools and other associated substances that were essential for maintaining the Aboriginal foraging lifestyle. Avian materials were also aesthetically prominent as decorations. The range of artefacts produced by Aboriginal toolmakers was essentially the same across the continent, albeit with some distinct regional styles.[1] As with all their cultural traditions, Aboriginal peoples ultimately referenced their artefact-making practices, along with the distribution of raw materials, to the order their spiritual ancestors established during the creation.

The focus of this chapter is to further explore how, back when Aboriginal peoples were chiefly living off the land, all parts of the bird's body had some potential use, if not for food then for making things. The leg bones from large flightless birds, the emu and southern cassowary, were robust enough to be used as tools and weapons, while the typically hollow wing bones from other smaller species, which are light-weight but strong, were once widely used in making ornaments. Leg sinews from emus supplied the strong cord for game nets and were used as stitching or binding for objects such as cloaks, spears and spearthrowers. Birdwings are used as fans and brushes, and in earlier times feathers were stuffed into skin bags and beaten to produce music. Bird oil served as a pigment fixative and protected artefacts from weathering. Many parts of the bird – including feather down, plumes, bones, beaks and claws – are still worn decoratively by Aboriginal men and women.

The parts of a bird's carcass that could be used to make artefacts were primarily determined by physical properties, but the choices were shaped by strong cultural perceptions of the ancestral powers associated with birds. Elaborate ceremonial decorations, often incorporating feathers, were produced not just for aesthetic reasons but to assist in drawing down the power of the ancestors, as discussed in Chapter 2. Transformed from their normal state, the decorated performers temporarily become the incarnation of their spirit ancestors. The decorations also protected the performers from being recognised by the malignant spirits who are believed to be attracted to such events.[2] The ritual objects that Aboriginal peoples used for healing, charming, fighting and rainmaking were seen as being spiritually powerful through their connection to the creation ancestors. In the case of bird ancestors, these links were often symbolised by decorations made from feathers and avian body parts extracted from relevant totemic species.[3]

The ancestors were credited with introducing the use of particular artefacts. For instance, Old Man Karramala the magpie goose is a major ancestor for Country on the lower reaches of the Finniss River.[4] Here, the contemporary Mak Mak Maranunggu people associate the honking call of *djulburr* (magpie geese) with the playing of the *kenbi* (didjeridu or drone-pipe), made from Arnhem Land bamboo of the same name, which Old Man Karramala did when coming down the Florence River to the Finniss River during the creation.[5] As discussed in Chapter 3, it was almost universally believed across Aboriginal Australia that during the creation bird ancestors caused the property of fire to escape into Country, from where people could obtain it by using firesticks.

In Aboriginal Australia, avian materials contain much symbolic power. Larger birds, which required high hunting skills to kill, were particularly important in this regard. Biologist/anthropologist Donald Thomson wrote about Arnhem Land culture and observed that:

> Certain materials, particularly the fat of *karritjambal* the Red Kangaroo, and of a few other animals, and the fat and feathers of some birds, chief of which are the Emu, Plain Turkey or [Australian] Bustard, the Jabiru [black-necked stork] and Native Companion [brolga], which are difficult to approach, have a high 'social value' and are greatly prized. These materials have a ritual significance and *must* be accumulated by the hunter. This fat, as well as the feathers of the birds mentioned above, are known as *tjarnbin* and play an important part in the ceremonies called *kunabibbi* [Gunapipi] and *ngulmark* [Ngulmarrk]. The long bones, bills and some of the flight feathers of these birds are also accumulated and wrapped with paperbark into neat parcels called *tabarr*.[6]

This chapter demonstrates how Aboriginal hunter-gatherers made objects from bird materials, and investigates the similarities and differences of such uses across Australia. It concerns material culture, which is the sum of beliefs and traditions concerning the tangible belongings of a people.[7] The information assembled here is primarily derived from the historical and anthropological literature, augmented by my own curatorial familiarity with artefacts in museum collections, along with my own field experience.[8] Another source of information drawn upon here are the images of Aboriginal artefacts in the paintings of early colonial artists.[9] This last source is particularly important when documenting Aboriginal life as it was at the frontier of British settlement, which for much of Australia occurred in an age before the advent of photography and before interest in documenting Aboriginal foraging practices had waned.[10] In spite of the delicate nature of avian bones, there is also some archaeological evidence for the butchering of bird carcasses at ancient campsites, albeit chiefly as food as discussed in Chapter 6. The past uses of relatively soft materials, such as feathers and sinews, are largely absent from the archaeological record.[11]

Feather objects

The bodies of most birds capable of flight have several types of feathers, namely the primaries, secondaries and tertiaries attached to the wing bones, the rectrices (tail-feathers), body contour feathers and the downy feathers mixed among the contour feathers.[12] For Aboriginal peoples, not all feathers are useful in the same way. The rectrices and pinions are typically used as head ornaments and the white bird down is particularly prized as a body and face decoration for both performers and objects during ceremonies. The other feathers are utilised for making a wide variety of objects for everyday use and for ceremonial purposes.

An advantage of using feathers was that they provided a greater range of colours, including blue and green, than could be produced from earthy materials such as charcoal (black), kaolin (white) and ochres (red, brown, yellow). In addition, the iridescent quality of feathers, which catches the rainbow, was perceived by Aboriginal peoples as being spiritually powerful.[13] As soft and aesthetically pleasing materials, feathers were in high demand for use in making decorations. When they were not readily available, the 'wool' from the dry flower heads of a wide variety of plants, such as daisies, slender pigweed and kapok-bush, was used instead.[14] Colonists also had uses for feather down. In South Australia, a settler suggested that breast down from the local pelican could be used in making clothes in place of swan down, presumably from the mute swan in Europe.[15]

Birdwing fans, brushes and whisks

Fans made from the wings of large birds are frequently reported in the literature, but in my curatorial experience there are surprisingly few of them in museum ethnographic collections. The Aboriginal need for them is apparent when camping in areas seasonally prone to flying insect vermin, such as march flies and mosquitoes. Fans had other uses too – as brushes to clean food and as devices to cool bodies and relight coals on the fire. The use of such objects in Australia can be documented from the pre-European period, as fans made from magpie goose wings, along with the specialised 'goose' spears, appear in western Arnhem Land rock paintings associated with what rock art experts call the 'Freshwater period' (1500 years ago to European colonisation).[16]

Birdwing fans/brushes were widely used in Aboriginal Australia. In the Lower Murray, a birdwing fan known as *tjelindjeri* was used to keep away flies.[17] Diyari people at Cooper Creek in the north-east of South Australia used emu tail-feathers to make fans.[18] At East Alligator River in western Arnhem Land, birdwings were habitually used to keep flies away.[19] Thomson reported that during the wet season, when Arnhem Land hunters entered the Arafura Swamp to hunt magpie geese and collect their eggs, the 'Goose wing fly or mosquito switches' were essential in order 'to drive away mosquitoes which, except when the wind is high, come in hordes'.[20] At Cobourg Peninsula in western Arnhem Land, a single goose-wing fan is made from parts of two wings belonging to the same bird; it is used to cool a person or to make a draught to rejuvenate a fading fire, as a brush for clearing ashes off cooked food and as a whisk for driving insects away.[21] Here, it was recorded that the expression *kunmanakaraka imajak*, which means 'you will fan with a goosewing', is used to describe the fanning action.[22]

In the Wadeye area south-west of Darwin, the pinion feathers from magpie geese and brolgas are cut off to make fans that are used to push away campfire smoke, sweep around the fire, keep flies away from food and babies, and slap at mosquitoes during the night.[23] They are also waved over children to help them go to sleep on hot nights. Here, to make a *nanhthi mangiu mirrirr* or goose-wing fan, local writer Jeff Hardwick explained that:

> The pinion feathers are cut from dead birds. Both Magpie goose and Brolga wings are used. The feathers from around the joint are plucked out and the flesh is removed. Placing small hot coals or ashes from the fire on the joint dries the remaining flesh. As the coals dry out the flesh, the wing is spread so that it remains open when dry. It is then hung up and left to dry for several days. String is tied around the joint and two, left and right wings are joined tighter to create a stronger draught of air fanning a fire or the face.[24]

In northern Queensland also, fans were objects with multiple uses. Among the Wik Mungkan people of western Cape York Peninsula, a goose-wing fan was used by a grandmother during the ceremonial presentation of a child to the father to drive off flies that might have followed the child from a now ritually prohibited place.[25] At Princess Charlotte Bay on the east coast of Cape York Peninsula, museum researchers Herbert M Hale and Norman B Tindale observed that:

> Wings of large birds provide fans with which flies and mosquitoes are warded off, and also form brushes with which the tidier members of a camp occasionally sweep out their huts; the wing of a wild [magpie] goose ... was secured from a Barunguan man who was suffering from large open sores, and was therefore particularly worried by the multitudinous flies and mosquitoes.[26]

A common feature of such fans is that they are made from the large feathers of big birds. In northern Queensland, early anthropologist/doctor Walter Edmund Roth stated that:

> In most camps, during the hotter months, the wing of some comparatively large bird, such as the 'Native Companion' [brolga] ... or 'Plain Turkey' [Australian bustard] ... is often to be seen employed as a fly-flick, possibly as a fan. At Cape Bedford, on the Palmer, Mitchell, Nassau, and Staaten Rivers, fly-flicks are manufactured with Emu-feathers, somewhat after the fashion of a feather-duster, the quills being bound tighter onto a short handle, the binding being strengthened with gum-cement ... these are made and used by men, generally the older ones, and known to the Koko-yimidir [Guugu-Yimidhirr] as wandaka, and to the Koko-minni [Kuku-Minni] as ata-angka (feather) or ariva- (emu) – ata-angka. On the Bloomfield, specimens of similar design are made of [southern] Cassowary feathers.[27]

Emu plumes tied together and mounted on sticks as a flywhisk or flyswitch were also used for mortuary practices. In the Lower Murray, when the smoke drying of a corpse on a raised platform was taking place, it was the responsibility of close female kin to keep flies away using whisks which were 2.4–2.7 m long, with a bunch of emu feathers attached to the end.[28] On the Yorke Peninsula in South Australia, flywhisks, called *gari wopa*, were made from emu feathers.[29] In the Museum of Victoria ethnographic collection there is a flyswitch collected by Donald Thomson in 1942, that was made by the Djabu people of Caledon Bay in north-east Arnhem Land from emu feathers mounted on the end of a stick, secured with string binding and wax or resin.[30] A smaller version was apparently used by children here when playing 'camp'.

Feather down decorations

Feather down was applied directly to human bodies for ceremonial purposes. By wearing the material in this way, during dances the performers appear to be emanating light and power. The down that is shaken loose during the performance enhances this effect.[31] Feather down is seen as ritually powerful material. For instance, in northern Australia anthropologists W Baldwin Spencer and Francis J Gillen noted that while plant-based down, such as from portulaca, is often used for decorations made for 'ordinary corroborees', 'the bird's down is used almost exclusively for sacred ceremonies'.[32] From an Aboriginal perspective, this power was derived from the creation. In the south-west Kimberley, artist Butcher Joe Nangan spoke of a creation myth whereby a female wedge-tailed eagle was killed and its feather down taken by the Australian owlet-nightjar and spotted nightjar in order to 'divine her place of origin in the spirit world'.[33] In this account, the excess down was caught in the air by two large willy-willies and taken up into the Skyworld, where it was transformed into the Southern Cross, seen as a wedge-tailed eagle.

For ceremonies across Aboriginal Australia, eagle down was widely used for decorative purposes on both bodies and objects, and the difficulties in procuring this material would have greatly added to its prestige. From Central Australia, Spencer and Gillen noted that 'In the Arunta [Arrernte] nation the down used in their ceremonies is obtained exclusively from birds – chiefly the eagle-hawk [wedge-tailed eagle]'.[34] The down was typically stuck on with human blood, which is an excellent fixative.[35] Doctor/biologist John B Cleland noted that:

> The down of birds lends itself to decorative purposes, especially that of the eagle. It is extensively used for ceremonial purposes being stuck on to the body by means of the sticky serum that exudes from the clotting of human blood. To obtain this blood, the men readily open the veins of the arm by means of a short longitudinal incision with a sharp pointed piece of stone, often prepared on the spot.[36]

During initiation ceremonies at the Macleay River in north-east New South Wales, Aboriginal men tied their hair in a knot and covered their heads with the snowy down of the cockatoo.[37] Similarly, in the Lower Murray region the white down of birds was combed into the hair of red-ochred initiates.[38] Here, the downy feathers of the 'goose' (probably the magpie goose) were used to make pendants.[39] In the Lower Murray and south-east of South Australia, colonial artist George French Angas recorded that during organised fights between opposing Aboriginal groups the men would bind the white down from musk ducks and black swans round their heads and twist it into fillets.[40] During the Mindari ceremony held in north-east South Australia, the Diyari people used the down from black swans and ducks for decorations.[41] Former colonist Thomas Worsnop noted that on the Eyre Peninsula in South Australia:

> The Port Lincoln tribes would catch birds, and from the white down they made a kind of chaplet ... The old men at Cooper's Creek [in north-east South Australia] used to wear a head-dress of feathers, but in widely different portions of the continent they wore their hair in the form known as a chignon, with the white downy feathers of the cockatoo fixed here and there in the coils.[42]

In many parts of Australia, small white downy bird feathers were used in the making of ceremonial string, being twined into plant or human hair fibres rolled on the thigh, giving the finished twine an attractive fluffy texture.[43] The fuzzy appearance of the down, coupled with its whiteness, imparts an aura of power. In north-east Arnhem Land, the feather down is woven into strings attached to the publicly seen *rangga* or ceremonial objects that are associated with the Barnumbirr Morning Star ceremonies.[44] In my experience, these are said by some artists to represent the rays of starlight. Similarly, in this region the ceremonial poles associated with Woial the honey man are typically decorated with feathered lengths of dangling string that represent lines of swarming bees.[45]

The Tiwi people on Bathurst and Melville Islands have a decoration known as a *tokwianga*, which is made from magpie goose feather down in the shape of a small ball attached to a tassel.[46] This object is worn round the neck or tied to the upper arm on all ceremonial occasions. Museum-based scholar Charles P Mountford noted that:

> These ornaments are made by embedding ends of the downy feathers of the pied goose [magpie goose] in a ball of wax. So much work was entailed in making a *tokwianga* that the aborigines were loath to part with them.[47]

Down-like material was also made by crushing large white feathers. Tindale was on Mornington Island of the Wellesley Islands in the Gulf of Carpentaria of Queensland for fieldwork in 1960 when he observed a group of Lardil men ceremoniously preparing themselves with decorations made from ochre and sulphur-crested cockatoo feathers. The feathers, he said, 'are pounded between stones and the matted tufts so formed stuck on using a native gum, applied to the body with a stick. In olden days blood would have been used'.[48] On Mornington Island, cylindrical-shaped hats were heavily decorated with bird down in white and red bands, the latter produced from down coloured with red ochre.[49] Down used in ceremonial decorations was often coloured by working in crushed mineral pigments, such as pipeclay and ochre.[50]

Head and body decorations

Bird plumage was regularly used across Aboriginal Australia for body and head ornamentation.[51] Primary feathers were often worn as plumes by men on the front of their head, by attaching them to a lock of hair.[52] Other plumes were a more integral part of an ornament, such as the emu feather plumes that hung gracefully from an apron cord made from kangaroo tail sinews, worn by young women.[53]

Apart from their spectacular appearance, the free movement of decorative plumes was also important. For instance, in 1940 Cleland noted in the southern Western Desert that 'Recently at Ooldea some of the younger men were wearing long feathers, such as those of the [Australian] bustard or native turkey, tucked into the hair arranged as a chignon at the back of the head. These plumes waving about had quite a picturesque effect'.[54] The Larrakia people in the Darwin area decorated their young men who had just come through initiations with emu or heron plumes attached to the head with a band.[55] In some cases, the feathers were modified to accentuate their free movement when worn. Australian bustard wing feathers were used in ceremonial dances held in the Wadeye area south-west of Darwin, being stripped and tied into a bundle for the purpose.[56] The stripping process involves splitting broad feathers, such as from a bustard or hawk, into the two vanes by tearing them along the rachis (rigid shaft) towards the calamus (hollow shaft section), which promotes movement when the feathers are tied together and shaken.[57]

In Aboriginal Australia, both men and women wore ornaments made from feathers, particularly during ceremonies. At other times, plumes were used to draw attention to the wearer's status. For example, in the reminiscences concerning local Aboriginal people, a former Adelaide Plains colonist said that 'Young men of 18 to 20 [years] who had been initiated who wished to show they desired a permanent wife wore a feather through their pierced nose'.[58] Cockatoo primary feathers as head ornaments were particularly favoured by

No. 6.—War Dance.
Drawing by an Aboriginal of the Kingston Tribe. [*From original in possession of E. Spiller, Esq.*]

Aboriginal men with feather plumes on their head. Drawn by an Aboriginal artist from Kingston, south-east of South Australia. Taplin G (1879) *Folklore, Manners, Customs, and Languages of the South Australian Aborigines.*

Aboriginal men. In Tasmania, the men inserted several cockatoo feathers into their hair for decoration.[59] Feathers were sometimes mixed with other materials for decorations. Worsnop noted that 'In the Tatiara country [the south-east of South Australia] the men would fix the crest of a cockatoo in the hair above the forehead, and sometimes a wild dog's [dingo's] tail or a bunch of pelicans' feathers was tied to the hair at the back of the head, and worn as a sort of pendant'.[60]

Plumes were also attached to a hairband, such as the pink feathers from the crest of the Major Mitchell cockatoo that were worn in arid areas,[61] and seen on female Warlpiri performers of the public Yawulyu dance.[62] From my own experience working with Aboriginal ethnographic collections in museums, the plumes to decorate the top of the head were often attached to a wooden or bone pin. According to colonist and writer Katherine Langloh Parker, in central New South Wales 'Feathers tied into little bunches and fastened on to small wooden skewers were stuck upright in the hair at corroborees, also swansdown fluffed in puff balls over the heads'.[63] The Warlpiri men of the Tanami Desert in northern Central Australia often wore an emu-feather pad, called a *wanya*, on their foreheads, chiefly as protection from the sun but also to hold small objects and ornaments.[64] This was made from a wad of short body feathers (also *wanya*) from the emu, tied together with hair-string, although more recently commercially coloured wool has been used.

Apart from being a decoration, many of the ornaments that the earliest Europeans noted being worn by Aboriginal peoples had important symbolic meanings and ritual uses. For instance, anthropologist Frederick D McCarthy observed that 'Ornaments such as chignons, head circlets and hair plumes were fashioned out of the feathers, which in western Queensland were wrapped in fur-twine or netting as amulets for the cure of ailments, or sent as an invitation to … assist in warfare'.[65] On ceremonial occasions, senior Tiwi men on Bathurst and Melville Islands wear a head ornament, known as *pimirtiki*, which is made from sulphur-crested cockatoo feathers fixed with beeswax to a wallaby bone pin.[66] Feather ornaments also play a symbolic role during performances. The white plumes of the Major Mitchell cockatoo worn by women dancers as a cockade during the Yawulyu in Central Australia are referred to as white clouds, in the accompanying singing.[67]

Spencer and Gillen provided a detailed account from Central Australia of Arrernte and Luritja men wearing a pad of emu feathers on their head, measuring 25 cm long by 13 cm wide and 5 cm in thickness.[68] Such pads formed a base to which decorative plumes could be attached. Spencer and Gillen said:

> The pad is made by stabbing the feathers together by means of bone pins and is called *Imampa*. It is worn on the back of the head and is fastened on, partly by fur-string which is wound round it and the hair beneath, and partly by means of bone pins. Into each of the upper corners is fixed a tuft of feathers of some bird such as the eagle-hawk [wedge-tailed eagle], owl or cockatoo, attached to a pointed stick about six or eight inches [15 or 20 cm] in length. Sometimes long white down is used, or tail-tips of the rabbit-kangaroo [greater bilby]. Very often the tufts of feathers, when no emu feather pad is used, are fixed into the matted locks.[69]

Doctor/anthropologist Herbert Basedow observed that for men, 'Very often, in the central as well as in the northern districts [of Australia], the hair thus tied back is worked up with a pad of emu feathers into a chignon, which is tied round and round with human hair-string'.[70] He also remarked that:

> The old Arunndta [Arrernte] men are very particular about their appearance. When one is stricken with baldness, he constructs a pad, resembling a skull cap, out of emu feathers, which he ties on top of his head with human hair-string and wears regularly to hide the bareness of his scalp. He refers to this feather-wig as 'memba.' Aluridja [Western Desert] men adopt a similar fashion, but call the article 'lorngai'.[71]

Explorers and settlers often encountered Aboriginal peoples wearing ornaments made from white cockatoo feathers, which indicated their senior status. For instance, Eliza Davies, who was part of explorer Charles Sturt's expedition of 1839 to the Murray River, noted that their guide Encounter Bay Bob, who was a prominent man in the local Aboriginal community, wore a white cockatoo feather in his hair.[72] Such ornaments were worn during formal occasions, such as for an organised fight. In 1873, the settler Mrs Dominic D Daly was out riding one day east of Darwin in the Top End with her father when they were surprised by a party of Wulna warriors, decorated with feathers.[73] She later wrote that each member of the Aboriginal party 'was armed to the teeth, carrying a full complement of spears, well burnished and freshly barbed', they were painted in white and yellow clay and 'on their heads were crowns of white cockatoo feathers which stood upright over their brows in true barbaric fashion'.[74]

Angas remarked after his 1844 trip with Governor George Grey through the Lower Murray and south-east of South Australia that 'On grand occasions – such as at a fight, or during a corrobbory or dance – the men adorn themselves with the feathers of the emu, the pelican, and the cockatoo, and ornament their bodies with strips and spots of red and white ochre'.[75] Here, emu feathers and bunches of leaves from gumtrees were part of the dancers' ornamentation,[76] the movement of which when performing would have added to the spectacle. Angas reported that when a group of Lower Murray men armed with spears and shields assembled on an open plain for a prearranged fight, 'a bunch of Emu feathers fastened at the end of a spear is sent as a challenge to the opposite party'.[77] In this region, small feather bunches were tied to sticks that were then attached to the hair, so that they would dangle and quiver with every head movement.[78] In the Lower Murray, girdles worn by young women were decorated with emu feathers.[79] On the Eyre Peninsula in South Australia, emu feathers were woven into string to make waist belts as girdles.[80] Adnyamathanha people of the Flinders Ranges made a similar object to cover a man's backside, using a hair string belt and a bunch of emu feathers threaded together with sinews.[81]

In 1860, naturalist George Bennett described the superb lyrebird living in the mountains of eastern New South Wales, and remarked that 'the natives also use the feathers, as well as those of the Emeu [emu], as ornaments in their hair'.[82] Plumes were often used ceremonially, such as by the Tiwi people in the Northern Territory who incorporated seabird feathers into their headdresses.[83] In 1951, travel writer Julitha Walsh described an Aboriginal ceremonial dance at Noonkanbah in the west Kimberley, noting that 'For the "Devil Dance" the men wear their tall headdresses made from the tail-feathers of turkeys [Australian bustards] and emus, their bodies are greased with emu oil and decorated with white ochre and down from the breasts of wild ducks'.[84] Across Aboriginal Australia, emu plumes were widely worn as decorations, particularly by dancers during ceremonies.[85] The decorative use of bird plumes has ancient origins, as shown by the existence of rock paintings of people wearing feathered objects such as headdresses and tassels.[86]

During the Wilyaru ceremonies held by the Arabana people of Lake Eyre, the initiates were painted with the wedge-tailed eagle design and the ceremonial leader carried a long spear decorated on the end with feathers from the same bird.[87] In the Mindari ceremony of the Wangkangurru people, whose Country lay north of the Arabana, the plumes worn on the head were composed of mixed feathers from black cockatoos and barn owls, or from those of white cockatoos.[88] For dances, the Diyari people of Cooper Creek in eastern Central Australia made decorative bunches, termed *kootchar*, from the feathers of hawks, crows and eagles that were tied together with emu sinews.[89] In this area, emu feathers were also heavily used as dancing paraphernalia, inserted into headbands[90] and stuffed into a net for wearing on the head.[91] Among the Diyari, budgerigar feathers were woven into a long girdle comprising string made from fur or human hair, which was worn by initiated men.[92] The Diyari used a message-stick decorated with emu feathers when calling people from neighbouring areas together for the Wilyaru and Mindari ceremonies,[93] and wrappers were made from emu feathers to hold teeth removed from initiates during the Chirrinchirrie tooth evulsion ceremony.[94] In the Flinders Ranges, sulphur-crested cockatoo feathers were chosen as ceremonial decorations. Linguist Dorothy Tunbridge and Adnyamathanha woman Annie Coulthard described them as:

> ... a special white feather worn on the forehead by the *vardnapa* [initiates] and *wilyaru* [senior initiated men] when they returned to a camp after going through the rule. The mother also wore it when meeting the boy coming back.[95]

Gamebirds of high value to hunters were sources of both meat and decorative material. In the North Wellesley Islands of the Gulf of Carpentaria, the *thaankur wangal* or 'comeback boomerangs' were used to hunt brolgas, cranes, Australian pelicans, corellas, pigeons, ducks, white-bellied sea eagles, Australian bustards and other species, not just for meat but for their feathers to be worn as body decorations.[96] According to anthropologist Paul Memmott, feathers were also used to decorate message-sticks[97] and the:

> Bunched feathers and leaves were also employed as pubic tassels. Brolga and cockatoo feathers were bound with grass string into a bunch to form the *wuulbuul* or feather tassel. Feathers were occasionally attached to the side of the belt.[98]

Bird feathers were fashioned into ornaments that were worn to indicate a person's totemic affiliations. For instance, Lower Murray people at trade fairs would wear ornaments that

During the public ritual house opening at Aurukun for the late Victor Wolmby (b.1905, d.1976), George Sydney Yunkaporta (front), Bernard Pootchemunka and the man behind him (possibly Francis Yunkaporta) wore feather plumes as head ornaments. The leader was Clive Yunkaporta (left, obscured). Sculptures in the foreground are the Estuarine Shark and the Two Young Women of Cape Keerweer. Peter Sutton, Aurukun, northern Queensland, 1976.

signified their *ngaitji* (totemic ancestors), such as feathers from a relevant bird species in their headbands and hair.[99] In 1930, Milerum (Clarence Long), who was a Tangani-speaker from the Coorong, made a sedge 'war basket' for Tindale at the South Australian Museum that was decorated with feathers belonging to the boobook owl, which was one of his *ngaitji*.[100] According to anthropologist Alfred W Howitt, the holding of trade fairs and ceremonies was communicated by messengers carrying message-sticks that were generally a carved piece of wood. In the Lower Murray 'a messenger is called *Brigge* [*prigi*]. When on a mission, he carries some part of his totem as an emblem. For instance, a messenger of the Tanganarin [Tangani] carried a pelican feather'.[101] During formal exchange rituals in the Murrumbidgee River region of New South Wales, Aboriginal people reportedly traded things such as 'coloured clays' (ochre), animal skins, string, spears and the 'richly coloured feathers of rare birds'.[102]

Feather decorations added to the gravity of formal occasions. In the Lower Murray, the formal establishment of a *ngengampi*, a ceremonial trading relationship between distant partners, required a *pulanggi-kalduki* (human navel string [umbilical cord] with feathers attached) to be sent to the proposed partners.[103] Navel strings were also kept for ritual purposes by the Lardil people on Mornington Island in the Gulf of Carpentaria, who wrapped them in a bundle tied up with grass string and decorated with feathers.[104] For tribal government meetings in the Lower Murray, the *mungkumbuli* leader wore a bunch of feathers, *kalduki*, on his forehead to signify his authority.[105] Angas painted a *kalduki* made from emu feathers, which was worn by young initiated men in the Lower Murray.[106] In western Victoria, colonist James Dawson recorded that when Tuurap warneen, who was 'chief of the Mount Kolor [Mt Rouse] tribe', attended:

> … korroboraes [ceremonies] and tribal meetings he was distinguished from the common people by having his face painted red, with white streaks under the eyes, and his brow-band adorned with a quill feather of the turkey bustard, or with the crest of a white [sulphur-crested] cockatoo.[107]

Lures

Feathers were components of the lures made for hunting, as described in Chapter 6. In some cases, the noise of the feathers as hunters swished them through the air elicited the desired effect from the game animals. Basedow explained that:

> North of the Great Australian Bight the small wallaby is captured as follows: The hunter ties a bundle of feathers to the top of a long pole, up to twenty feet [6 m] in length, and this he whirls around his head, high in the air, as he walks across the tussocky plains known to harbour the game they call 'wilpa.' The wallabies, apparently taking the whizzing feathers to be an eagle hawk [wedge-tailed eagle], squat in fear, and, for the moment, do not attempt to escape from the native. Before the animal recognizes the fraud, the treacherous spear of the hunter has pinned it to earth.[108]

Music bags

The Diyari people at Cooper Creek in eastern Central Australia stuffed downy feathers into a small skin bag, a bit larger than a hand, that was used on top of a wooden drum made from a hollow eucalypt log and beaten during dancing.[109] Adnyamathanha people used the same object, which they called *wandatha varlka* (literally 'feather-down drum'), stuffed with emu or cockatoo down.[110] These bags protected the player's hands, and initially produced a mellow sound before the feathers became compacted. The same object was used on the Adelaide Plains in South Australia, where a colonist reported in 1844 that:

> A stuffed opossum skin is used as a kind of bass or drum. This is named *Tapuroo*. It is placed in the laps of the women singers, and on it they beat a stunning dull noise, keeping time with their voices.[111]

Needles for cord

In the Lower Murray, a quill made from a Eurasian coot feather was used with a spiny-headed sedge stem as cord to sew shut the apertures of a human corpse in order to keep flies out when preparing it for smoking on a platform.[112] From the inland regions of Queensland, it was recorded that the mid-rib cut out from a primary wing feather was used as a needle when making bags,[113] such as those used to transport pituri.[114] Roth explained that:

> In the Boulia district [of south-west Queensland], and beyond it on the Georgina, the needle employed in weaving the pituri-bags and other than netting-stitch bags was formerly made from the mid-rib of a Plain-Turkey's [Australian bustard's] wing-feather, with a piece of twine attached to its proximal extremity … Some of the older Glenormiston blacks told me last year that such a needle, the so-called tatti, might also be made out of other birds' wing-feathers.[115]

Ritual materials

Feathers were essential parts of many ritual objects, for both aesthetic and ritual reasons. The Adnyamathanha people of the Flinders Ranges in South Australia placed a red-ochred feather, *urdaki*, on the point of a spear used during initiation ceremonies.[116] In the Lower Murray, healers known as *putari* utilised a bunch of feathers as a ritual tool for breaking up sickness and drawing out blood from their patients.[117] Blood extracted from a sufferer was buried so that sorcery birds, particularly Australia magpies as described in Chapter 4, would not take it and cause sickness to reappear. For Lower Murray rituals aimed at improving the wellbeing of people, two special long heavy spears, *parmuri*, were made from native pine timber and decorated with emu feathers.[118] As described above, in Aboriginal Australia down-like materials were routinely used to decorate ritually powerful objects and the people who used them. According to Spencer and Gillen, across northern Australia down was collected from birds and plants, then 'Designs are drawn with this material upon the body and also on ceremonial and magical objects'.[119]

In the Lower Murray region, an emu feather object played a major role in the account of a boy being rescued from the camp of the *mulgyewonk* (*muljuwonk*, a 'bunyip' river spirit), which was a hole in the bed of the Murray River at Brinkley near Wellington.[120] The boy's father had drawn a bunch of emu feathers across his mouth before diving in, which gave him the power to render the *mulgyewonk* senseless when he encountered it deep beneath the water surface. In this region, emu feathers that had been soaked in dead body fat were used a wrapping for a type of sorcery object, the *ngildjeri*, which was a wooden point.[121] Elsewhere in Aboriginal Australia clusters of feathers were utilised to draw out sickness from a patient's body. In the Gulf Country of north-west Queensland, Roth recorded that the *mul-ta-ra* was a 'roll of emu feathers worn over portion of body wherever pain is'.[122] Emu feathers were also used to wrap a range of ritually powerful objects.[123] From my own curatorial experience in museums, the coarse emu tail-feathers are generally used for this purpose.

The highly ritualised use of feathers was ordained by the ancestral events of the creation itself. For instance, the Diyari people in the north-east of South Australia had a tradition of how two of their creation ancestors, the Mura-mura, created rain at Mungaranie after dipping their eagle feather headdresses into a wooden bowl of water then sprinkling it all round them. McCarthy wrote that since then:

> The Dieri [Diyari] rain-maker in this ceremony decorates his body with a feather-down design stuck on with human blood, and wears a feathered hoop on his head to make himself invisible. Other men sit around bowls of water and chant rain songs while he dances over his bowl and sprinkles the water about, his head-dress making a swishing noise like the falling of rain.[124]

In northern Australia, Aboriginal peoples used a variety of colourful feather plumes from the red-tailed black cockatoo, sulphur-crested cockatoo, little corella, red-collared lorikeet, galah, red-winged parrot and other parrots as decorations for dancers.[125] In some cases, it was culturally important which bird species provided the material from which the decoration was made. For instance, in a Central Australian rainmaking ceremony, a Southern Arrernte man would wear a headdress decorated with a small tuft of barred feathers from the red-tailed black cockatoo and white feather down from another bird, with the red bars representing the rippled water and the white the clouds.[126] Spencer and Gillen explained that its significance was based on the tradition that during the creation the 'black cockatoo brought rain down from the north'.[127]

In the Wadeye area south-west of Darwin, tufted yellow and white feathers from the sulphur-crested cockatoo were used to decorate the head of a short ceremonial spear, which was thrown into camp to announce the return of an initiate.[128] Here, dancing whisks held between the hands were made with split brolga feathers tied by string into a bunch, a handgrip covered with beeswax, and strings with brolga feather tassels.[129] At Cobourg Peninsula in western Arnhem Land, dancing whisks were made from feathers of the brolga, emu and Australian bustard, with small 'flowers' made from plumes attached to ceremonial strings.[130]

In central and eastern Arnhem Land, Yolngu people store ritual items in baskets decorated with tassels made from the bright feathers of birds, such as the red-collared lorikeet.[131] In this region, the feather-decorated tassels are attached to the top of Morning Star dancing poles.[132] They are also worn by initiates and performers in ceremonies, and decorate the sacred objects that are publicly displayed at certain times to represent the ancestors.[133] Necklaces include feathers from red-collared lorikeets and red-winged parrots, often along with fish vertebrae.[134] McCarthy observed that here:

Among the ornaments might be mentioned the beautiful multicoloured strings of parakeet [red-collared lorikeet] and other bird feathers made into armlets, chest-bands, and girdles, by the *dua* [Dhuwa] moiety for the *jirritja* [Yirritja] men to wear in ceremonies and to bind the *rangga* [sacred objects].[135]

The feathered string from which the tassels are made have high ritual value. They are reused after each occasion in a variety of ways to help maintain the Yolngu connection with their ancestral world. Anthropologist Howard Morphy explained that:

As well as its symbolic meaning (referring to attributes of Ancestral Beings) feather string gains its power through association with particular ritual contexts and by being connected with stages in the life history of individual clan members. The same length of string may have begun life as part of a longer skein of string unravelled at ceremonies, draped over and stretched between objects and actors. It may subsequently have been used as a headdress worn by an initiate at his circumcision ceremony, and later on wound around a restricted sacred object then finally used to form one of the tassels of a clan member's dilly bag.[136]

In eastern Arnhem Land, the brilliant orange-red feathers from the red-collared lorikeet are mainly used by clans from the Dhuwa moiety to ornament the sacred *rangga* objects.[137] They are termed *jukurr*, meaning 'fat' and a reference to its perceived spiritual power.[138] In north-eastern Arnhem Land, the decorative pendants that are hung from sacred dilly bags belonging to each clan, and publicly displayed during ceremonies such as those associated with the Djang'awu ancestors of the Dhuwa moiety, have powerful symbolic meanings.[139] For instance, anthropologist Nancy Williams described the pendants on a dilly bag from the Rirratjingu clan. She recorded that 'the green feathers [of the red-collared lorikeet] that protrude at intervals from the red feathers represent casuarina trees along the beach at Yalangbara [Port Bradshaw]. Tufts of white feathers at the ends of the pendants represent sea foam along the shore at Yalangbara'.[140] On a similar pendant hanging on a dilly bag from the Ngaymil clan, the same white feather tufts 'represent bubbles in the water coming up from the well made by Djang'awu's walking stick'.[141] A *rangga* object photographed by Ronald M Berndt and Catherine H Berndt, that represented the ancestral goanna tail, was decorated with the same lorikeet pendants terminating in white feather tufts.[142]

Use of bird materials is perceived by Aboriginal peoples to give the wearers access to ancestral powers. Women among the Walmajarri people of the Great Sandy Desert of Western Australia wear the feathers of the whistling kite as a head decoration, enabling them to sing love songs to attract a partner.[143] There is a Walmajarri love song associated with the nankeen kestrel and the manner in which it hovers.[144] In north-eastern Arnhem Land, small human figures made from beeswax were covered with white-bellied sea-eagle or silver gull feathers for use in sorcery.[145] Here, men ritually used 'love magic' objects, which were attached to long lengths of string decorated with white down and red-collared lorikeet breast feathers, in order to obtain women.[146]

Sorcerers' slippers

Emu feathers are widely associated with powerful sorcery objects. An Aboriginal English name for sorcerers, 'feather-foots', refers to them typically wearing what is popularly known as 'kadaitcha shoes', which are ritual footwear or 'slippers' generally made from the emu's coarse tail-feathers, knotted together with human hair string and decorated with bird down.[147] When worn, the shoes are said to make the sorcerer invisible, so that he can sneak about to inflict harm upon someone without leaving a track. From my own field experience across the Western Desert and Central Australia, it is believed that sorcerers can also project their spirits into the form of animals, particularly large black dogs or eagles.

In the Lower Murray, feathers were used in the making of *karaigatatami*, which were 'magic shoes' used to practise *mantalanganar* (magical disappearance).[148] A sorcerer would attempt to kill people with a mixture of sorcery and trickery. During my 1980s fieldwork in the south-east of South Australia, Moandik man Ron Bonney claimed that a *kuratji* or sorcerer would cut the feet off an emu carcass and tie them to the underside of his own feet, in order to leave a bird track in the sand as he walked.[149] Apparently, the sorcerer would then wait with a spear in hand to kill whoever came tracking the presumed emu. McCarthy claimed that 'Among the central Australian tribes the Kurdaitja [kadaitcha] shoes worn by a sorcerer and the avenger of a crime were made of emu feathers stuck together with human blood'.[150]

There are many historical accounts of ritually powerful footwear that incorporates emu feathers.[151] For instance, in 1882 surveyor T Brown was in the Musgrave Ranges in Central Australia when he found a pair of kadaitcha shoes. He claimed that he:

> … was disturbed one night in camp by the near presence of some natives; and in the early morning, on approaching the place from which he imagined the sound came, he found, not the natives, but a pair of native shoes, or sandals, made of emu feathers and human hair stuck together with blood (which is taken from the arm of the maker). The soles were about 1 1/2in. [3.8 cm] thick, very soft, and of even breadth. The upper portions were nets made of human hair. These shoes are only used by the blacks at night when on an expedition to attack their enemies. The object they serve is to prevent the wearers being pursued after a murderous night attack. It is only on the softest ground they leave any mark, and even then it is impossible to say in what direction the enemy came or returned, as the toe or heel cannot be distinguished in the track; so that the natives say they are able to track anything that walks except the wearer of the kooditcha [kadaitcha], the name by which the shoe is known.[152]

Sinews, skin and intestines

The more flexible parts of the bird's body that Aboriginal peoples used to make artefacts were the sinews, skin and intestines. Typically, the presence of such useful materials for making tools for foraging was credited to events of the creation. Museum curator Aldo

Giuseppe Massola wrote that the Wotjobaluk people of western Victoria had a myth whereby:

> Werimul, the Emu, at one time was not as swift of foot as he is now. One day he wanted to go hunting, but having no energy to do so asked Wallup, the Stump-tailed Lizard, who was then a very energetic individual, for a loan of his sinews. He promised to return them together with some spoils of the chase; but he did not keep his word. Wallup was left helpless, and from that day spends most of his time basking in the sun or sluggishly crawling about.[153]

Cordage

Emu leg sinew was widely used across Aboriginal Australia as a tough cord with high tensile strength for fastening together components of artefacts such as spears, spearthrowers, feather decorations, cloaks and bark canoes.[154] For instance, Basedow remarked that 'A straight, single-piece, hard-wood spear is made more effective by splicing a barb on to the point with kangaroo or emu sinew'.[155] Aboriginal toolmakers in the Murray River region used emu sinew as a strong cord for game nets and as thread to sew mammal skins together when making cloaks, with a sharpened emu fibula used as an awl.[156] Similarly, in the Lower Murray the tendons from emu legs were used as cord.[157] Bird sinews in general were highly resistant to breaking. Former colonist Molineux, who was searching for local sources of medical thread, noted that:

> There is one very valuable item in the anatomy of the pelican, which should be useful to the surgeon. The sinews of the neck and wings are extra long, fine, and strong, and will split, so that for sewing up wounds made in operations these fibres ought to be of value.[158]

Emu intestines were used in a manner similar to sinews. In central Victoria, in the Goulburn River district north of Melbourne, colonist WH Baylie reported in 1843 that local Aboriginal people:

> … employ as a means for their relief a number of tightened cords above or over the parts affected, thereby checking the circulation and allaying the symptoms or paroxysm of pain; these cords are made from the intestines of the kangaroo or emu, neatly twisted like our fine twine, and possess a great degree of strengthened elasticity; in choice or spasmodic action of the intestines, they apply this cord and check the violence of the pain by pressure.[159]

Ritual materials

As discussed in Chapter 2, in the Lower Murray the *ngaitji* or totemic species associated with the clans were seen as a source of spiritual and physical power. At trade fairs, people wore a piece of bird skin to signify their *ngaitji*.[160] These materials needed to be kept guarded when fresh, as the sinews and skin from a bird just eaten by an intended victim could be used to make powerful sorcery. The Berndts explained that in the Lower Murray:

> Depending on the fragment procured, the sorcery practised on it resulted in the victim experiencing a particular ailment that could lead to death. For instance, a splinter of bone from a duck's head could cause serious headaches; skin from its wing, a diseased arm; skin from its body, an internal disease. The sinew of a bald coot [Eurasian coot] treated in the appropriate way could cripple a victim.[161]

Some bird intestines were used for sinister purposes. In the Lower Murray, a way for a sorcerer to kill was to put *ngruwi* (dead body fat) in a black swan's gut before the gut was cooked and eaten by the intended victim.[162]

Egg shells, bone, beaks and claws

This section discusses the hard parts from the bird carcass that Aboriginal peoples used to make tools and as source material for ritual purposes. For Aboriginal hunters, most parts of the bird could be used for either food or artefact-making.

Containers

Since colonial times, Aboriginal peoples have carved designs into emu eggs for sale to Europeans.[163] Emu egg shells are larger and thicker than those of all other Australian birds, save the elusive southern cassowary,[164] and so in earlier times Aboriginal foragers used them as water carriers.[165] Basedow remarked that 'now and then the broken shell of the emu egg also makes a very serviceable cup'.[166] In Western Australia, a newspaper columnist claimed that:

> Emu eggs, too, were eaten, the shells being carefully handled in order that they could be used as water vessels. The shell is capable of holding a substantial drink and was one of the most popular water carriers used by the aborigines.[167]

Gouges

In northern Queensland, the *tawat*, or gouges, for carving the wooden trough-like containers for carrying water and for food preparation were made from the tibia of emus and kangaroos. Donald Thomson observed that many of those in use by the Koko Taiyuri people on the Edward River were old and had been used many times. He said:

> They were known generally as *nampi* (emu), or, more specifically, by the name of the principal bone from which they were made – *yan'ka nampi* (*nampi*, emu, *yan'ka* tibia) or *min nampi kummandonon*. The medullary cavity of the bone is generally filled with a plug of bark or wood to prevent the entry of chips and splinters. A typical example is twelve and a half inches [32 cm] in length. The shaft is cut away for a distance of nearly two and a half inches [6 cm] in and is

sharpened to a chisel edge. In use, this gouge is … always worked towards the
body by the user, never away from it.[168]

Necklaces

Apart from feathers, Aboriginal peoples used a variety of more robust bird materials to make
decorations. This was apparent to former colonist Thomas Worsnop, who reminisced that:

> On the Lower Murray I saw a necklace made of very fine reeds, with a curious
> pendant, greatly prized by the owner. The string was made of very fine fibre, and
> about 18ft. [5.5 m] in length, making, as the wearer chose, several circlets round
> the neck. On the twine were strung very short pieces of fine thin reed, colored
> alternately white and red; the pendant being composed also of twine, on which
> were fastened downy feathers of the goose, shells, the mandible of a duck, the
> upper mandible of a black swan, and tufts of human hair. Necklaces were also
> made of red berries and eagles' claws, and had a very pleasing effect.[169]

Among the Arrernte people of Central Australia, eagle talons were made into ornaments[170]
and sorcery charms.[171] These objects were generally composed of one or more pairs of claws,
their bases mounted in resin or wax on a hair string cord and the tips placed opposite to each
other.[172] These cords, wound round the head or neck, served either as a headband with the
claws hanging from it or as a necklace with the claw pendant resting on the chest. Similarly,
Waramungu people in northern Central Australia had a neck-band that:

> … consists of a thin strand of human hair-string, to either end of which a little
> lump of resin is attached, and each of them carries a pair of eagle-hawk [probably
> wedge-tailed eagle] claws. The strand is tied so that the claws hang down the
> back.[173]

At West Point in Tasmania, the claws of a hawk that archaeologists found at a possible
cremation site may have been part of a necklet.[174] Aboriginal people in the Sydney area also
used bird talons and feathers as ornaments, sometimes gummed directly into the hair.[175] The
Wardaman people in the Katherine region of the Northern Territory keep wedge-tailed eagle
claws, stuffed with fibre, as good luck charms.[176] Due to a combination of their aesthetic appeal
and totemic associations, bird parts were widely used to make ornaments. Anthropologist/
archaeologist Kim Akerman recorded that:

> The *ngorrekanarra* pendants made by the Djinang of Castlereagh Bay, Arnhem
> Land, consisted of a beeswax-mounted beak of a spoonbill (either *Platalea
> regia* or *P. flavipes*), suspended from a twine neck cord. These were worn in the
> *gamboi* ceremony. A similar necklace known as *lenderra*, made by the Burrada
> of Kupanga in Arnhem Land, consisted of up to ten spoonbill beaks fixed with
> resin to a fibre cord.[177]

The bony bird materials worn by an individual gave power from their close association with a particular totemic ancestor. Angas observed this in 1844 at the Murray River where it enters Lake Alexandrina, where he met a woman with a boy aged about four and named Rimmelliperingery. He was described as 'the pride of his tribe, and wears the upper mandible of the black swan round his neck; which is regarded as a *gunwarrie* or wizard charm'.[178] Here, the Berndts recorded the same object as a necklet charm known as *kurrindjeri*, which was worn to signify the *ngaitji*:

> Being made up of actual parts of the *ngatji* [*ngaitji*] animal, bird, and so forth, it was said to be imbued with its own power that could be released or communicated with only by the person who wore the necklet and whose *ngatji* [*ngaitji*] it was ... Moreover, a necklet of bird's claws (according to one example given) gripped or clung to a wearer at particular times. This grip stimulated a wearer's nerves and transfused the power that protected him from illness during his boyhood.[179]

In the Lower Murray, the *kurrindjeri* necklets were kept wrapped up and hidden in the shelter as heirlooms.[180] On some occasions, male children would wear those associated with the clan of their maternal relatives, although their main connection to Country came from the father. The wearer nonetheless received the ancestor's power through such ornaments. For instance, if a Turiorn clan member had trouble with their stomach, a necklet made from their *ngaitji*, the *turi* (Eurasian coot), was taken and tied around their body so that its claws could grip the stomach and remove the pain.[181] The Berndts recorded that 'The youth wears the *ngatji* [*ngaitji*] claws until he is a young man. The necklet is then passed on to a younger brother'.[182]

Many of the ornaments that European colonists observed on Aboriginal peoples would have been seen by the wearers as drawing power from totemic ancestors. Worsnop observed an Aboriginal person in the Lower Murray wearing a necklace fashioned out of fine reeds that was 'greatly prized by the owner'; attached to it was a pendant made from twine that incorporated tufts of human hair and goose down, and held shells, a duck mandible and the upper beak of a black swan.[183]

Nose-bones

Bird bones, being typically hollow, are light in weight in comparison with equivalent sized mammal bones but are nonetheless strong. In Aboriginal Australia, this attribute makes avian leg bones an excellent choice for decorative nosepegs.[184] In New South Wales, Bennett remarked that:

> Both sexes have the *septum naris* [nasal septum] perforated, in which a piece of straw, stick, or emu-bone is worn, looking like what Jack would term a 'spritsail yard;' this practice is universal among the whole of the tribes seen in the colony, and is regarded as highly ornamental.[185]

In eastern Central Australia, the sharpened radius of an emu bone was used for piercing the nasal septum.[186] Across Central Australia, the piercing was performed during an early stage of initiation for men, and after marriage for women.[187] The most common bones selected as nosepegs are the ulnae and radii of larger bird species like the Australian bustard, wedge-tailed eagle and Australian pelican.[188] The bones generally have the exposed ends plugged with plant resin. Examination of artefacts in museum collections indicates that the bones are sometimes decorated with finely incised lines and have feathers, such as from the red-tailed black cockatoo, inserted in the ends.

Points

The literature is generally inconsistent in descriptions of bone points, which are variously called awls, borers, drills, pegs, pins, needles, daggers, charms and death-pointers.[189] The intended use for each point is often unclear from the context in which they were collected. They are on occasion found in archaeological excavations, such as the 'very sharp and highly polished awls made from bird bone' 3–10 cm long that archaeologist D John Mulvaney uncovered in the late 1950s at Fromm's Landing, south of Walker Flat on the Murray River in South Australia.[190] It is likely that as tools the bone points were multipurpose. As weathered bones are brittle, these implements were probably made from the bones of recently dead animals. McCarthy stated that:

> The rib, leg and wing bones of many kinds of birds and mammals were used throughout the continent as awls. The joint was left on one end as a rule, the other end broken off at the required length and then rubbed on a stone to a sharp point. They are from 5–30cm long. In the southern half of Australia they were used to peg out possum and other skins on bark and wood to dry them for cloaks and rugs and to perforate their edges in sewing them together.[191]

In some parts of Australia, a sharpened section of emu bone was used as the peg or spur for holding the spear on the spearthrower,[192] and they were more extensively utilised as bone points and needles.[193] Emu bones were widely used as shoulder pins for skin cloaks and as drills when boring holes through animal skin and into wood.[194] On the Adelaide Plains in South Australia, a dagger used by assassins and known as a *werpoo* was made from a curved emu bone.[195] In the Lower Murray, emu bones were fashioned into needles and awls.[196] According to archaeologist Graeme L Pretty, burials at Roonka on the Murray River in South Australia contain scatters and bands of small bones that suggest that birdwings were incorporated into the manufacture of a marsupial skin cloak.[197] In the Murray River region, a sharpened emu fibula was used as an awl when sewing mammal skins together for cloaks.[198] According to Langloh Parker, in central New South Wales an Aboriginal 'woman's needle was a little bone from the leg of an emu, pointed'.[199] The Diyari people of Cooper Creek in eastern Central Australia had an object termed *wanapanyi*, which was 'The polished and pointed radius of an emeu [emu], used for piercing a hole through the nose'.[200] Southern cassowary bones were similarly used as borers in the rainforest area north of Cairns in Queensland.[201]

Large sharpened avian bones were widely employed as an awl when drilling into spear shafts to make a mortice for the point.[202]

Ritual materials

Avian bones also had important ritual uses. In the Lower Murray, sorcery bone objects known as *ngadhungi*, or 'skewers' in Aboriginal English, were hidden within an emu feather pouch or inside the hollow of a duck bone.[203] Bones from a person's meal were reportedly stolen and used for these 'magical purposes',[204] with grey teal bones being considered particularly potent.[205] After a collection of sorcery objects was seized in the Lower Lakes in 1862, the missionary George Taplin noted that 'I found today that the particular bones used for making Ngathungi [*ngadhungi*] are the leg bones of [grey] teal, water-hens and coots, head bones of [Pacific] black ducks and of ponde [*pondi*, Murray cod]'.[206] He also observed that 'It appears that ngadhungi is made with the eggs and red bills of bald coots [sic; purple swamphens] as well as by other bones',[207] and 'Sometimes they put in ngadhungi a ponde's eye or emu feathers or duck bone'.[208] During my Lower Murray fieldwork, it was claimed that it was a sorcery practice to take an individual's duck bones and other food remains, wrap them in hair then put them into the fire.[209] For this reason, in the past parents told their children to carefully dispose of any food scraps from a meal.[210]

The *ngadhungi* were said to work by causing the fat covering the bone or sinew to melt when placed somewhere very warm, like in the roof over a shelter near the campfire.[211] The action of the *ngadhungi* could be stopped by submerging it in water, where it cooled off. The Lower Murray and Murray River people greatly feared this kind of sorcery attack from their eastern neighbours, the Ngarkat, who lived in the semi-arid mallee country to the east. According to Tindale, the Ngarkat were believed to have 'great magic powers and could kill one if even one piece of bone belonging to your food or a piece of hair fell into their hands'.[212] The intended action was the transference of a property from the animal to the human who consumed it. The Berndts explained that a sharp fragment of bone taken by a sorcerer from a discarded duck's head could give a serious headache to the person who earlier ate the bird.[213]

Emu bones served as pointing bones for sorcery in the Lake Eyre district of Central Australia.[214] Here, during ceremonies to increase the local waterfowl population and to make rain, senior men pushed emu bone skewers through skin folds on their arms, thighs and scrotum, then drank the blood that flowed from the wounds.[215] The same bone, sharpened at each end, could be used as the support structure for a headdress. In this district, the sharpened emu bones were also tools during subincision rituals.[216] In the Central Australian region, paired eagle talons were used as sorcery objects, and for this purpose they were attached by hair string to a set of pointing bones.[217] Used in this way, the claws were seen as ripping apart the intended victim's bowels and intestines. In the North Wellesley Islands in the Gulf of Carpentaria, a pair of pointers used for sorcery was made from black-necked stork bones.[218]

The Dua moiety men of north-eastern Arnhem Land used 'love magic' objects attached to feathered strings which had silver gull heads, or carvings of them with yellow beaks, often with food like a small mouse, worm or fish wedged in the mouth to symbolise either the spirit of the intended woman or her 'desire'.[219] Other 'love magic' objects used here had bird heads that represented the pelican, magpie goose and royal spoonbill.[220] Anthropologists Adolphus P Elkin, Catherine H Berndt and Ronald M Berndt explained that 'The beaks suggest the swooping of the birds for food, and symbolically represent, for the purpose of love magic, the "catching" of a woman'.[221] For members of the silver gull totemic clan, a wooden object, representing a silver gull's head, was tied to a string and used by kin of the deceased during mortuary rituals. In this case it was said that the 'bird's head represents a small gull which has just emerged from its egg, like the spirit of the dead person which has departed from the body'.[222]

Whistles

In the Normanton area of the Gulf Country in northern Queensland, whistles were made from bird bones, with one end sealed with resin and the other left open.[223] Similarly, among the Yandruwandha people of north-east South Australia, bird bone whistles were used to lure emus towards concealed hunters.[224] On Cape York Peninsula, hunters blew through magpie geese trachea to decoy geese in the open swamps to come in within striking range.[225]

Oil

Avian oil was used on both people and artefacts. Emu oil is used to fix red ochre onto human bodies for ceremonial occasions, such as during a warrior's preparations or when a novice is being made ready for initiation.[226] It was widely applied to dancers as a fixative for ochre, charcoal and pipeclay paints.[227] Aboriginal peoples in south-eastern Australia considered emu oil to be a ritually powerful substance,[228] and for this reason men rubbed emu oil on their bodies before a fight.[229] On the Adelaide Plains in South Australia, the term *marnitti* referred to 'grease; a boy greased and painted with red ochre' in preparation for the first stage of initiation.[230] In the Lower Murray, for initiation ceremonies novices were covered with red ochre mixed with emu oil to make them *narambi* (sacred), as Waijungari the 'red-ochre-man' ancestor had done during the creation.[231] According to the Berndts, the oil for initiations was ritually gifted, 'contributions of fat or oil came from all directions – from Piltangk [near Poltalloch homestead on the southern shore of Lake Alexandrina] and Ngulunmunangk [Ngulun-malang on the south-eastern shore of Lake Alexandrina] where large pelicans were caught, and from Mulbarapa clansmen in whose territory [surrounding Mason Lookout] many emus were caught'.[232] A herb, *papuri*, was mixed with *tjelnggai*, pelican oil, for rubbing onto a corpse being prepared for smoking on burial platforms.[233]

As described in Chapter 8, the oil from animals, such as birds, was used widely as medicine. In the Lower Murray, oil for rubbing onto the body was prepared from sources such as whale blubber, fish and birds.[234] The fat was heated to produce oil, which was collected in a vessel such as a human skull cap.[235] Body oil was used decoratively and for warmth and cleanliness. The main bird species used in the Lower Murray for oil production were the Cape Barren goose, Eurasian coot, emu, Australian pelican and black swan.[236] In Tasmania, animal oil, in the form of grease, was mixed with ochre and worked into the ringlets of hair until it became stiff and mop-like.[237]

Avian oil was used by Tasmanian Aboriginal peoples to polish shells used for necklaces; *mariner* shells in particular had their rainbow hues brought out by being rubbed with fat.[238]

Across northern Australia, emu oil was also used to make supple the sticks being bent over the fire for spear shafts.[239] Here, when mixed with red ochre it was utilised for preserving wooden objects and for painting in rock shelters.[240] In Aboriginal Australia, emu oil was widely used for the maintenance of sorcery paraphernalia, such as the special clothing and 'slippers' made from human hair, emu feathers and bark, in order to keep them soft and pliable.[241]

It can be reasonably assumed that Europeans learnt about the uses of emu oil, particularly those relating to medicine as discussed in Chapter 8, from Aboriginal peoples on the colonial frontier. Ron Bonney described emu oil as excellent for rejuvenating old and brittle leather,[242] which is a fact supported by commercial accounts of its use.[243] Emu oil was reportedly used to treat a leather saddle and bridle.[244] A former colonist reminisced that in 1860 along the Murray River in northern Victoria:

> The Murray blacks used to drive a good trade with the steamers in emu oil, for a quart bottle of which they received 1s. [shilling] or its equivalent. Most of the birds were caught by the leg in looped snares, to which they were decoyed by the blacks imitating their peculiar call or whistle, which they can mimic to perfection. If the line is sufficiently strong, when once snared there is no get-away, for their legs are the only weapons of defence the emus are furnished with. A kick from one of them is forcibly expressive, and many a horse has been crippled by the bird empanelled in Australia's coat of arms. In skinning them particular care is taken to leave all the fat on the skin, which is pegged out, and hung up in front of a fire; the oil then drips into wooden vessels placed underneath for that purpose. The fat from the intestines is obtained in the usual way, an old fryingpan being generally used for that purpose. Before bottling, the oil requires straining to free it from impurities.[245]

*****

In Aboriginal Australia, a wide range of avian materials has been used to make artefacts. The bones, talons, sinews, skins and oils were valuable items for toolmakers. The incorporation of feathers into ornaments enabled the use of a range of colours beyond the earthy set available from mineral sources alone, and in some cases had the much-valued quality of iridescence. The physical properties of the bird parts selected for incorporation into objects are important, as are the cultural connections between the bird species and the intended users. Aboriginal peoples perceived that the ancestor's power in an artefact is greatly enhanced through the use of materials obtained from the owner's animal kin. Since many of the spirit ancestors are birds, the use of avian materials for making and decorating Aboriginal objects is understandably high. Apart from these totemic connections, materials from certain birds such as eagles, emus and cockatoos were more widely used across Australia, particularly as ornaments for men. Many of the species used to make artefacts are also eaten as gamebirds, providing the hunters with a ready supply of valued material.

Endnotes

[1] For broad descriptions of Australian Aboriginal material culture, refer to Worsnop (1897), Spencer (1922), Davidson (1933, 1934, 1935a, 1935b, 1936, 1937, 1941, 1947a, 1947b), McCarthy (1940, 1961, 1974, 1976, 1978), Mitchell (1949), McCourt (1975), Sutton (1988a), Akerman (1990, 1992, 1995, 2005, 2011b, 2018), Haagen (1994), Jones (1996a, 1996b, 2007), Morphy (1998), Berndt *et al.* (1999), Clarke (2000, 2001b, 2012), Davies (2002) and Nicholls (2008). Examples of regional studies include the following: Western Australia – Akerman (2011a, 2017); south-eastern Australia – Bonwick (1870), Sculthorpe *et al.* (1990), Hemming *et al.* (2000) and Taçon *et al.* (2003); Central Australia – Spencer & Gillen (1899), Horne & Aiston (1924), Mountford (1976a), Tunbridge & Coulthard (1985) and Jones & Sutton (1986); Top End – Elkin *et al.* (1950), Taçon (1991), Barnes (1999), Hamby & Young (2001), Hamby (2005) and Akerman *et al.* (2014); and northern Queensland – Roth (1901a, 1904), Thomson (1936b), McConnel (1953), Anderson (1996) and Sutton (2003b).

[2] Mountford (1958:92).

[3] For instance, see figures in Spencer & Gillen (1899, 1904).

[4] Bird Rose *et al.* (2002:10–13,89).

[5] Refer to the 'Map of MakMak Dreamings' in Bird Rose *et al.* (2002:xiv).

[6] Thomson (1949:31–32). For accounts of the Gunapipi (Kunapipi) and Ngulmarrk (Ngurlmak) ceremonies, refer to Berndt (1951) and Keen (1994:142,159,165,257–262).

[7] Barfield (1997:311). See also Hicks & Beaudry (2010).

[8] For an account of my curatorial background and field experience, see Clarke (2014a). Examples of studies concerning collections of material culture held in the South Australian Museum include Cooper (1948), Edwards (1966, 1972), Sutton (1988a), Hemming & Clarke (1989), Jones (1996a, 1996b, 2007), Clarke (2000, 2001b, 2012) and Hemming *et al.* (2000).

[9] See Lycett (1820–22) and Angas (1847a).

[10] Urry (1985).

[11] Attenbrow (2010:110).

[12] Chandler (1916) and Dove & Koch (2011).

[13] Sutton & Snow (2015).

[14] Spencer & Gillen (1904:723) and Clarke (2012:213–215).

[15] Molineux (1905:6).

[16] Chaloupka (1993:11,66,185,188–189). See also Mountford (1976c:9).

[17] Berndt *et al.* (1993:190). The term *tjelindjeri* is also written as *djalindjeri*.

[18] Gason (1879:289).

[19] Spencer (1913:8).

[20] Thomson (1996:51). Originally written in 1937.

[21] Akerman *et al.* (2014:201–202).

[22] Akerman *et al.* (2014:202).

[23] Hardwick (2019b:55,62, 2019c:20,29).

[24] Hardwick (2019c:20).

[25] Thomson (1983b:34).

[26] Hale & Tindale (1934:133).

[27] Roth (1904:26).

[28] Angas (1847b:95) and Berndt *et al.* (1993:273).

[29] Tindale (1936c:61).

[30] Haagen (1994:19).

[31] K Akerman (pers. comm.).

[32] Spencer & Gillen (1904:718, see also 722).

[33] BJ Nangan (cited in Akerman 2020:118).

[34] Spencer & Gillen (1904:723).

35 Spencer & Gillen (1904:723), Johnston (1943:279–280) and Cleland (1966:116).

36 Cleland (1940:10).

37 Worsnop (1897:54).

38 Mathews (1898a:341, 1900:632).

39 Worsnop (1897:158).

40 Angas (1847b:92–93).

41 Gason (1879:272), Howitt (1904:662) and Johnston (1943:280).

42 Worsnop (1897:157–158, see also 54).

43 Clarke (2012:171,186).

44 For accounts of the Barnumbirr Morning Star ceremony refer to Johnson (1998:18), Morphy (1998:227–230,242,259) and Berndt *et al.* (1999:105).

45 Berndt *et al.* (1999:105).

46 Mountford (1958:97–98,Plate 31B).

47 Mountford (1958:97).

48 Tindale (1960:317).

49 Memmott (2010:Fig.69).

50 Spencer & Gillen (1904:722–723).

51 Worsnop (1897:156–159), Mjöberg (1918:339,341–342), Cleland (1966:116), McCarthy (1974), Puruntatameri *et al.* (2001:97) and Akerman *et al.* (2014:187,201–203,206,210–211).

52 Worsnop (1897:50,52).

53 Worsnop (1897:52,Plate 27).

54 Cleland (1940:10).

55 Basedow (1925:249). The name for the Larrakia people is also written as Larrekiya.

56 Hardwick (2019b:63).

57 R Graham (pers. comm.).

58 Chalk, 7 November 1926 (Tindale 1935–69:168).

59 Bonwick (1870:27).

60 Worsnop (1897:157–158).

61 Cleland (1966:123).

62 M Laughren (pers. comm.). For descriptions of the Yawulyu, refer to Munn (1973:Ch.4), Barwick (2005), Barwick *et al.* (2013) and Laughren *et al.* (2016).

63 Langloh Parker (1905:129).

64 M Laughren (pers. comm.).

65 McCarthy (1965:17).

66 Smith (1990: Aloysius Puantulura plate).

67 Laughren *et al.* (2016:427).

68 Spencer & Gillen (1899:573, 1904:723–724).).

69 Spencer & Gillen (1899:573).

70 Basedow (1925:47, see Plate IX).

71 Basedow (1925:50, see Plate VIII).

72 Davies (1881:129).

73 Daly (1887:183). Wulna was written as Woolnah.

74 Daly (1887:183).

75 Angas (1847b:85).

76 Angas (1877).

77 Angas (1847b:88).

78 Angas (1847b:92–93).

79 Taplin (1879:Fig. opp.64) and Worsnop (1897:87).

80 Schürmann (1846:211–212,233) and Wilhelmi (1861:167–168).

[81] Tunbridge & Coulthard (1985:28).

[82] Bennett (1860:180).

[83] Worsnop (1897:53).

[84] Walsh (1951:16).

[85] Johnston (1943:271–272).

[86] For instance, see Chaloupka (1993:112,125 & Figs105,106,109,115,123,125–130,132–133,155,218) and Morwood (2002:158–159).

[87] Spencer & Gillen (1899:641) and Johnston (1943:280).

[88] Horne & Aiston (1924:45). The 'white owl' mentioned is most likely a reference to the barn owl.

[89] Gason (1879:289) and Johnston (1943:271).

[90] Horne & Aiston (1924:118), Fry (1937:201) and Johnston (1943:271–272).

[91] Horne & Aiston (1924:38).

[92] Horne & Aiston (1924:47).

[93] Howitt (1904:679) and Johnston (1943:280).

[94] Johnston (1943:280)

[95] Tunbridge & Coulthard (1985:14).

[96] Memmott (2010:14,103).

[97] Memmott (2010:113).

[98] Memmott (2010:52, see also 113–114).

[99] Berndt *et al.* (1993:198).

[100] South Australian Museum artefact (A67664, collected by NB Tindale 1930). See Hemming *et al.* (2000:25).

[101] Howitt (1904:706).

[102] Mathews (1897a:151).

[103] Berndt *et al.* (1993:119,211) and Gale (2009:118). Another spelling of *pulanggi-kalduki* is *bulangi-kalduki*.

[104] Memmott (2010:115).

[105] Berndt *et al.* (1993:62,Fig.4).

[106] Angas (1844a) and Hemming *et al.* (2000:18). Another spelling of *kalduki* is *kalduke*.

[107] Dawson (1881:58).

[108] Basedow (1925:141).

[109] Fry (1937:270) and Johnston (1943:270).

[110] Tunbridge & Coulthard (1985:44,53).

[111] Anonymous (1844:3).

[112] Berndt *et al.* (1993:271).

[113] Roth (1904:26). Note that McCarthy (1976:86) repeats some of the information but says 'the midrib of a Plains Turkey wing', omitting the reference to the feather in an apparent error.

[114] For a description of the pituri bags, refer to Horne & Aiston (1924:64–65,158–159), Watson (1983:31,37,40), Clarke (2007a:107–108, 2008a:137,139, 2012:48) and Silcock *et al.* (2012:39).

[115] Roth (1904:26). Note that McCarthy (1976:86) repeats some of the information but says 'the midrib of a Plains Turkey wing', omitting the reference to the feather in an apparent error.

[116] Tunbridge & Coulthard (1985:11,42).

[117] Berndt *et al.* (1993:193–195).

[118] Tindale (1941b:234).

[119] Spencer & Gillen (1904:700).

[120] Tindale (1930–52:269–270) and Harvey (1939). Ngarrindjeri man Henry J Rankine (1991:121) gave the same account, although he did not specify from which bird the magic feathers came.

[121] Berndt *et al.* (1993:261–262,493–494).

[122] Roth (1897:36).

[123] Vane-Millbank (1936:1).

[124] McCarthy (1953:76).

[125] Spencer & Gillen (1904:687), Raymond *et al.* (1999:106–108), Puruntatameri *et al.* (2001:96–97), Wiynjorrotj *et al.* (2005:134–135), Karadada *et al.* (2011:74–75), Doonday *et al.* (2013:141) and Hardwick (2019b:61, 2019d:49, 2019e:54).

[126] Spencer & Gillen (1904:727).

[127] Spencer & Gillen (1904:727).

[128] Hardwick (2019a:74).

[129] Hardwick (2019e:54).

[130] Akerman *et al.* (2014:202).

[131] Elkin *et al.* (1950:103–105), Morphy (1984:Plates 2–6,9,10,15), Tweedie (1998:26,144,146–147,151) and Berndt *et al.* (1999:Plate 103).

[132] Berndt *et al.* (1999:Plate 97) and Hamby & Young (2001:10,52).

[133] McCarthy (1974:36,Fig.28), Davis (1989:74–75) and Berndt *et al.* (1999:Plates 77,78A,78B,79).

[134] Hamby & Young (2001:8,82).

[135] McCarthy (1974:37).

[136] Morphy (1984:95).

[137] Elkin *et al.* (1950:35,39–41 & Plates 3A,4A).

[138] Thomson (1949:28).

[139] Kelly (2004:38) described the clan ownership of sacred dilly bags. The name of the Djang'awu has also been variously written as Djanggewul (Elkin *et al.* 1950:Map), Djanggawul (Berndt 1952; Berndt & Berndt 1954:9–12) and Djang'kawu (Morphy 1984:18, 1998:72–73,88–91,157; Keen 1994:48–61).

[140] Williams (1986:Fig.11).

[141] Williams (1986:Fig.11).

[142] Berndt (1952:Plate 8).

[143] Doonday *et al.* (2013:134).

[144] Doonday *et al.* (2013:135).

[145] Elkin *et al.* (1950:106–108,Plate 21). The authors referred to the 'seagull', which from the plates in the book can be identified as the silver gull. The 'white eagle-hawk' can reasonably be assumed to be the white-bellied sea-eagle. See also Berndt *et al.* (1999:109).

[146] Elkin *et al.* (1950:84–91,Plate 18A). The authors referred to 'red parakeet' feathers, which from the plates in the book can be identified as being from red-collared lorikeet.

[147] There is an extensive literature on the kadaitcha (kurdaitcha) sorcerers who used these ritual shoes or 'slippers' made from emu feathers and human hair string, much of it localised to Central Australia (Magarey 1899:122; Spencer & Gillen 1899:477,650–651, 1927:Vol.2:454–461; Horne & Aiston 1924:138; Basedow 1925:188; Sullivan 1928:168; Ramson 1988:354–355; Dixon *et al.* 1992:58,156–157; Akerman 2005; Doonday *et al.* 2013:129).

[148] NB Tindale (cited in Gale 2009:12,45).

[149] Clarke (2018i:42).

[150] McCarthy (1965:17).

[151] Refer to Akerman (2005).

[152] T Brown 1882 (cited in Worsnop 1897:52–53). See also Curr (1886:Vol.1:148).

[153] Massola (1968:10).

[154] Gason (1879:289), Roth (1901a:7–8), Langloh Parker (1905:128), Horne & Aiston (1924:79), Anonymous (1940), McCarthy (1965:17) and Gould (1969a:62).

[155] Basedow (1925:192).

[156] Eyre (1845:Vol.2:166–167).

[157] Ramsay Smith (1909:12).

[158] Molineux (1905).

[159] Baylie (1843:88).

[160] Berndt *et al.* (1993:198).

[161] Berndt *et al.* (1993:258–259).

[162] Clarke (2018i:41).

[163] Sutton (1988a:Fig.234), Hemming & Clarke (1989:14), DETE (1998:86–87), Clarke (2003b:100, 2018:18), Scott & Laurie (2007:52–53,57) and Fisher (2012:253).

[164] Beruldsen (1980:129–130).

[165] Wiynjorrotj *et al.* (2005:123) and Cocker & Tipling (2013:25).

[166] Basedow (1925:95).

[167] Anonymous (1940).

[168] Thomson (1936b:73).

[169] Worsnop (1897:158).

[170] Roth (1897:112,Fig.270) and Chewings (1936,Plate opp. 66).

[171] Spencer (1928:253,Fig.148).

[172] Akerman (2018:201).

[173] Spencer & Gillen (1904:692).

[174] Jones (1966:8) and Akerman (2018:202).

[175] Attenbrow (2010:117).

[176] Raymond *et al.* (1999:111).

[177] Akerman (2018:201).

[178] Angas (1847b:60).

[179] Berndt *et al.* (1993:150). The term *kurrindjeri* can also be written as *gurindjeri*.

[180] Berndt *et al.* (1993:150). The term *kurrindjeri* can also be written as *gurindjeri*.

[181] Berndt *et al.* (1993:150).

[182] Berndt *et al.* (1993:359).

[183] Worsnop (1897:158).

[184] Cleland (1966:116).

[185] Bennett (1834:Vol.1:176).

[186] Sanger (1883:1223) and Johnston (1943:271).

[187] Spencer & Gillen (1899:216).

[188] Sanger (1883:1224), Roth (1897:110), Horne & Aiston (1924:41,Fig.32) and Walshe (2008:182).

[189] Walshe (2008).

[190] Mulvaney (1960:61,66–67). The 'awls' found during the excavation were placed in the South Australian Museum archaeology collection (A52005 from Level 0, A52059–60 from Level 6, A52105 from Level 7).

[191] McCarthy (1976:86).

[192] Worsnop (1897:124). Such examples in the South Australian Museum ethnographic collection are from the Lower Murray region.

[193] Anonymous (1940) and Langloh Parker (1953:35).

[194] McCarthy (1965:17) and Roth (1904:25).

[195] McCarthy (1952).

[196] Ramsay Smith (1909:12).

[197] Pretty (1977:288–331).

[198] Eyre (1845:Vol.2:166–167).

[199] Langloh Parker (1905:128).

[200] Sanger (1883:1223).

[201] Tindale (1938–60:68,72).

[202] Roth (1904:25).

[203] Taplin (1859–79:9 November 1861).

[204] Tindale (1930–52:267). See also Meyer (1846:196–197) and Elkin (1977:43–46).

[205] Atkins (1911). The 'teal duck' has been identified here as the grey teal.

[206] Taplin (1859–79:5 February 1862).

[207] Taplin (1859–79:5 December 1862).

[208] Taplin (1859–79:9 November 1861).

[209] Clarke (2018i:41).

[210] Clarke (2018i:41).

[211] Berndt *et al.* (1993:490).

[212] Tindale (1953:1).

[213] Berndt *et al.* (1993:258–259).

[214] McCarthy (1965:17, see also 1953:74).

[215] Gason (1879:278), Horne & Aiston (1924:44), Johnston (1943:271) and McCarthy (1965:17, 1953:74).

[216] Horne & Aiston (1924:161).

[217] Akerman (2018:201).

[218] Memmott (2010:113).

[219] Elkin *et al.* (1950:85,89–91 & Plates 18A,19A,19B) and Berndt *et al.* (1999:108,Plate 98).

[220] Elkin *et al.* (1950:91,Plate 19B). The authors' reference to a 'goose' can be interpreted as the magpie goose in keeping with other uses of the term within the text, while the head of the 'spoon-billed duck' in the plate can be readily identified as that of a royal spoonbill.

[221] Elkin *et al.* (1950:85).

[222] Elkin *et al.* (1950:88).

[223] Akerman (2018:202).

[224] Kerwin (1986:22,33).

[225] Kennedy (1933:154) and Akerman (2018:202–203).

[226] Basedow (1925:184,255).

[227] Silverleaf (1883).

[228] Smyth (1878:Vol.1:450).

[229] Tindale (1937a:115).

[230] Teichelmann & Schürmann (1840:20).

[231] Tindale (1935:267,270–271). See also Tellurian (1935). Waijungari was also written as Wyoongurrie, and for an account of this mythology refer to Clarke (1999b).

[232] Berndt *et al.* (1993:351).

[233] Gale (2009:97,146).

[234] Amateur Naturalist (1886), Harris (1894–1895:1 June 1894) and Berndt *et al.* (1993:115–117,175,349–351,486–487).

[235] Berndt *et al.* (1993:115).

[236] Clarke (2018i:29–32).

[237] Bonwick (1870:25) and Cleland (1940:10).

[238] Bonwick (1870:26).

[239] Basedow (1925:190).

[240] Basedow (1925:275,342).

[241] Dawson (1881:54), Tindale (1937a:110) and Berndt *et al.* (1993:485).

[242] Clarke (2018i:32).

[243] Anonymous (1885b, 1955).

[244] Anonymous (1955:30).

[245] EKV (1884).

10
Conclusion

Aboriginal relationships with organisms, which Western Europeans have collectively defined as birds, are broad. They extend from the cultural perceptions of birds as totemic species, ancestral creators and spirit beings, to the physical uses of their bodies as foods, medicines and artefact-making materials. Holistically, beliefs and perceptions of birds are so heavily integrated into the culture that it is possible to use the study of ethno-ornithology as a window into the whole of Aboriginal Australia. In this chapter, some of the major themes and subthemes that have emerged in the book will be further discussed with the aim of summarising Aboriginal relationships with Australian birds.

In an Aboriginal world view, the physical form of each bird species is widely seen as tangible 'proof' of the ancestral events that took place in the creation, with the origin of particular behaviours, plumage colour and shape being major elements. Aggressiveness is a prominent theme in many Australian bird myths, such as with the wedge-tailed eagle and crow/raven, and this reflects the observed characteristics of living species. Avian habitat preferences are explained, such as the brolga living in the wetlands and emus ranging across inland areas. The mythology blurs the distinction between people and animals, giving them shared origins as reflected in Aboriginal totemic systems and in the traditions concerning the origin of places. The Aboriginal perception of time is cyclical, and therefore not as deep as the linear view of time held by Westerners. This means that the events of the creation are seen as just beyond living memory; for instance, an Aboriginal person with totemic links to the emu could, in some situations, claim that their great-grandfather was the ancestral emu man.

Aboriginal peoples believe that they co-occupy the landscape with a diverse range of spirits, some of which may take human-like forms while others adopt the shape of animals, particularly birds. For this reason, it is not always clear whether a sighted bird is a spirit being or just a wild animal. Birds that are observed acting strangely or that appear at critical times in certain places are often treated as spirits. Malignant spirits are believed capable of wearing several 'cloaks' or 'skins', including those of birds. Most Australian birds possess the power of flight, which enables them to traverse the connected regions of the Skyworld, Earth and Underworld. For this reason, bird spirits are seen as capable of gaining greater insights into the spirit world than people are generally able to.

To determine the place of birds in Aboriginal culture, it is convenient to consider the nature of the corpus of environmental knowledge that Aboriginal peoples have held.[1] Applied ecologist Fikret Berkes defined 'traditional ecological knowledge' as 'a cumulative body of knowledge, practice, and belief, evolving by adaptive processes and handed down through generations by cultural transmission, about the relationship of living beings (including humans) with one another and with their environment'.[2] Following on from this, I would add that the learners of environmental knowledge are like apprentices who, while on the job, absorb all the insights that are relevant to their ongoing experiences with their environment.

The knowledge so gained comprises relevant facts about the local ecology that are put into the practice of environmental use. Due to the complexities of human relationships with the environment, Berkes has insightfully recommended that 'traditional knowledge as process, rather than as content, is perhaps what we should be examining'.[3]

Aboriginal environmental knowledge encompasses all the cultural values, ethics and philosophies that define humans within what Westerners have described as the 'natural world'.[4] For example, as shown in this book the Aboriginal knowledge about brolgas is expressed in a manner that interweaves both the ecological and cultural aspects of the species – as a major source of protein and desired materials for making artefacts there is detailed knowledge about brolga behaviour for hunting, and yet they are universally regarded as major ancestral beings who were involved in the creation and as such are celebrated in ceremony. In Aboriginal Australia, certain birds, such as magpies, eagles, owls, willie-wagtails and many totemic species, are able to move freely between the worlds of spirits and people, and because of this they have transformative powers. Beliefs in certain malignant spirit beings, often described as targeting children, are used to spatially define dangerous places, such as certain deep waterholes or large trees prone to lightning. They also dictate appropriate behaviours, such as when outsiders are visiting a strange area. As monsters, they are seen as a physical amalgam of human and animal traits, including those of birds. In Aboriginal perception, birds are like other people, to the extent that those that are encountered may be seen as friend or foe, depending on a range of factors.

For Aboriginal peoples, the environmental knowledge they hold is part of their corpus of information regarding the total landscape, and imbues it with a range of spiritual, ethical and community values. This knowledge contains a wealth of facts about ecological components and processes based upon observation over time. For example, in birds it will explain such things as migration, nesting and feeding behaviours, and be used to determine appropriate foraging methods. Knowledge accumulation is the result of the process whereby oral traditions are constantly augmented through lived experiences. The loss of environmental data may be extreme at certain times, such as through the impact of European colonisation upon Aboriginal living styles, although the framework within which it is organised by the community persists for as long as their world view remains largely intact. An example is the retention of Aboriginal knowledge of bird spirits, even in areas where the ceremonial life has severely declined. It is therefore the structure of the environmental knowledge that is the community asset, not its content as that varies widely among members within the group and changes through time.

Two key features of Aboriginal environmental knowledge are that it is oral in nature and it does not follow the empirical systems of Western science, in terms of cause and effect. It is based in a more qualitative system that is highly adaptive to the seasonal cyclic changes that can be observed within Country and the local culture.[5] An example from Chapter 7 is the knowledge that the flowers appearing on calendar plants, which are often yellow, signal that gamebirds, such as swans and terns, are laying eggs with yellow yolks – these are phenomena that Western scientists would regard as independent events. The Aboriginal holders of accumulated environmental knowledge make sense of it through its relevance to their beliefs in spirits and ancestors, their ongoing relationship with the natural environment and the use of its resources, and the connections between people.[6] Based on the richness of bird knowledge as discussed within this book, it is apparent that for a non-literate society the system has ably supported the development and maintenance of extensive understanding of how the cultural and physical elements within the Aboriginal world are intermeshed.

Aboriginal myth narratives gain much of their substance from environmental knowledge, which is why the discussion of mythology has been given so much attention in this book. The dynamism of Aboriginal myth is demonstrated by anthropological ethnographic accounts that stress the seemingly endless variations that speakers weave into their narratives from the corpus of their myth knowledge. Anthropologists Ronald M Berndt and Catherine H Berndt, whose detailed fieldwork experiences covered much of Australia, remarked that when Aboriginal people are telling stories 'The material may have a direct bearing on the main theme or plot or none at all. It can be elaborated or cut short or even omitted altogether, depending on the speaker, the listeners and the situation'.[7] Given the importance in Aboriginal Australia of acknowledging that the creation ancestors set the scene for all the people who were to follow, it is consistent that myths also contain much practical data. The Berndts explained that:

> The stories centring on a cluster of sites or a strip of territory are like a loosely organized collection of guide books, where basic information is embedded in quasi-factual accounts of local history. Without giving a complete picture, they indicate what visitors can expect to find there: what kinds of terrain, vegetation and food, what difficulties or dangers, whether insects are troublesome, what surface waters will last through the dry season or whether water must be dug for, and so on. Geared to the demands of semi-nomadic living, the stories are a convenient medium for transmitting a series of 'hints to travellers'.[8]

Scholars have recovered elements of the Aboriginal environmental knowledge from south-eastern Australia that was in existence when Europeans first arrived, primarily from historical sources.[9] Analysis of this has determined that the knowledge was gained via the intimate relationships that hunter-gatherers had with the whole of the environment with which they interacted, and that it was interpreted through the lens of Aboriginal tradition. It is reasonable to conclude that there was much passing on of this knowledge between generations, but learning it would have required reinforcement through the constant observation of actual processes. Since so much of this environmental knowledge was built upon the wealth of lived experience that foragers gained through daily subsistence activities, it can be deduced that Aboriginal peoples who have largely grown up in European-controlled situations, such as on cattle stations and mission reserves, will have a different body of knowledge in terms of content. The fundamental relationships with the biota that Aboriginal peoples throughout the continent now possess has been irreversibly altered due to changing living patterns, although elements of the earlier forms of knowledge about the environment persist as new knowledges are created.

The majority of the data on Aboriginal relationships with birds compiled for the writing of this book has come from historical sources concerning landscapes where both the biota and the people who lived within it have since undergone varying degrees of change.[10] Aboriginal peoples living at the time of first European settlement realised that there was a different source of powerful knowledge about the world, shown with the arrival of strange 'new' ideas and technologies, and they needed to incorporate it into their universe.[11] For contemporary Aboriginal peoples who are still embedded in their traditional culture, this means that the facts learnt from the realms of both 'Indigenous belief' and 'science' must be made to address each other. In relation to the colonial encounter between Aboriginal peoples and the first Europeans, anthropologist Jeremy Beckett wrote that in this context:

> Indigenous knowledge is no longer self-evident; it must be assessed relative to the knowledge of the colonizers, [and] if it is not to be abandoned as worthless, it must either be consigned to a separate domain or made commensurate with the knowledge of the other through some kinds of articulation.[12]

The study of Aboriginal environmental knowledge is not just of antiquarian interest. The records of the early forms of this knowledge provide evidence of the depth and extent of past relationships to Country. It is for this reason, in my experience, that the connection reports written by anthropologists for land rights and native title claims typically include accounts of such things as Aboriginal mythology and hunter-gatherer subsistence – and birds generally feature prominently in both. This environmental knowledge is also incorporated into museum and art gallery displays of the unique and diverse material culture that was produced by Aboriginal peoples, both past and present. For instance, as discussed in Chapter 3, Aboriginal relationships with birds form major themes in Aboriginal art, embracing both the classical pre-European culture and its modern forms.

Non-Aboriginal people have embraced the study of Aboriginal environmental knowledge as a means of potential insights into the culture of the knowledge-holders and the physical environment that they and their ancestors have occupied for tens of thousands of years. Academically trained ecologists, who study the relationships that organisms have with the broader environment, are particularly well placed to consider Aboriginal environmental knowledge and its wider context. Applied ecological research incorporating an Aboriginal perspective on the environment informs the development of conservation programs run by Indigenous ranger groups on Indigenous Protected Areas and the reinstatement of 'firestick farming' practices in remote areas.[13] Biologists have also been at the forefront of programs that involve working alongside senior Aboriginal people to document their detailed knowledge of the fauna and flora of their Country, so it can be passed on to future generations.[14] This book draws heavily upon the records such researchers have compiled.

Across the world, it is widely recognised that indigenous systems of environmental knowledge, along with the associated practices of local communities, are critical in maintaining the biological and cultural diversity of the planet.[15] The continuity and dynamism of this knowledge is threatened by ongoing globalisation, harsh government policies, capitalism and

Rock engravings. Dated to many thousands of years ago, they contain Australian bustard and emu prints, alongside what are possibly emu eggs or stars – or even both? Philip A Clarke, Panaramittee, near Yunta, South Australia, 2004.

colonialism. Rapid environmental changes interfere with the relationships that indigenous communities have historically possessed with their landscape. According to an international group of ethnobiologists, 'the foundations of these knowledge systems are compromised by ongoing suppression, misrepresentation, appropriation, assimilation, disconnection, and destruction of biocultural heritage'.[16] It is largely in response to this situation that I wrote this book on Aboriginal relationships with birds.

Palaeontologists have been active in embracing aspects of Aboriginal culture in publication of their research; for example, shown by their choice of scientific names when describing newly discovered fossil species, as discussed in Chapter 5. In the context of formally recognising the presence of Aboriginal peoples and their ancestors upon the continent, along with their custodianship of it since ancient times, the use of Indigenous-derived names for genera and species is, in my opinion, highly commendable. The problem for science occurs when the naming is interpreted by others to mean that the ancestral beings of Aboriginal myth were actual species that vanished many thousands or even millions of years ago. Aboriginal mythology, and the environmental knowledge encoded within it, is dynamic and not static, so logically it should not be taken unquestioningly as a scientific record of ancient fossil species. This is not to say that in the past Aboriginal peoples did not explain the presence of fossils that they found, some of them millions of years old, in terms of their ancestors' feats during the creation. Aboriginal peoples have consistently stated that during the creation period their spiritual ancestors existed as an amalgam of animal and human characteristics, and were often seen as more powerful and as a consequence morphologically larger than species that are present today. More recently, Aboriginal peoples have been drawn towards making sense of the public discourse concerning the scientific discoveries of fossils, in terms of their own creation traditions. Syncretism is to be expected in these situations, although there remains tension between Indigenous and scientific explanations of the past over what is a 'fact' in both systems.

Outside of ecology and palaeontology, certain scholars have taken from recorded Aboriginal traditions what has been termed the 'racial memories' of past landscape-transforming events, such as volcanic eruptions, tektite showers and rapid sea level changes.[17] As discussed in Chapters 2, 3 and 7, these researchers have connected physical cataclysmic events, that occurred many thousands of years previously, with particular Aboriginal myths, many of which include bird ancestors. A recent example of what can be termed the 'racial memory' literature, written by co-authors comprising two astronomers, two historians, two physicists and a geographer, makes the fantastic claim that Tasmanian Aboriginal oral traditions, formed between 16 300 and 11 800 years BP, recorded the sea level rises separating Tasmania from Victoria in the late Pleistocene.[18] Running against contemporary anthropological understandings of the dynamism of culture, these authors are in effect supporting the early 20th century views, as espoused by Norman B Tindale, that Aboriginal culture was 'primitive' and that the content of mythological traditions did not change before European influences.

In my opinion, it is an intellectually poor approach for researchers, most of them non-Aboriginal, to take myth as an account from which scientific facts about ancient times can be derived, even though contemporary Aboriginal peoples with complex cultural backgrounds may refer to their mythologies as 'history'. As discussed earlier in this book, the Aboriginal notion of cyclical time stresses the phases of the environment as in tune with the cultural life with which it is intrinsically connected. The 'facts' from Aboriginal myths, as described by the Berndts, are therefore those that are most pertinent to the lived present. They should not be selectively extracted by scientists for interpretation as a linear record of the deep past. There is, however, compelling ethnographic evidence demonstrating that more recent events, such as the arrival of Europeans from the late 18th century, are incorporated into myth when they are considered relevant to the speaker's community.

The recordings of Aboriginal environmental knowledge from the hunter-gatherer period are of immense heritage value, but it is unreasonable to expect today that, as an asset, the knowledge can be reclaimed and then relearnt in its original form. In many areas the Indigenous languages, within which this knowledge was orally recorded, are no longer spoken. Anthropologist Nicolas Peterson questioned whether the recording of ecological knowledge, mythology and other aspects of Australian Indigenous culture would help to maintain an Aboriginal culture in the modern world for communities of former hunter-gatherers. He came to the conclusion that 'doing so turns knowledge and belief into information, inevitably distanced from the contexts in which such knowledge and belief were a part of lived experience, and literally disembodies them'.[19] I agree with this in relation to the environmental knowledge gained and held by former hunter-gatherers, but it is also apparent that contemporary Aboriginal peoples have created new bodies of environmental knowledge, incorporated from a variety of sources including those external to Aboriginal culture. These contemporary knowledges are relevant to their own world views and have been distilled from their still often-unique relationships with the modern environment.

In the present day, investigations of the earliest historical records of Aboriginal environmental knowledge, derived from foragers who had lived at the time of first European settlement, have much potential for informing contemporary Aboriginal peoples who are looking back into the past as inspiration when developing a modern non-Western perspective of Country.[20] The Aboriginal community's development of new forms of the environmental knowledge possessed by earlier generations, with some inclusion of past elements, is supported through the Indigenous ranger programs that employ many young people within the remote Indigenous Protected Areas, under the guidance of Elders in the community.[21] In many regions, the national parks are under a co-management regime, whereby environmental managers work alongside Traditional Owners.[22] Indigenous-run wildlife tourism relies heavily upon the ability of guides to give tourist clients a meaningful overview of their environmental knowledge relevant to the local area.[23] Conservation of bird species in areas where Indigenous peoples maintain some foraging activities is dependent upon the authorities developing a working understanding of the local cultural perspective of Country, as has been shown in other countries.[24] As a Gunditjmara man remarked in 2018 in relation to Aboriginal environmental knowledge being an important asset for his community, 'you use it or lose it'.[25]

The availability of wild food sources, such as gamebirds, remains essential for Aboriginal peoples living on outstations in remote parts of Australia.[26] Here, the procurement of game has been a highly valued activity for hunters, with the meat so produced adding to their

Mornington Island dance group, performing in capital cities of eastern Australia, 1972. The dancers are decorated with bird down and are wearing ceremonial paraphernalia made from feathers. John Bissell Collection, Mirndiyan Gununa Aboriginal Corporation, Mornington Island Art.

prestige.[27] Seasonal foraging has added to the food supply of many Aboriginal households, although its dietary value overall was diminished by the late 20th century.[28] When assessing the importance of hunter-gatherer traditions for contemporary Aboriginal communities, it is wrong to assume that they were simply food-producing activities conducted solely in response to economic and recreational needs. For many Aboriginal peoples, foraging is an expression of their local identity – a case of 'you are what you eat'. For instance, linguist Joseph Blythe and botanist Glen Wightman noted that among the Kija and Jaru people of the east Kimberley region of Western Australia:

> Animals and plants figure among the favourite topics of conversation; many stories on a variety of themes include details about animals and plants encountered in travels and what people ate when in the bush. This traditional knowledge is part of what it means to know the country.[29]

Today, the emerging bushfood industry – which at the moment is largely controlled by non-Aboriginal agents – has seen renewed interest in traditional Aboriginal food sources, particularly from plants, and the culinary history of the ingredients is a major marketing tool.[30] The hunt for new ingredients to develop a truly Australian cuisine is gaining momentum.

In recent years there has been a growing market for magpie goose as a commercial meat,[31] sourced from wild northern Australian populations because the species' range has contracted from southern regions since European settlement.[32] The Bass Strait Aboriginal tradition of muttonbirding continues in Tasmania, where it is a distinctive aspect of the local identity.[33] In addition to the harvesting of wild foods, there is more intensive food production, such as Aboriginal people involved with emu-farming enterprises for the production of meat, leather, oil, decorative feathers and engraved egg shells.[34] Aboriginal communities who are living on and managing their traditional Country potentially have a wide range of wildlife, including indigenous and feral species, available for internal use and for developing commercial ventures.[35]

Through native title determinations and land rights grants, Indigenous peoples have gained legal recognition for their rights and interests to many parts of Australia; however, in terms of non-Indigenous concepts of property this is not considered to equate to the total 'ownership' of the wildlife.[36] Therefore, Aboriginal rights to subsistence foraging for wild animals and plants is problematic in many areas, with overall poor prospects for commercial harvests by Traditional Owners.[37] There is a role for Indigenous peoples in the management of wild animal resources,[38] although it requires greater understanding of the complex hybrid economies operating in many Aboriginal communities.[39]

Birds are prominent ancestral beings to the extent that they have been a major element in the expression of Aboriginal identity. In Aboriginal Australia, it is a commonly held belief that during the creation the bird ancestors possessed transformative powers that gave meaning to the landscape, both on land and in the heavens. Importantly, their actions also established the basis for Aboriginal social structure and seasonal life. The incorporation of observations concerning the physical form and behaviour of individual bird species within the narratives of Aboriginal myths demonstrates the existence of a body of ornithological knowledge outside of Western science. Findings from ethno-ornithological research therefore have the potential to inform how land management agencies can perceive the landscape and its rich avifauna, and in doing so recognise its cultural values. Through legal processes beginning in the late 20th century, such as land rights and native title, Indigenous communities are today managing the natural resources over vast areas of Australia.[40] This has created a need to bring together insights derived from the knowledge systems of both Westerners and Indigenous Australians, in order to work together. The deep understanding of the ethno-ornithology of Aboriginal Australia provides unique insights into the distinctive regional cultures.

Internationally, it has been recognised that there is an imperative for traditional forms of environmental knowledge to be incorporated into the contemporary understanding and management of natural and cultural landscapes.[41] Ethnobiologist Dana Lepofsky remarked that:

> We live in a world where biological and cultural diversity are being lost at dizzying rates. As ethnobiologists we know that losses of diversity in both of these realms are inextricably intertwined.[42]

Endnotes

[1] Refer to overviews of Aboriginal environmental knowledge by Ens *et al.* (2015) and Pert *et al.* (2015).

[2] Berkes *et al.* (2000:1252), in relation to Traditional Ecological Knowledge.

[3] Berkes (2009:151), in relation to Traditional Ecological Knowledge.

[4] Thomas (1991).

[5] De Guchteneire *et al.* (2002), in relation to Indigenous Knowledge.

[6] Andrews & Buggey (2008), in relation to Traditional Ecological Knowledge.

[7] Berndt & Berndt (1970:41).

[8] Berndt & Berndt (1970:41).

[9] Cahir *et al.* (2018a).

[10] For changes in relation to Aboriginal peoples and the environment, refer to Rose (1987:Ch.1), Clarke (2003a:Pt4, 2007a:Pt4, 2012:Pt3) and Zeanah *et al.* (2015).

[11] Mulvaney (1989).

[12] Beckett (1993:691).

[13] For example, refer to overview publications by EJ Ens, such as Ens *et al.* (2012, 2014, 2016), Pert *et al.* (2015) and McKemey *et al.* (2020). See also the specific publications of J Davies and FJ Walsh, such as Davies (1998), Walsh & Mitchell (2002), Davies *et al.* (2013, 2018) and Hill *et al.* (2013). Andersen (1999), Russell-Smith *et al.* (2009) and Petty *et al.* (2015) have investigated the contemporary application of Indigenous fire management practices.

[14] For example, refer to the publications to which biologists G Wightman and NM Smith contributed, such as Barr *et al.* (1993), Smith *et al.* (1993), Yunupingu *et al.* (1995), Raymond *et al.* (1999), Puruntatameri *et al.* (2001), Wiynjorrotj *et al.* (2005), Karadada *et al.* (2011) and NM Smith (2013).

[15] Whyte (2013), Berkes (2017), McGregor *et al.* (2018) and Fernández-Llamazares *et al.* (2021).

[16] Fernández-Llamazares *et al.* (2021:144).

[17] For 'racial memories' and related concepts, see Tindale (1959:46–47, 1974:50,119).

[18] Hamacher *et al.* (2021).

[19] Peterson (2017:236).

[20] Cahir *et al.* (2018b, 2018c).

[21] Davies *et al.* (2016), Godden & Cowell (2016) and McKemey *et al.* (2020).

[22] Young *et al.* (1991), Walsh (2008) and Hill *et al.* (2013).

[23] For sources on Aboriginal-run ecotourism, refer to Muloin *et al.* (2001) and Zander *et al.* (2014).

[24] Muiruri & Maundu (2010), Sault (2010) and Thomas (2010).

[25] Author (unpublished).

[26] Young (1983:Ch.5), Altman (1987, 2001), Altman *et al.* (1995, 1997), Bomford & Caughley (1996), Barber (2005:Ch.5) and Humphries (2007).

[27] Sackett (1979).

[28] O'Dea *et al.* (1980, 1982) and O'Dea (1984).

[29] Blythe & Wightman (2003:69).

[30] Cherikoff & Brand (1987), Davies (1998), Parliament of the Commonwealth of Australia (1998), Gorman *et al.* (2008) and Jones & Clarke (2018).

[31] J Zonfrillo (pers. comm.). For contemporary accounts of cooking magpie goose refer to https://northernterritory.com/things-to-do/art-and-culture/aboriginal-culture/bush-food and https://www.abc.net.au/news/2018-05-10/indigenous-bush-foods-on-rise-michelin-chefs-in-yirrkala/9742480.

[32] Blakers *et al.* (1984:68,662).

[33] Smith (1965) and Skira (1996).

[34] Anonymous (1985:2, 1994:5), Altman *et al.* (1997:5,9), O'Malley (1998) and Sales (2007).

[35] Wilson *et al.* (1992).

[36] Davies (1998).

[37] Resource Assessment Commission (1993:166–189), Smyth (1993:211–225), Altman *et al.* (1995), Peterson & Rigsby (1998), Gorman *et al.* (2008) and Jones & Clarke (2018:42–44).

[38] Wilson *et al.* (1992), Bomford & Caughley (1996), Altman *et al.* (1997), Parliament of the Commonwealth of Australia (1998) and Zander *et al.* (2014).

[39] Altman (1987, 2001).

[40] Smyth *et al.* (2004).

[41] Lertzman (2009) and Nabhan (2009).

[42] Lepofsky (2009:161).

References

Abbott I (2009) Aboriginal names of bird species in south-west Western Australia, with suggestions for their adoption into common usage. *Conservation Science Western Australia* **7**(2), 213–278.

Abbott I (2013) Extending the application of Aboriginal names to Australian biota: *Dasyurus* (Marsupiala: Dasyuridae) species. *Victorian Naturalist* **130**(3), 109–126.

Abdulla I (1993) *As I Grew Older: The Life and Times of a Nunga Growing Up Along the River Murray.* Omnibus Books, Adelaide.

Abdulla I (1994) *Tucker.* Omnibus Books, Adelaide.

Abimosleh SM, Tran CD, Howarth GS (2012) Emu oil: a novel therapeutic for disorders of the gastrointestinal tract? *Journal of Gastroenterology and Hepatology* **27**, 857–861. doi:10.1111/j.1440-1746.2012.07098.x

Agrawal A (2002) Indigenous knowledge and the politics of classification. *International Social Science Journal* **54**(173), 287–297. doi:10.1111/1468-2451.00382

Akerman K (1979) Contemporary Aboriginal healers in the south Kimberley. *Oceania* **50**(1), 23–30. doi:10.1002/j.1834-4461.1979.tb01928.x

Akerman K (1990) *Tools, Weapons and Utensils.* Culture and Society Series. Aboriginal and Torres Strait Islander Commission, Canberra.

Akerman K (1992) *Carving and Sculpture.* Culture and Society Series. Aboriginal and Torres Strait Islander Commission, Canberra.

Akerman K (1995) Tradition and change in aspects of contemporary Australian Aboriginal religious objects. In *Politics of the Secret.* (Ed. C Anderson) pp. 23–30. Oceania Monographs No. 45. Oceania, Sydney.

Akerman K (2005) Shoes of invisibility and invisible shoes: Australian hunters and gatherers and ideas on the origins of footwear. *Australian Aboriginal Studies* **2**, 55–64.

Akerman K (2011a) Some aspects of the material culture of the Aborigines of the Canning Stock Route, Western Australia. In *Ngurra Kuju Walyja: One Country, One People. Stories from the Canning Stock Route.* (Eds M La Fontaine and J Carty) pp. 354–363. Macmillan, Melbourne.

Akerman K (2011b) The use of animal and human teeth in the material culture of Aboriginal Australians. *Ethnozootechnie* **89**, 133–142.

Akerman K (2016) *Wanjina: Notes on Some Iconic Ancestral Beings of the Northern Kimberley.* Hesperian Press, Perth.

Akerman K (2017) The material culture of Western Australia. In *Australia: The Vatican Museums Indigenous Collections.* (Ed. K Aigner) pp. 124–131. Aboriginal Studies Press and Edizioni Musei Vaticani, Canberra.

Akerman K (2018) The esoteric and decorative use of bone, shell and teeth in Australia. In *The Archaeology of Portable Art: Southeast Asian, Pacific and Australian Perspectives.* (Eds MC Langley, D Wright, M Litster and SK May) pp. 199–219. Routledge, London.

Akerman K (2020) *From the Bukarikara: The Lore of the Southwest Kimberley Through the Art of Butcher Joe Nangan.* UWA Publishing, Perth.

Akerman K, Birch B, Evans N (2014) Notes on the contemporary knowledge of traditional material culture among the Iwaidja: Cobourg Peninsula, Arnhem Land, Northern Territory, 2005–2006. *Transactions of the Royal Society of South Australia* **138**(2), 181–213. doi:10.1080/03721426.2014.11649008

Albrecht FW (1959) *The Natural Food Supply of the Australian Aborigines.* Aborigines' Friends' Association, Adelaide.

Altman JC (1987) *Hunter-gatherers Today: An Aboriginal Economy in North Australia.* Australian Institute of Aboriginal Studies, Canberra.

Altman JC (2001) *Sustainable Development Options on Aboriginal Land: The Hybrid Economy in the Twenty-first Century*. Centre for Aboriginal Economic Policy Research, Australian National University, Canberra.

Altman JC, Bek HJ, Roach LM (1995) *Native Title and Indigenous Australian Utilisation of Wildlife: Policy Perspectives*. Centre for Aboriginal Economic Policy Research, Australian National University, Canberra.

Altman JC, Roach LM, Liddle LE (1997) *Utilisation of Native Wildlife by Indigenous Australians: Commercial Considerations*. Centre for Aboriginal Economic Policy Research, Australian National University, Canberra.

Amery R (2016) *Warraparna Kaurna! Reclaiming an Australian Language*. University of Adelaide Press, Adelaide.

Amundson R (1982) Science, ethnoscience and ethnocentrism. *Philosophy of Science* **49**(2), 236–250. doi:10.1086/289052

Andersen A (1999) Cross-cultural conflicts in fire management in northern Australia: not so black and white. *Conservation Ecology* **3**(1), 1–12. doi:10.5751/ES-00093-030106

Anderson A (1989) *Prodigious Birds: Moas and Moa-hunting in Prehistoric New Zealand*. Cambridge University Press, Cambridge.

Anderson C (1996) Traditional material culture of the Kuku-Yalanji of Bloomfield River, north Queensland. *Records of the South Australian Museum* **29**(1), 63–83.

Andrews TD, Buggey S (2008) Authenticity in Aboriginal cultural landscapes. *APT Bulletin* **39**(2/3), 63–71.

Anell B (1960) *Hunting and Trapping Methods in Australia and Oceania*. Lund: Hakan Ohlssons Boktryckeri, Uppsala.

Angas GF (1847a) *South Australia Illustrated*. T McLean, London.

Angas GF (1847b) *Savage Life and Scenes in Australia*. Smith, Elder & Co., London.

Angas GF (1877) Amongst the blacks on the Murray River. *Colonies and India*, London. 13 January, p. 2.

Anonymous (1844) The native corroboree. *South Australian Register*, South Australia. 16 March, p. 3.

Anonymous (1845) Wonderful discovery of a new animal. *Port Phillip Patriot and Melbourne Advertiser*, Victoria. 5 July, p. 2.

Anonymous (1857) The bunyip. *Argus*, Victoria. 14 March, p. 6.

Anonymous (1885a) The pelican. *Border Watch*, South Australia. 4 February, p. 4.

Anonymous (1885b) Colonial and Indian exhibition. *Albany Mail & King George's Sound Advertiser*, Western Australia. 26 May, p. 4.

Anonymous (1888) The history of Victoria. *Illustrated Australian News*, Victoria. 1 August, pp. 2–8.

Anonymous (1889) On Australian Aborigines. *Northern Argus*, South Australia. 13 September, p. 4.

Anonymous (1893) The mutton bird on the Furneaux Islands. *Geelong Advertiser*, Victoria. 13 May, p. 4.

Anonymous (1895) The mutton bird of the Furneaux Islands. *Petersburg Times*, South Australia. 26 July, p. 2.

Anonymous (1904) Red ochre. *Adelaide Observer*, South Australia. 24 December, p. 43.

Anonymous (1906) Aboriginal witchcraft. *Port Pirie Recorder and North Western Mail*, South Australia. 31 March, p. 3.

Anonymous (1911a) Native companions and emus. *The Register*, South Australia. 19 June, p. 6.

Anonymous (1911b) Cheery bush birds. *Sydney Mail and New South Wales Advertiser*, New South Wales. 18 October, p. 32.

Anonymous (1913a) In the Never-Never. *Victor Harbor Times and Encounter Bay and Lower Murray Pilot*, South Australia. 7 March, p. 4.

Anonymous (1913b) The native companion. *Maryborough Chronicle, Wide Bay and Burnett Advertiser*, Queensland. 19 April, p. 5.

Anonymous (1914) How names of places came to be bestowed. *Dandenong Advertiser and Cranbourne Berwick and Oakleigh Advocate*, Victoria. 8 January, p. 3.

Anonymous (1915) Remedies. *Sun*, New South Wales. 10 March, p. 2.

Anonymous (1917) Bird nomenclature. *The Queenslander*, Queensland. 12 May, p. 8.

Anonymous (1918) Memories of Pantonie. *The Register*, South Australia. 21 January, p. 4.

Anonymous (1922) Concerning people. *The Register*, South Australia. 19 June, p. 6.

Anonymous (1925) Australian myths and legends. *Murray Pioneer and Australian River Record*, South Australia. 27 June, p. 8.

Anonymous (1930a) Murray Bridge diggers. Visit by Col. Dollman. *Mount Barker Courier and Onkaparinga and Gumeracha Advertiser*, South Australia. 23 May, p. 4.

Anonymous (1930b) Well-remembered blacks. *Observer*, South Australia. 23 October, p. 54.

Anonymous (1931) Price on emu heads. *News*, South Australia. 20 June, p. 2.

Anonymous (1932) Towns, people and things we ought to know. *Chronicle*, South Australia. 8 September, p. 44.

Anonymous (1940) Speed birds of the open plains. *Daily News*, Western Australia. 30 December, p. 5.

Anonymous (1944) Bittern maybe bunyip. *The Advertiser*, South Australia. 4 August, p. 7.

Anonymous (1952) Aboriginal legends. The parrot. *Argus*, Victoria. 24 October, p. 23.

Anonymous (1955) The battling. *World's News*, New South Wales. 26 March, p. 30.

Anonymous (1985) Playing the game. *Canberra Times*, Australian Capital Territory. 21 July, p. 2.

Anonymous (1994) Emu edibles part of an export push into the US. *Canberra Times*, Australian Capital Territory. 30 April, p. 5.

Archer M (1976) Miocene marsupiacarnivores (Marsupialia) from central South Australia *Ankotarinja tirarensis* gen. et sp. nov, *Keeuna woodburnei* gen. et. sp. nov. and their significance in terms of early marsupial radiations. *Transactions of the Royal Society of South Australia* **100**(2), 53–73.

Argus M (1947) Booming bitterns and bunyips. *Examiner*, Tasmania. 12 April, p. 6.

Arthur JM (1996) *Aboriginal English: A Cultural Study*. Oxford University Press, Melbourne.

Ash A, Giacon J, Lissarrague A (2003) *Gamilaraay, Yuwaalaraay and Yuwaalayaay Dictionary*. Institute for Aboriginal Development Press, Alice Springs.

AT (1950) Carved emu eggs. *World's News*, New South Wales. 11 March, p. 21.

Atkins B (1911) River Murray Aborigines: early reminiscences. *The Advertiser*, South Australia. 29 June, p. 12.

Attenbrow V (2010) *Sydney's Aboriginal Past: Investigating the Archaeological and Historical Records*, 2nd edn. UNSW Press, Sydney.

Austin P, Tindale NB (1985) The brolga and emu myth. *Aboriginal History* **9**(1), 8–21.

Bailey FM (1880) Medicinal plants of Queensland. *Proceedings of the Linnean Society of New South Wales* **5**(1), 1–29. doi:10.5962/bhl.part.15864

Baker G (1957) The role of australites in Aboriginal customs. *Memoirs of the National Museum of Victoria* **22**(8), 1–26. doi:10.24199/j.mmv.1957.22.08

Baker G (1959) Tektites. *Memoirs of the National Museum of Victoria* **23**, 1–313. doi:10.24199/j.mmv.1959.23.01

Baker RM (1999) *Land is Life: From Bush to Town – The Story of the Yanyuwa People*. Allen & Unwin, Sydney.

Barber M (2005) Where the clouds stand: Australian Aboriginal relationships to water, place and the marine environment in Blue Mud Bay, Northern Territory. PhD thesis. School of Archaeology and Anthropology, Australian National University, Canberra.

Barber M, Shellberg J, Jackson S, Sinnamon V (2012) *Working Knowledge: Local Ecological and Hydrological Knowledge about the Flooded Forest Country of Oriners Station Cape York*. National Research Flagships, Water for a Healthy Country, CSIRO, Darwin.

Barfield T (Ed.) (1997) *The Dictionary of Anthropology*. Blackwell, Oxford.

Barnes K (1999) *Kiripapurajuwi: Skills of Our Hands – Good Craftsmen and Tiwi Art*. The author, Darwin.

Barr A, Chapman J, Smith N, Beveridge M (1988) *Traditional Bush Medicines: An Aboriginal Pharmacopoeia*. Greenhouse Publications, Melbourne.

Barrett JW (1909) The mutton bird industry. *Mercury*, Tasmania. 17 February, p. 7.

Barrett CL (1932) Our national bird: epic of the emu. *Herald*, Victoria. 3 November, p. 6.

Barrett CL (1938) Mystery of the australites. *Weekly Times*, Victoria. 16 April, p. 42.

Bartholomaeus E (2012) *Indigenous Food Garden and Native Riverina Birds*. 2012 Green Steps Internship Report. School of Education, Charles Sturt University, Wagga Wagga.

Bartholomai A (2008) New lizard-like reptiles for the Early Triassic of Queensland. *Alcheringa: An Australasian Journal of Palaeontology* **3**(5), 225–234.

Barwick L (2005) Performance aesthetics, experience: thoughts on *Yawulyu mungamunga* songs. In *Aesthetics and Experience in Music Performance*. (Eds E Mackinlay, S Owens and D Collins) pp. 1–18. Cambridge Scholars Press, Newcastle, UK.

Barwick L, Laughren M, Turpin M (2013) Sustaining women's Yawulyu/Awelye: Some practitioners' and learners' perspectives. *Musicology Australia* **35**(2), 191–220. doi:10.1080/08145857.2013.844 491

Basedow H (1905) Geological report on the country traversed by the South Australian Government North-west Prospecting Expedition, 1903. *Transactions of the Royal Society of South Australia* **29**, 57–102.

Basedow H (1907) Anthropological notes on the western coastal tribes of the Northern Territory of South Australia. *Transactions of the Royal Society of South Australia* **31**, 1–61.

Basedow H (1914) Aboriginal rock carvings of great antiquity in South Australia. *Journal of the Royal Anthropological Institute* **44**, 195–211.

Basedow H (1925) *The Australian Aboriginal*. Preece, Adelaide.

Basedow H (1935) *Knights of the Boomerang: Episodes from a Life Spent Among the Native Tribes of Australia*. Endeavour Press, Sydney.

Bassani P, Lakefield A, Popp T (2006) *Lamalama Country: Our Country, Our Culture-way*. Yintjingga Land Trust, Queensland.

Bates DM (1906) The West Australian Aborigines: marriage laws and some customs. *Western Mail*, Western Australia. 28 April, p. 44.

Bates DM (1912) Aboriginal nomenclature. *The West Australian*, Western Australia. 10 February, p. 8.

Bates DM (1918a) Wirilya life and legend. No. I. *The Australasian*, Victoria. 7 December, p. 53.

Bates DM (1918b) Wirilya life and legend. No. II. *The Australasian*, Victoria. 14 December, p. 58.

Bates DM (1918c) Aborigines of the west coast of South Australia: vocabularies and ethnographical notes. *Transactions of the Royal Society of South Australia* **42**, 152–167.

Bates DM (1925) The serpent cult of Aborigines. *The Australasian*, Victoria. 29 August, p. 64.

Bates DM (1928) Central Australian birds' names. *The Australasian*, Victoria. 1 December, p. 56.

Bates DM (1947) *The Passing of the Aborigines: A Lifetime Spent Among the Natives of Australia*, 2nd edn. John Murray, London.

Bates DM (1901–14) *The Native Tribes of Western Australia*. (Ed. I White, 1985). National Library of Australia, Canberra.

Bates DM (1992) *Aboriginal Perth: Bibbulmun Biographies and Legends*. (Ed. PJ Bridge). Hesperian Press, Perth.

Bauer A (2014) *The Use of Signing Space in a Shared Sign Language of Australia*. Mouton De Gruyter, Berlin.

Baylie WH (1843) On the Aborigines of the Goulburn district. *Port Phillip Magazine* **1**, 86–92.

Bayly IAE (1999) Review of how Indigenous people managed for water in desert regions of Australia. *Journal of the Royal Society of Western Australia* **82**, 17–25.

Beckett J (1993) Walter Newton's history of the world – or Australia. *American Ethnologist* **20**, 675–695. doi:10.1525/ae.1993.20.4.02a00010

Beckett J (1994) Aboriginal histories Aboriginal myths: an introduction. *Oceania* **65**(2), 97–115. doi:10.1002/j.1834-4461.1994.tb02493.x

Bednarik RG (2013) Megafauna depictions in Australian rock art. *Rock Art Research* **30**(2), 197–215.

Bell D (1998) *Ngarrindjeri Wurruwarrin: A World That Is, Was and Will Be*. Spinifex, Melbourne.

Bellchambers TP (1931) *A Nature Lovers' Notebook*. (Ed. H Basedow). Nature Lovers' League, Adelaide.

Bennett G (1834) *Wanderings in New South Wales Batavia, Pedir Coast, Singapore and China: Being the Journal of a Naturalist in Those Countries, during 1832, 1833 and 1834*. 2 volumes. Richard Bentley, London.

Bennett G (1860) *Gatherings of a Naturalist in Australasia: Being Observations Principally on the Animal and Vegetable Productions of New South Wales, New Zealand and Some of the Austral Islands*. John Van Voorst, London.

Bennett DC, Code WE, Godin DV, Cheng KM (2008) Comparison of the antioxidant properties of emu oil with other avian oils. *Australian Journal of Experimental Agriculture* **48**, 1345–1350. doi:10.1071/EA08134

Berkes F (2009) Indigenous ways of knowing and the study of environmental change. *Journal of the Royal Society of New Zealand* **39**(4), 151–156. doi:10.1080/03014220909510568

Berkes F (2017) *Sacred Ecology*, 4th edn. Routledge, London.

Berkes F, Colding J, Folke C (2000) Rediscovery of traditional ecological knowledge as adaptive management. *Ecological Applications* **10**(5), 1251–1262. doi:10.1890/1051-0761(2000)010[1251:ROTEKA]2.0.CO;2

Berlin B (1992) *Ethnobiological Classification: Principles of Categorization of Plants and Animals in Traditional Societies*. Princeton University Press, Princeton.

Berlin B, O'Neill JP (1981) The pervasiveness of onomatopoeia in Aguaruna and Huambisa bird names. *Journal of Ethnobiology* **1**(2), 238–261.

Berndt RM (1940a) Some aspects of Jaraldi culture, South Australia. *Oceania* **11**(2), 164–185. doi:10.1002/j.1834-4461.1940.tb00283.x

Berndt RM (1940b) Aboriginal sleeping customs and dreams, Ooldea, South Australia. *Oceania* **10**(3), 286–294. doi:10.1002/j.1834-4461.1940.tb00294.x

Berndt RM (1940c) A curlew and owl legend from the Narunga tribe, South Australia. *Oceania* **10**(4), 456–462. doi:10.1002/j.1834-4461.1940.tb00306.x

Berndt RM (1940d) Some Aboriginal children's games. *Mankind* **2**(9), 289–293.

Berndt RM (1947) Wuradjeri magic and 'clever men'. *Oceania* **17**(4), 327–365; **18**(1), 60–86.

Berndt RM (1951) *Kunapipi: A Study of an Australian Aboriginal Religious Cult*. FW Cheshire, Melbourne.

Berndt RM (1952) *Djanggawul: An Aboriginal Religious Cult of North-eastern Arnhem Land*. FW Cheshire, Melbourne.

Berndt RM (1974) *Australian Aboriginal Religion*. EJ Brill, Leiden.

Berndt CH (1981) Interpretations and 'facts' in Aboriginal Australia. In *Woman the Gatherer*. (Ed. F Dahlberg) pp. 153–203. Yale University Press, New Haven.

Berndt CH (1982) Sickness and health in western Arnhem Land: a traditional perspective. In *Body, Land and Spirit: Health and Healing in Aboriginal Society*. (Ed. J Reid) pp. 121–138. University of Queensland Press, Brisbane.

Berndt CH (1988) *When the World Was New in Rainbow Snake Land*. Bookshelf Publishing, Gosford.

Berndt RM, Berndt CH (1942) Preliminary report of fieldwork in the Ooldea region, western South Australia (continued). *Oceania* **13**(2), 143–169. doi:10.1002/j.1834-4461.1942.tb00375.x

Berndt RM, Berndt CH (1943) Preliminary report of fieldwork in the Ooldea region, western South Australia (continued). *Oceania* **14**(1), 30–66. doi:10.1002/j.1834-4461.1943.tb00397.x

Berndt RM, Berndt CH (1954) *Arnhem Land: Its History and its People*. FW Cheshire, Melbourne.

Berndt RM, Berndt CH (1970) *Man, Land and Myth in North Australia: The Gunwinggu People*. Ure Smith, Sydney.

Berndt RM, Berndt CH (1989) *The Speaking Land: Myth and Story in Aboriginal Australia*. Penguin, Melbourne.

Berndt RM, Berndt CH (1999) *The World of the First Australians: Aboriginal Traditional Life – Past and Present*. Aboriginal Studies Press, Canberra.

Berndt RM, Vogelsang T (1941) Comparative vocabularies of the Ngadjuri and Dieri tribes, South Australia. *Transactions of the Royal Society of South Australia* **65**(1), 3–10.

Berndt RM, Berndt CH, Stanton JE (1993) *A World That Was: The Yaraldi of the Murray River and the Lakes, South Australia.* Melbourne University Press, Melbourne.

Berndt RM, Berndt CH, Stanton JE (1999) *Aboriginal Australian Art.* Reed New Holland, Sydney.

Berry KA, Jackson SE, Saito L, Forline L (2018) Reconceptualising water quality governance to incorporate knowledge and values: case studies from Australian and Brazilian indigenous communities. *Water Alternatives* **11**(1), 40–60.

Beruldsen G (1980) *A Field Guide to Nests and Eggs of Australian Birds.* Rigby, Adelaide.

Beveridge P (1883) Of the Aborigines inhabiting the Great Lacustrine and Riverine Depression of the Lower Murray, Lower Murrumbidgee, Lower Lachlan and Lower Darling. *Journal and Proceedings of the Royal Society of New South Wales* **17**, 19–74.

Biernoff D (1982) Psychiatric and anthropological interpretations of 'aberrant' behaviour in an Aboriginal community. In *Body, Land and Spirit: Health and Healing in Aboriginal Society.* (Ed. JS Reid) pp. 139–153. University of Queensland Press, Brisbane.

Bindon P (1996) *Useful Bush Plants.* Western Australian Museum, Perth.

Bindon P, Lofgren M (1982) Walled rock shelters and a cached spear in the Pilbara region, Western Australia. *Records of the Western Australian Museum* **10**(2), 111–126.

Bird DW, Bird RB, Parker CH (2004) Women who hunt with fire: Aboriginal resource use and fire regimes in Australia's Western Desert. *Australian Aboriginal Studies* **1**, 90–97.

Bird DW, Bird RB, Codding BF (2009) In pursuit of mobile prey: Martu hunting strategies and archaeofaunal interpretation. *American Antiquity* **74**(1), 3–29. doi:10.1017/S000273160004748X

Bird Rose D (1988) Jesus and the dingo. In *Aboriginal Australians and Christian Missions.* (Eds T Swain and D Bird Rose) pp. 361–375. Australian Association for the Study of Religions, Adelaide.

Bird Rose D, Swain T (1988) Introduction. In *Aboriginal Australians and Christian Missions.* (Eds T Swain and D Bird Rose) pp. 1–8. Australian Association for the Study of Religions, Adelaide.

Bird Rose D (1992) *Dingo Makes Us Human: Life and Land in an Aboriginal Australian Culture.* Cambridge University Press, Cambridge.

Rose D Bird (Ed.) (1995) *Country in Flames: Proceedings of the 1994 Symposium on Biodiversity and Fire in North Australia.* Biodiversity Unit, Department of the Environment, Sport and Territories; North Australia Research Unit, Canberra and Darwin.

Bird Rose D (1996) *Nourishing Terrains: Australian Aboriginal Views of Landscape and Wilderness.* Australian Heritage Commission, Canberra.

Bird Rose D, D'Amico S, Daiyi N, Deveraux K, Daiyi M, *et al.* (2002) *Country of the Heart: An Indigenous Australian Homeland.* Aboriginal Studies Press, Canberra.

Bird Rose D (2005) Rhythms, patterns, connectivities: Indigenous concepts of seasons and change. In *A Change in the Weather: Climate and Culture in Australia.* (Eds T Sherratt, T Griffiths and L Robin) pp. 32–41. National Museum of Australia Press, Canberra.

Bishop IM (Koormundum) (2000) *Ngun Koongurrukun: Speak Koongurrukun.* The author, Perth.

Black JM (1917) Vocabularies of three South Australian native languages: Wirrung, Narrinyeri and Wongaidya. *Transactions of the Royal Society of South Australia* **41**, 1–13.

Black A, McEntee J, Sutton P, Breen G (2018) The pre-European distribution of the galah, *Eolophus roseicapilla* Vieillot: reconciling scientific, historical and ethno-linguistic evidence. *South Australian Ornithologist* **42**(2), 37–57.

Blake BJ (1981) *Australian Aboriginal Languages: A General Introduction.* Angus & Robertson, Sydney.

Blake BJ (Ed.) (1998) *Wathawurrung and the Colac Language of Southern Victoria.* Pacific Linguistics Series C, No. 147. Australian National University, Canberra.

Blakers M, Davies SJJF, Reilly PN (1984) *The Atlas of Australian Birds.* Melbourne University Press, Melbourne.

Blandowski W (1858) Recent discoveries in natural history on the Lower Murray. *Transactions of the Philosophical Society of Victoria* **2**, 124–137.

Blows M (1975) Eaglehawk and crow: birds, myths and moieties in south-east Australia. In *Australian Aboriginal Mythology.* (Ed. LR Hiatt) pp. 24–45. Australian Institute of Aboriginal Studies, Canberra.

Blythe J, Wightman G (2003) The role of animals and plants in maintaining the links. *Proceedings of the 7th Foundation for Endangered Languages Conference.* Broome, Western Australia, 22–24 September 2003. (Eds J Blythe and RM Brown) pp. 69–77. Foundation for Endangered Languages, Bristol.

Boles WE (1992) Revision of *Dromaius gidju* Patterson and Rich, 1987 from Riversleigh, northwestern Queensland, Australia, with a reassessment of its generic position. *Los Angeles County Museum Science Series* **36**, 195–208.

Boles WE, Longmore NW (2014) A 'new' night parrot specimen? *Australian Field Ornithology* **31**, 141–149.

Bomford M, Caughley J (Eds) (1996) *Sustainable Use of Wildlife by Aboriginal Peoples and Torres Strait Islanders.* Australian Government Publishing Service, Canberra.

Bonta M (2010a) Ethno-ornithology and biological conservation. In *Ethno-ornithology: Birds, Indigenous Peoples, Culture and Society.* (Eds S Tidemann and A Gosler) pp. 13–29. Earthscan, London.

Bonta M (2010b) Transmutation of human knowledge about birds in 16th-century Honduras. In *Ethno-ornithology: Birds, Indigenous Peoples, Culture and Society.* (Eds S Tidemann and A Gosler), pp. 89–109. Earthscan, London.

Bonta M, Gosford R, Eussen D, Ferguson N, Loveless E, Witwer M (2017) Intentional fire-spreading by 'firehawk' raptors in northern Australia. *Journal of Ethnobiology* **37**(4), 700–718. doi:10.2993/0278-0771-37.4.700

Bonwick J (1870) *Daily Life and Origin of the Tasmanians.* Sampson, Low, Son & Marston, London.

Bonwick J (1884) *The Lost Tasmanian Race.* Sampson, Low, Marston, Searle & Rivington, London.

Bowyang B (1929) The native companion. *The Queenslander*, Queensland. 7 March, p. 63.

Bradley J (1988) *Yanyuwa Country: The Yanyuwa People of Borroloola Tell the History of Their Land.* Greenhouse Publications, Melbourne.

Bradley J, Holmes M, Norman D, Isaac A, Miller J, Ninganga I (2006) *Yumbulyumbulmantha ki-Awarawu: All Kinds of Things from Country – Yanyuwa Ethnobiological Classification.* Research Report Series 6. Aboriginal and Torres Strait Islander Studies Unit, University of Queensland, Brisbane.

Bradley J with Yanyuwa Families (2010) *Singing Saltwater Country: Journey to the Songlines of Carpentaria.* Allen & Unwin, Sydney.

Bradshaw CJA, Norman K, Ulm S, Williams AN, Clarkson C *et al.* (2021) Stochastic models support rapid peopling of Late Pleistocene Sahul. *Nature Communications* **12**(1), 1–11. doi:10.1038/s41467-021-21551-3

Brand Miller J, James KW, Maggiore P (1993) *Tables of Composition of Australian Aboriginal Foods.* Aboriginal Studies Press, Canberra.

Brown AR (1918) Notes on the social organisation of Australian tribes. *Journal of the Anthropological Institute of Great Britain and Ireland* **48**, 222–253.

Brown CH (1986) The growth of ethnobiological nomenclature. *Current Anthropology* **27**(1), 1–19. doi:10.1086/203375

Brown IB, Naessan PA (2014) *Birds from the Country: Tjurlpu Tjurta Ngurraritja. Olden Time Anthikirrinya Bird Knowledge.* Indigenous Language Support Report. Linguistics Discipline, University of Adelaide, Adelaide.

Browne HYL (1895) Government geologist's report on explorations in the Northern Territory. *South Australian Parliamentary Papers* **82**, 3–13.

Bruce R (1902) *Reminiscences of an Old Squatter.* WK Thomas, Adelaide.

Brusnahan M (1992) *Raukkan and Other Poems.* Magabala Books, Broome.

Buchler IR, Maddock K (Eds) (1978) *The Rainbow Serpent: A Chromatic Piece.* Mouton Publishers, The Hague.

Bukowick KA (2004) *Truth and Symbolism: Mythological Perspectives of the Wolf and Crow*. Boston College University Libraries, Boston. http://hdl.handle.net/2345/489

Bulmer J (1887) Some account of the Aborigines of the Lower Murray, Wimmera, Gippsland and Maneroo. *Transactions and Proceedings of the Royal Geographical Society of Australasia, Victorian Branch* **5**(1), 15–43.

Bulmer R (1967) Why is the cassowary not a bird? A problem of zoological taxonomy among the Karam of the New Guinea Highlands. *Man* **2**, 5–25. doi:10.2307/2798651

Bulmer R (1978) Totems and taxonomy. In *Australian Aboriginal Concepts*. (Ed. LR Hiatt) pp. 1–19. Australian Institute of Aboriginal Studies, Canberra.

Bulmer J (1855–1908) *John Bulmer's Recollections of Victorian Aboriginal Life, 1855–1908*. (Ed. R Vanderwal, 1999). Museum Victoria, Melbourne.

Burr T, Grey G (1845) Account of Governor G. Grey's exploratory journey along the south-eastern sea-board of South Australia. *Journal of the Royal Geographical Society of London* **15**, 160–184. doi:10.2307/1797905

Byard R (1988) Traditional medicine of Aboriginal Australia. *Canadian Medical Association Journal* **139**(8), 792–794.

Cahir F, Clark I, Clarke PA (Eds) (2018a) *The Indigenous Bio-cultural Knowledge of Southeastern Australia*. CSIRO Publishing, Melbourne.

Cahir F, Clark I, Clarke PA (2018b) Introduction. In *The Indigenous Bio-cultural Knowledge of Southeastern Australia*. (Eds F Cahir, I Clark and PA Clarke) pp. xv–xxiv. CSIRO Publishing, Melbourne.

Cahir F, Clark I, Clarke PA (2018c) Conclusion: the future of Aboriginal biocultural knowledge. In *The Indigenous Bio-cultural Knowledge of Southeastern Australia* (Eds F Cahir, I Clark and PA Clarke) pp. 281–284. CSIRO Publishing, Melbourne.

Callomon P (2016) The nature of names: Japanese vernacular nomenclature in natural science. MSc thesis. Department of Science, Technology and Society, Drexel University, Philadelphia.

Cameron ALP (1893) The Aborigines of New South Wales. *The Riverine Grazier*, New South Wales. 27 January, p. 4.

Cameron-Bonney L (1990) *Out of the Dreaming*. South East Kingston Leader, Kingston.

Campbell AJ (1893) Some Australian birds. *The Australasian*, Victoria. 2 December, p. 29.

Campbell TG (1926) Insect foods of the Aborigines. *Australian Museum Magazine* **2**(12), 407–410.

Cane S (1987) Australian Aboriginal subsistence in the Western Desert. *Human Ecology* **15**(4), 391–434. doi:10.1007/BF00887998

Cane S (2002) *Pila Nguru: The Spinifex People*. Fremantle Arts Centre Press, Fremantle.

Cataldi L (2004) *A Dictionary of Ngardi*. Ngardi Elders – Mungkirna Napaljarri Balgo, Western Australia.

Cawte J (1974) *Medicine is the Law*. University Press of Hawaii, Honolulu.

Cawte J (1996) *Healers of Arnhem Land*. UNSW Press, Sydney.

Cawthorne WA (1868) Natives killing parrots. *Illustrated Adelaide Post*, South Australia. 23 January, pp. 1, 5.

Cayley NW (1931) *What Bird is That?* Rev. TR Lindsay, signature edn, 2011. Australia's Heritage Publishing, Sydney.

Chaloupka G (1993) *Journey in Time: The World's Longest Continuing Art Tradition – the 50,000 Year Story of the Australian Aboriginal Rock Art of Arnhem Land*. Reed, Sydney.

Chamberlain AF (1902) Algonkian words in American English: a study in the contact of the white man and the Indian. *Journal of American Folklore* **15**(59), 240–267. doi:10.2307/533199

Chandler AC (1916) *A Study of the Structure of Feathers, with Reference to their Taxonomic Significance*. University of California Press, Berkeley.

Chapman A, Hadfield B, Douglas W, Gardner J, Hutchinson P, Rolland G (1995) Aboriginal names for some biota from the Great Victoria Desert, Western Australia. *CALMscience* **1**(3), 349–361.

Chase A, Sutton P (1981) Hunter-gatherers in a rich environment: Aboriginal coastal exploitation in Cape York Peninsula. In *Ecological Biogeography of Australia*. (Ed. A Keast) pp. 1819–1852. Junk, The Hague.

Chatwin B (1988) *The Songlines*. Picador, London.

Cherikoff V, Brand JC (1987) Is there a trend towards Indigenous foods in Australia? *Food Habits in Australia*. Rene Gordon, Melbourne.

Chewings C (1936) *Back in the Stone Age*. Angus & Robertson, Sydney.

Christidis L, Boles WE (2008) *The Systematics and Taxonomy of Australian Birds*. CSIRO Publishing, Melbourne.

Clark ID (2010) Colonial tourism in Victoria Australia, in the 1840s: George Augustus Robinson as a nascent tourist. *International Journal of Tourism Research* **12**(5), 561–573. doi:10.1002/jtr.775

Clarke PA (1986a) The study of ethnobotany in southern South Australia. *Australian Aboriginal Studies* **1986**(2), 40–47.

Clarke PA (1986b) Aboriginal use of plant exudates, foliage and fungi as food and water sources in southern South Australia. *Journal of the Anthropological Society of South Australia* **24**(3), 3–18.

Clarke PA (1987) Aboriginal uses of plants as medicines, narcotics and poisons in southern South Australia. *Journal of the Anthropological Society of South Australia* **25**(5), 3–23.

Clarke PA (1988) Aboriginal use of subterranean plant parts in southern South Australia. *Records of the South Australian Museum* **22**(1), 73–86.

Clarke PA (1989) Aboriginal non-plant medicines in southern South Australia and western Victoria. *Journal of the Anthropological Society of South Australia* **27**(5), 1–10.

Clarke PA (1990) Adelaide Aboriginal cosmology. *Journal of the Anthropological Society of South Australia* **28**, 1–10.

Clarke PA (1994) Contact conflict and regeneration. Aboriginal cultural geography of the Lower Murray, South Australia. PhD thesis. Anthropology Discipline and Department of Geography, University of Adelaide, Adelaide.

Clarke PA (1995) Myth as history: the Ngurunderi mythology of the Lower Murray, South Australia. *Records of the South Australian Museum* **28**(2), 143–157.

Clarke PA (1996a) Early European interaction with Aboriginal hunters and gatherers on Kangaroo Island, South Australia. *Aboriginal History* **20**(1), 51–81.

Clarke PA (1996b) Adelaide as an Aboriginal landscape. In *Terrible Hard Biscuits. A Reader in Aboriginal History*. (Eds V Chapman and P Read) pp. 69–93. Allen & Unwin, Sydney.

Clarke PA (1997) The Aboriginal cosmic landscape of southern South Australia. *Records of the South Australian Museum* **29**(2), 125–145.

Clarke PA (1999a) Spirit beings of the Lower Murray, South Australia. *Records of the South Australian Museum* **31**(2), 149–163.

Clarke PA (1999b) Waiyungari and his role in the mythology of the Lower Murray, South Australia. *Records of the South Australian Museum* **32**(1), 51–67.

Clarke PA (2000) *The Australian Aboriginal Cultures Gallery*. South Australian Museum, Adelaide.

Clarke PA (2001a) The significance of whales to the Aboriginal people of southern South Australia. *Records of the South Australian Museum* **34**(1), 19–35.

Clarke PA (2001b) The South Australian Museum, Adelaide. In *Aboriginal Art Collections. Highlights from Australia's Public Museums and Galleries*. (Ed. S Cochrane) pp. 108–113. Craftsman House, Sydney.

Clarke PA (2002) Early Aboriginal fishing technology in the Lower Murray, South Australia. *Records of the South Australian Museum* **35**(2), 147–167.

Clarke PA (2003a) *Where the Ancestors Walked. Australia as an Aboriginal Landscape*. Allen & Unwin, Sydney.

Clarke PA (2003b) Twentieth-century Aboriginal harvesting practices in the rural landscape of the Lower Murray, South Australia. *Records of the South Australian Museum* **36**(1), 83–107.

Clarke PA (2003c) Australian ethnobotany: an overview. *Australian Aboriginal Studies* **2003**(2), 21–38.

Clarke PA (2003d) Australian Aboriginal mythology. In *Mythology. Myths, Legends, and Fantasies.* (Eds J Parker and J Stanton) pp. 382–401. Global Book Publishing, Sydney.

Clarke PA (2007a) *Aboriginal People and Their Plants.* Rosenberg Publishing, Dural Delivery Centre, New South Wales.

Clarke PA (2007b) Indigenous spirit and ghost folklore of 'settled' Australia. *Folklore* **118**(2), 141–161. doi:10.1080/00155870701337346

Clarke PA (2008a) *Aboriginal Plant Collectors. Botanists and Aboriginal People in the Nineteenth Century.* Rosenberg Publishing, Dural Delivery Centre, New South Wales.

Clarke PA (2008b) Aboriginal healing practices and Australian bush medicine. *Journal of the Anthropological Society of South Australia* **33**, 3–38.

Clarke PA (2009a) Aboriginal culture and the Riverine environment. In *The Natural History of the Riverland and Murraylands.* (Ed. JT Jennings) pp. 142–161. Royal Society of South Australia, Adelaide.

Clarke PA (2009b) An overview of Australian Aboriginal ethnoastronomy. *Archaeoastronomy. The Journal of Astronomy in Culture* **21**, 39–58.

Clarke PA (2009c) Australian Aboriginal ethnometeorology and seasonal calendars. *History and Anthropology* **20**(2), 79–106. doi:10.1080/02757200902867677

Clarke PA (2012) *Australian Plants as Aboriginal Tools.* Rosenberg Publishing, Dural Delivery Centre, New South Wales.

Clarke PA (2013a) The Aboriginal ethnobotany of the Adelaide region, South Australia. *Transactions of the Royal Society of South Australia* **137**(1), 97–126. doi:10.1080/3721426.2013.10887175

Clarke PA (2013b) The use and abuse of Aboriginal ecological knowledge. In *The Aboriginal Story of Burke and Wills: Forgotten Narratives.* (Eds I Clark and F Cahir) pp. 61–79. CSIRO Publishing, Melbourne.

Clarke PA (2014a) *Discovering Aboriginal Plant Use. Journeys of an Australian Anthropologist.* Rosenberg Publishing, Dural Delivery Centre, New South Wales.

Clarke PA (2014b) Australian Aboriginal astronomy and cosmology. In *Handbook of Archaeoastronomy and Ethnoastronomy.* (Ed. CLN Ruggles) pp. 2223–2230. Springer, New York.

Clarke PA (2014c) The Ethnobotany of the Skyworld, Part 1: The Flora and the Aesthetics of the Heavens in Aboriginal Australia. *Journal of Astronomical History and Heritage* **17**(3), 307–325.

Clarke PA (2014d) Review of 'Science and Sustainability. Learning from Indigenous Wisdom' by Joy Hendry. *Times Higher Education*, London. 27 November, p. 48.

Clarke PA (2015a) The Aboriginal Australian cosmic landscape. Part 2: Plant connections with the Skyworld. *Journal of Astronomical History and Heritage* **18**(1), 23–37.

Clarke PA (2015b) The Aboriginal ethnobotany of the South East of South Australia region. Part 1: seasonal life and material culture. *Transactions of the Royal Society of South Australia* **139**(2), 216–246 doi:10.1080/03721426.2015.1073415.

Clarke PA (2015c) The Aboriginal ethnobotany of the South East of South Australia region. Part 2: foods, medicines and narcotics. *Transactions of the Royal Society of South Australia* **139**(2), 247–272 doi:10.1080/03721426.2015.1074339.

Clarke PA (2015d) The Aboriginal ethnobotany of the South East of South Australia region. Part 3: mythology and language. *Transactions of the Royal Society of South Australia* **139**(2), 273–305 doi:10.1080/03721426.2015.1074340.

Clarke PA (2016a) Birds as totemic beings and creators in the Lower Murray, South Australia. *Journal of Ethnobiology* **36**(2), 277–293. doi:10.2993/0278-0771-36.2.277

Clarke PA (2016b) Birds and the spirit world of the Lower Murray, South Australia. *Journal of Ethnobiology* **36**(4), 746–764. doi:10.2993/0278-0771-36.4.746

Clarke PA (2017) Early Indigenous practices of bird foraging in the Lower Murray, South Australia. *Transactions of the Royal Society of South Australia* **141**(1), 26–47 doi:10.1080/03721426.2016.12 66571.

Clarke PA (2018a) Totemic life. In *The Indigenous Bio-cultural Knowledge of Southeastern Australia.* (Eds F Cahir, I Clark and PA Clarke) pp. 1–18. CSIRO Publishing, Melbourne.

Clarke PA (2018b) Terrestrial spirit beings. In *The Indigenous Bio-cultural Knowledge of Southeastern Australia.* (Eds F Cahir, I Clark and PA Clarke) pp. 19–34. CSIRO Publishing, Melbourne.

Clarke PA (2018c) Water spirits. In *The Indigenous Bio-cultural Knowledge of Southeastern Australia.* (Eds F Cahir, I Clark and PA Clarke) pp. 35–53. CSIRO Publishing, Melbourne.

Clarke PA (2018d) Plant food. In *The Indigenous Bio-cultural Knowledge of Southeastern Australia.* (Eds F Cahir, I Clark and PA Clarke) pp. 55–71. CSIRO Publishing, Melbourne.

Clarke PA (2018e) Animal food. In *The Indigenous Bio-cultural Knowledge of Southeastern Australia.* (Eds F Cahir, I Clark and PA Clarke) pp. 73–93. CSIRO Publishing, Melbourne.

Clarke PA (2018f) Space. In *The Indigenous Bio-cultural Knowledge of Southeastern Australia.* (Eds F Cahir, I Clark and PA Clarke) pp. 247–263. CSIRO Publishing, Melbourne.

Clarke PA (2018g) Time. In *The Indigenous Bio-cultural Knowledge of Southeastern Australia.* (Eds F Cahir, I Clark and PA Clarke) pp. 265–280. CSIRO Publishing, Melbourne.

Clarke PA (2018h) Aboriginal foraging practices and crafts involving birds in the post-European period of the Lower Murray, South Australia. *Transactions of the Royal Society of South Australia* **142**(1), 1–26 doi:10.1080/03721426.2017.1415588.

Clarke PA (2018i) A review of early Indigenous artefacts incorporating bird materials in the Lower Murray River region, South Australia. *Transactions of the Royal Society of South Australia* **142**(1), 27–48 doi:10.1080/03721426.2018.1424505.

Clarke PA (2018j) Australites. Part 1: Aboriginal involvement in their discovery. *Journal of Astronomical History and Heritage* **21**(2–3), 115–133.

Clarke PA (2019a) The Ngarrindjeri nomenclature of birds in the Lower Murray River region, South Australia. *Transactions of the Royal Society of South Australia* **143**(1), 118–146 doi:10.1080/037214 26.2018.1534530.

Clarke PA (2019b) Australites. Part 2: Early Aboriginal perception and use. *Journal of Astronomical History and Heritage* **22**(1), 155–178.

Clarkson C, Jacobs Z, Marwick B, Fullagar R, Wallis L *et al.* (2017) Human occupation of northern Australia by 65,000 years ago. *Nature* **547**(7663), 306–310. doi:10.1038/nature22968

Cleland JB (1940) Some aspects of the ecology of the Aboriginal inhabitants of Tasmania and southern Australia. *Papers and Proceedings of the Royal Society of Tasmania* 1–18.

Cleland JB (1953) The healing art in primitive society. *Mankind* **4**(10), 395–411.

Cleland JB (1957) Ethno-ecology: our natives and the vegetation of southern Australia. *Mankind* **5**(4), 149–162.

Cleland JB (1966) The ecology of the Aboriginal in south and central Australia. In *Aboriginal Man in South and Central Australia. Part 1* (Ed. BC Cotton) pp. 111–158. Government Printer, Adelaide.

Cleland JB, Johnston TH (1933) The ecology of the Aborigines of Central Australia; botanical notes. *Transactions of the Royal Society of South Australia* **57**, 113–124.

Cleland JB, Johnston TH (1937–38) Notes on native names and uses of plants in the Musgrave Ranges region. *Oceania* **8**(2), 208–215; **8**(3), 328–342.

Cleland JB, Tindale NB (1954) The ecological surroundings of the Ngalia natives in Central Australia and native names and uses of plants. *Transactions of the Royal Society of South Australia* **77**, 81–86.

Clements FE (1932) *Primitive Concepts of Disease.* University of California Press, Berkeley.

Cocker M, Tipling D (2013) *Birds and People.* Jonathan Cape, London.

Collins D (1798–1802) *An Account of the English Colony in New South Wales: With Remarks on the Dispositions Customs, Manners, &c. of the Native Inhabitants of that Country.* Cadell & W Davies, London. Republished in 1975, AH & AW Reed, Sydney.

Condon HT (1941) The stone plover (*Burhinus magnirostris*). *South Australian Naturalist* **21**(1), 6–7.

Condon HT (1955a) Aboriginal bird names: South Australia. Part 1. *South Australian Ornithologist* **21**(6–7), 74–88.

Condon HT (1955b) Aboriginal bird names: South Australia. Part 2. *South Australian Ornithologist* **21**(8), 91–98.

Cooper HM (1948) Examples of native material culture from South Australia. *South Australian Naturalist* **25**(1), 1–8.

Cooper HM (1952) The burning of the crows, or why the crows' feathers are now black. *South Australian Ornithologist* **20**(7), 80.

Cooper HM (1962) *Australian Aboriginal Words. 3000 Examples and Their Meaning*, 4th edn. South Australian Museum, Adelaide.

Cooper HM, Condon HT (1947) On some fragments of emu egg-shell from an ancient camp-site on Kangaroo Island. *South Australian Ornithologist* **18**(7), 66–68.

Cosgrove R, Allen J, Marshall B (1998) Palaeoecology and Pleistocene occupation in south central Tasmania. In *Archaeology of Aboriginal Australia: A Reader*. (Ed. T Murray) pp. 235–256. Allen & Unwin, Sydney.

Curr EM (1886–87) *The Australian Race: Its Origins, Languages, Customs, Place of Landing in Australia and the Routes by which it Spread Itself over the Continent*. 4 volumes. Trubner, London.

Curran G, Barwick L, Turpin M, Walsh F, Laughren M (2019) Central Australian Aboriginal songs and biocultural knowledge: evidence from women's ceremonies relating to edible seeds. *Journal of Ethnobiology* **39**(3), 354–370. doi:10.2993/0278-0771-39.3.354

Daly DD (1887) *Digging, Squatting and Pioneering Life in the Northern Territory of South Australia*. Sampson Low, Marston, Searle & Rivington, London.

Davidson J (1898) Language of the Pinejunga people. In *Corartwalla: A History of Penola, the Land and its People*. (Ed. C Hanna, 2000) pp. 328–333. Magill Publications, South Australia.

Davidson DS (1933) Australian netting and basketry techniques. *Journal of the Polynesian Society (N. Z.)* **42**, 257–299.

Davidson DS (1934) Australian spear-traits and their derivations. *Journal of the Polynesian Society (N. Z.)* **43**, 41–72, 143–162.

Davidson DS (1935a) The chronology of Australian watercraft. *Journal of the Polynesian Society (N. Z.)* **44**, 1–16, 69–84, 137–152, 193–207.

Davidson DS (1935b) Knotless netting in America and Oceania. *American Anthropologist* **37**, 117–134. doi:10.1525/aa.1935.37.1.02a00110

Davidson DS (1936) Australian throwing-sticks, throwing clubs and boomerangs. *American Anthropologist* **38**, 76–100. doi:10.1525/aa.1936.38.1.02a00080

Davidson DS (1937) A preliminary consideration of Aboriginal Australian decorative art. *Memoirs of the American Philosophical Society* **9**, 1–147.

Davidson DS (1941) Aboriginal Australian string figures. *Proceedings of the American Philosophical Society* **84**(6), 763–901.

Davidson DS (1947a) Footwear of the Australian Aboriginal: environmental vs cultural determination. *Southwestern Journal of Anthropology* **3**, 114–123. doi:10.1086/soutjanth.3.2.3628727

Davidson DS (1947b) Fire-making in Australia. *American Anthropologist* **49**, 426–437. doi:10.1525/aa.1947.49.3.02a00040

Davies E (1881) *The Story of an Earnest Life. Adventures in Australia and in Two Voyages Around the World*. Central Book, Cincinnati.

Davies J (1998) Who owns the animals? Sustainable commercial use of wildlife and Indigenous rights in Australia. In *Crossing Borders: Seventh Annual Conference of the International Association for the Study of Common Property*. Vancouver, British Columbia, 10–14 June 1998. Environmental Science and Management Department, University of Adelaide, Adelaide.

Davies SM (2002) *Collected: 150 Years of Aboriginal Art and Artifacts at the Macleay Museum*. Macleay Museum, University of Sydney, Sydney.

Davies J, Hill R, Walsh FJ, Sandford M, Smyth D, Holmes MC (2013) Innovation in management plans for community conserved areas: experiences from Australian Indigenous protected areas. *Ecology and Society* **18**(2), 14. doi:10.5751/ES-05404-180214

Davies J, Walker J, Maru YT (2018) Warlpiri experiences highlight challenges and opportunities for gender equity in Indigenous conservation management in arid Australia. *Journal of Arid Environments* **149**, 40–52. doi:10.1016/j.jaridenv.2017.10.002

Davis S (1989) *Man of All Seasons. An Aboriginal Perspective of the Natural Environment.* Angus & Robertson, Sydney.

Dawson J (1881) *Australian Aborigines.* Robertson, Melbourne.

De Guchteneire P, Krukkert I, von Liebenstein G (Eds) (2002) *Best Practices on Indigenous Knowledge.* NUFFIC, Amsterdam.

DETE [Department of Education, Training and Employment] (1998) *Aboriginal Artists in South Australia.* Hyde Park Press, Adelaide.

Devanesen D (1985) *Traditional Aboriginal Medicine and Bicultural Approach to Healthcare in Australia's Northern Territory.* In *Alcohol and Drug Use in a Changing Society.* (Eds KP Larkins, D McDonald and C Watson). Alcohol and Drug Foundation, Canberra.

Devanesen D (2000) Traditional Aboriginal medicine practice in the Northern Territory. In *Proceedings of the International Symposium on Traditional Medicine: Better Science, Policy and Services for Health Development.* 11–13 September 2000. Centre for Health Development, World Health Organization, Kobe.

Diamond JM (1991) Interview techniques in ethnobiology. In *Man and a Half: Essays in Pacific Anthropology and Ethnobiology in Honour of Ralph Bulmer.* (Ed. A Pawley) pp. 83–86. Memoirs of the Polynesian Society, No. 48. Polynesian Society, Auckland.

Dixon RMW (1972) *The Dyirbal Language of North Queensland.* Cambridge University Press, Cambridge.

Dixon RMW (1991) *Words of Our Country: Stories, Place Names and Vocabulary in Yidiny, the Aboriginal Language of the Cairns–Yarrabah Region.* University of Queensland Press, Brisbane.

Dixon RMW, Ramson WS, Thomas M (1992) *Australian Aboriginal Words in English: Their Origin and Meaning.* Oxford University Press, Melbourne.

Doonday B, Samuels C, Clancy M, Milner J, Chungulla R *et al.* (2013) *Walmajarri Plants and Animals: Aboriginal Biocultural Knowledge from the Paruku Indigenous Protected Area, Southern Kimberley Australia.* Paruku IPA, Mulan Aboriginal Community, Western Australia.

Doran EW (1903) The vernacular names of birds. *The Auk* **20**(1), 38–42. doi:10.2307/4070096

Douglas MT (1982) *In the Active Voice.* Routledge & Kegan Paul, London.

Dousset L (1997) Naming and personal names of Ngaatjatjarra-speaking people, Western Desert: some questions related to research. *Australian Aboriginal Studies* **2**, 50–55.

Dove CJ, Koch SL (2011) Microscopy of feathers: a practical guide for forensic feather identification. *Microscope* **59**(2), 51–71.

Dow DD (1975) Display of the honeyeater, *Manorina melanocephala. Ethology* **38**(1), 70–96.

Duncan-Kemp AM (1933) *Our Sandhill Country.* Angus & Robertson, Sydney.

Duranti A (Ed.) (2001) *Linguistic Anthropology: A Reader.* Blackwell Publishers, Malden, MA.

Durkheim E (1915) *The Elementary Forms of the Religious Life.* Translated from the French by JW Swain, 1964. George Allen & Unwin, London.

Dwyer PD (2005) Ethnoclassification, ethnoecology and the imagination. *Journal de la Société des Océanistes* **120**, 11–25. doi:10.4000/jso.321

Eastwell HD (1973a) Co-operating with the medicine man. *Health.* First quarter, 12–14.

Eastwell HD (1973b) The traditional healer in modern Arnhem Land. *Medical Journal of Australia* **2**, 1011–1017. doi:10.5694/j.1326-5377.1973.tb129909.x

ECS (1935) Aboriginal anaesthetic. *Queenslander*, Queensland. 25 April, p. 2.

Edwards R (1966) Australites used for Aboriginal implements in South Australia. *Records of the South Australian Museum* **15**, 243–251.

Edwards R (1972) *Aboriginal Bark Canoes of the Murray Valley.* Rigby, Adelaide.

Eickelkamp U (2014) Specters of reality: Mamu in the eastern Western Desert of Australia. In *Monster Anthropology in Australasia and Beyond.* (Eds Y Musharbash and GH Presterudstuen) pp. 57–73. Springer, The Hague.

Eisenmann E, Poor HH (1946) Suggested principles for vernacular nomenclature. *Wilson Bulletin* **58**, 210–215.

EKV (1884) Colonial fragments: overland from Victoria to Queensland, 1860. *Queenslander*, Queensland. 1 November, p. 709.

Elkin AP (1931) The social organization of South Australian tribes. *Oceania* **2**(1), 44–73. doi:10.1002/j.1834-4461.1931.tb00022.x

Elkin AP (1933) Studies in Australian totemism: the nature of Australian totemism. *Oceania* **4**(2), 113–131. doi:10.1002/j.1834-4461.1933.tb00096.x

Elkin AP (1934) Cult-totemism and mythology in northern South Australia. *Oceania* **5**(2), 171–192. doi:10.1002/j.1834-4461.1934.tb00139.x

Elkin AP (1964) *The Australian Aborigines: How to Understand Them*, 4th edn. Angus & Robertson, Sydney.

Elkin AP (1977) *Aboriginal Men of High Degree*, 2nd edn. University of Queensland Press, Brisbane.

Elkin AP, Berndt CH, Berndt RM (1950) *Art in Arnhem Land*. FW Cheshire, Melbourne.

Endacott SJ (1944) *Australian Aboriginal Native Words and Their Meanings*. National Handbook No. 20. Robertson & Mullens, Melbourne.

Ens EJ, Finlayson M, Preuss K, Jackson S, Holcombe S (2012) Australian approaches for managing 'Country' using Indigenous and non-Indigenous knowledge. *Ecological Management & Restoration* **13**(1), 100–107. doi:10.1111/j.1442-8903.2011.00634.x

Ens EJ, Pert P, Clarke PA, Budden M, Clubb L *et al.* (2015) Indigenous biocultural knowledge in ecosystem science and management: review and insight from Australia. *Biological Conservation* **181**, 133–149. doi:10.1016/j.biocon.2014.11.008

Ens EJ, Scott ML, Yugul Mangi Rangers, Moritz C, Pirzl R (2016) Putting indigenous conservation policy into practice delivers biodiversity and cultural benefits. *Biodiversity and Conservation* **25**(14), 2889–2906. doi:10.1007/s10531-016-1207-6

Eureka (1938) So they say: a famous gathering. *Queenslander*, Queensland. 16 March, p. 2.

Evans N (1992) *Kayardild Dictionary and Thesaurus*. Department of Linguistics and Language Studies, University of Melbourne, Melbourne.

Everist SL (1981) *Poisonous Plants of Australia*, rev. edn. Angus & Robertson, Sydney.

Eyre EJ (1845) *Journals of Expeditions of Discovery into Central Australia and Overland from Adelaide to King George's Sound in the Years 1840–1 … Including an Account of the Manners and Customs of the Aborigines and the State of Their Relations with Europeans*. 2 volumes. T & W Boone, London.

Faast R, Clarke PA, Taylor GS, Salagaras RL, Weinstein P (2020) Indigenous use of lerps in Australia: so much more than a sweet treat. *Journal of Ethnobiology* **40**(3), 328–347; Supplement, 1–19. doi:10.2993/0278-0771-40.3.328

Fabian (1931) The bushlover. *Brisbane Courier*, Queensland. 9 May, p. 19.

Farb P, Armelagos G (1980) *Consuming Passions: The Anthropology of Eating*. Houghton Mifflin, Boston.

Fenner C (1946) *Gathered Moss*. Georgian House, Melbourne.

Fernández-Llamazares Á, Lepofsky D, Lertzman K, Armstrong CG, Brondizio ES *et al.* (2021) Scientists' warning to humanity on threats to Indigenous and local knowledge systems. *Journal of Ethnobiology* **41**(2), 144–169. doi:10.2993/0278-0771-41.2.144

Fields CV (1951) The bunyip. *Western Mail*, Western Australia. 9 August, p. 13.

Fisher L (2012) The art/ethnography binary: post-colonial tensions within the field of Australian Aboriginal art. *Cultural Sociology* **6**(2), 251–270. doi:10.1177/1749975512440224

Fison L, Howitt AW (1880) *Kamilaroi and Kurnai: Group-marriage and Relationship and Marriage by Elopement. Drawn Chiefly from the Usage of the Australian Aborigines. Also the Kurnai Tribe. Their Customs in Peace and War*. George Robertson, Melbourne.

Flannery T (1994) *The Future Eaters: An Ecological History of the Australasian Lands and People*. Reed Books, Melbourne.

Fleay D (1940) Brown bittern our real 'bunyip'. *The Australasian*, Victoria. 18 May, p. 36.

Fleck DW (2007) Field linguistics meets biology: how to obtain scientific designations for plant and animal names. *STUF-Sprachtypologie und Universalienforschung* **60**(1), 81–91. doi:10.1524/stuf.2007.60.1.81

Fletcher TG (2007) *Thanakupi's Guide to Language and Culture: A Thaynakwith Dictionary*. Jennifer Isaacs Arts & Publishing, Sydney.

Flood J (1983) *Archaeology of the Dreamtime: The Story of Prehistoric Australia and Her People*. Collins, Sydney.

Foley WA (1997) *Anthropological Linguistics: An Introduction*. Blackwell Publishers, Cambridge, MA.

Folkmanova V (2015) The oil of the dugong: towards a history of an Indigenous medicine. *History Australia* **12**(3), 97–112. doi:10.1080/14490854.2015.11668588

Forbes C, Owen T, Veale S (2019) Risk and resilience: Baiame's Cave and creation landscape, NSW Australia. In *8th ICBR Lisbon Book of Papers*. (Eds AN Martins, L Hobeica, A Hobeica, PP Santos, N Eltinay and JM Mendes) pp. 219–228. ICBR, Lisbon.

Forth G (2010a) What's in a bird's name: relationships among ethno-ornithological terms in Nage and other Malayo-Polynesian languages. In *Ethno-ornithology. Birds, Indigenous Peoples, Culture and Society*. (Eds S Tidemann and A Gosler) pp. 223–237. Earthscan, London.

Forth G (2010b) Symbolic birds and ironic bats: varieties of classification in Nage folk ornithology. *Ethnology* **48**(2), 139–159.

Foster GM (1976) Disease etiologies in non-western medical systems. *American Anthropologist* **78**(4), 773–782. doi:10.1525/aa.1976.78.4.02a00030

Frazer JG (1890) *The Golden Bough: A Study in Comparative Religion*. Republished in 1933. Gramercy Books, New Jersey.

Frazer JG (1910) *Totemism and Exogamy. A Treatise on Certain Early Forms of Superstition and Society*. Macmillan, London.

Fred S (1893) Some southern beasts: under which head are included the Aborigines of this country. *Adelaide Observer*, South Australia. 2 December, p. 33.

Froggatt W (1932) Mind of an emu. *Sydney Morning Herald*, New South Wales. 27 December, p. 3.

Fruits of the Earth (1999) Relieve aches and pains with the new medicated range. *Riverine Herald*, Victoria. 12 April, p. 9.

Fry HK (1937) Dieri legends. *Folklore* **48**(2), 187–206, 269–287. doi:10.1080/0015587X.1937.9718686

Fry HK (1950) Aboriginal social systems. *Transactions of the Royal Society of South Australia* **73**(2), 282–294.

Fuller RS, Anderson MG, Norris RP, Trudgett M (2014) The emu sky knowledge of the Kamilaroi and Euahlayi peoples. *Journal of Astronomical History and Heritage* **17**(2), 171–179.

Gale M (2000) Poor bugger whitefella got no dreaming: the representation and appropriation of published Dreaming narratives with special reference to David Unaipon's writings. PhD thesis. European Studies, University of Adelaide, Adelaide.

Gale M (2009) *Ngarrindjeri Dictionary*. Raukkan Community Council, Raukkan, South Australia and Department of Environment, Water, Heritage and the Arts Australian Government, Canberra.

Garvey J, Cochrane B, Field J, Boney C (2011) Modern emu (*Dromaius novaehollandiae*) butchery, economic utility and analogues for the Australian archaeological record. *Environmental Archaeology* **16**(2), 97–112. doi:10.1179/174963111X13110803260840

Gason S (1879) The manners and customs of the Dieyeri tribe of Australian Aborigines. In *The Native Tribes of South Australia*. (Ed. JD Woods) pp. 253–307. ES Wiggs, Adelaide.

Giacon J (2010) Etymology of Yuwaalaraay Gamilaraay bird names. In *Lexical and Structural Etymology: Beyond Word Histories*. (Ed. R Mailhammer) pp. 251–291. De Gruyter, Berlin.

Giles E (1875) *Geographic Travels in Central Australia from 1872 to 1874*. M'Carron Bird & Co., Melbourne.

Giles T (1887) Old time memories: the blacks in the south. *South Australian Register*, South Australia. 5 October, p. 3.

Gill AM, Woinarski JCZ, York A (1999) *Australia's Biodiversity – Responses to Fire: Plants Birds and Invertebrates*. Biodiversity Technical Paper. Environment Australia, Canberra.

Gillen FJ (1896) Notes on some manners and customs of the Aborigines of the McDonnell Ranges belonging to the Arunta tribe. In *Report on the Work of the Horn Expedition to Central Australia*, vol. 4. (Ed. WB Spencer) pp. 161–186. Dulau & Co., London.

Glass A, Hackett D (2003) *Ngaanyatjarra and Ngaatjatjarra to English Dictionary*. IAD Press, Alice Springs.

Goddard C (1992) *Pitjantjatjara/Yankunytjatjara to English Dictionary*. IAD Press, Alice Springs.

Goddard C, Kalotas A (1988) *Punu: Yankunytjatjara Plant Use*. Angus & Robertson, Sydney.

Godden L, Cowell S (2016) Conservation planning and Indigenous governance in Australia's Indigenous Protected Areas. *Restoration Ecology* **24**(5), 692–697. doi:10.1111/rec.12394

Gonzalez JCT (2011) Enumerating the ethno-ornithological importance of Philippine hornbills. *Raffles Bulletin of Zoology* **24**, 149–161.

Gordon T (1979) *Milbi: Aboriginal Stories from Queensland's Endeavour River*. Republished in 2021. Aboriginal Studies Press, Canberra.

Gorman JT, Whitehead PJ, Griffiths AD, Petheram L (2008) Production from marginal lands: Indigenous commercial use of wild animals in northern Australia. *International Journal of Sustainable Development and World Ecology* **15**(3), 240–250. doi:10.3843/SusDev.15.3:7

Gosford R (2003) *Towards an (Australian) Indigenous Ornithology: Is Australia an Ornithological Terra Nullius?* Paper presented at Australasian Ornithological Congress. December 2003, Australian National University, Canberra.

Gosford R (2009) Luurnpa: the red-backed kingfisher and Aboriginal knowledge and significance of birds in the central Australian deserts. *Northern Myth*. 4 January https://blogs.crikey.com.au/northern/2009/01/04/bird-of-the-week-red-backed-kingfisher/

Gosford R (2010) The bush stone curlew as a harbinger of death … and more. *Northern Myth*. 27 September. https://blogs.crikey.com.au/northern/2010/09/27/bird-of-the-week-the-bush-stone-curlew-as-a-harbinger-of-death-and-more/

Gosford R (2016) 'Troublemakers for fire': raptors spreading fire in Australian savanna woodlands. *Northern Myth*. 1 October. https://blogs.crikey.com.au/northern/2016/10/01/troublemakers-fireraptors-spreading-fire-australian-savannawoodlands/

Gott B (2005) Aboriginal fire management in south-eastern Australia: aims and frequency. *Journal of Biogeography* **32**, 1203–1208. doi:10.1111/j.1365-2699.2004.01233.x

Gott B (2008) Indigenous use of plants in south-eastern Australia. *Telopea* **12**(2), 215–226. doi:10.7751/telopea20085811

Gould RA (1969a) *Yiwara: Foragers of the Australian Desert*. Collins, London.

Gould RA (1969b) Subsistence behaviour among the Western Desert Aborigines of Australia. *Oceania* **39**(4), 253–274. doi:10.1002/j.1834-4461.1969.tb01026.x

Gould RA (1971) Uses and effects of fire among the Western Desert Aborigines of Australia. *Australian Journal of Anthropology* **8**, 14–24. doi:10.1111/j.1835-9310.1971.tb01436.x

Grant C (2012) Analogies and links between cultural and biological diversity. *Journal of Cultural Heritage Management and Sustainable Development* **2**(2), 153–163. doi:10.1108/20441261211273644

Grassie J (1876) Notes by the way: mustering 'the big paddock'. *Border Watch*, South Australia. 9 December, p. 4.

Gray J, Fraser I (2013) *Australian Bird Names: A Complete Guide*. CSIRO Publishing, Melbourne.

Green J (2010) *Central and Eastern Anmatyerr to English Dictionary*. IAD Press, Alice Springs.

Green D, Jackson S, Morrison J (Eds) (2009) *Risks from Climate Change to Indigenous Communities in the Tropical North of Australia*. Department of Climate Change and Energy Efficiency, Canberra.

Green R, Green J, Hamilton-Hollaway A, Meakins F, Osgarby D *et al.* (2019) *Mudburra to English Dictionary*. Aboriginal Studies Press, Canberra.

Gregory JW (1906) *The Dead Heart of Australia. A Journey Around Lake Eyre in the Summer of 1901–1902, with Some Account of the Lake Eyre Basin and the Flowing Wells of Central Australia*. John Murray, London.

Grey G (1841) *Journals of Two Expeditions of Discovery in North-west and Western Australia During the Years 1837, 38 and 39, Under the Authority of Her Majesty's Government Describing Many Newly Discovered, Important and Fertile Districts, with Observations on the Moral and Physical Condition of the Aboriginal Inhabitants, &c. &c*. 2 volumes. Boone, London.

Griffin T, McCaskill M (Eds) (1986) *Atlas of South Australia*. Government Printer, Adelaide.

Griffiths T (1996) *Hunters and Collectors: The Antiquarian Imagination in Australia*. Cambridge University Press, Cambridge.

Gunn RC (1847) On the bunyip of Australia Felix. *Tasmanian Journal of Natural Science* **3**(2), 147–149.

Gunn RG, Douglas LC, Whear RL (2011) What bird is that? Identifying a probable painting of *Genyornis newtoni* in western Arnhem Land. *Australian Archaeology* **73**, 1–12. doi:10.1080/031224 17.2011.11961918

Haagen C (1994) *Bush Toys: Aboriginal Children at Play*. Aboriginal Studies Press, Canberra.

Hagger J (1979) *Australian Colonial Medicine*. Rigby, Adelaide.

Hahn DM (1838–39) Extracts from the 'Reminiscences of Captain Dirk Meinertz Hahn, 1838–1839.' Translated by FJH Blaess and LA Triebel, 1964. *South Australiana* **3**(2), 97–134.

Hale HM, Tindale NB (1934) Aborigines of Princess Charlotte Bay, North Queensland. Parts I–II. *Records of the South Australian Museum* **5**(1), 63–172.

Hall FJ, McGowan RG, Guleksen GF (1951) Aboriginal rock carvings: a locality near Pimba SA. *Records of the South Australian Museum* **9**, 375–380.

Hallam SJ (1975) *Fire and Hearth: A Study of Aboriginal Usage and European Usurpation in South-western Australia*. Australian Institute of Aboriginal Studies, Canberra.

Hamacher DW (2012) On the astronomical knowledge and traditions of Aboriginal Australians. PhD thesis. Department of Indigenous Studies, Macquarie University, Sydney.

Hamacher DW (2015) Identifying seasonal stars in Kaurna astronomical traditions. *Journal of Astronomical History and Heritage* **18**(1), 39–52.

Hamacher DW, Norris R (2011) Eclipses in Australian Aboriginal astronomy. *Journal of Astronomical History and Heritage* **14**(2), 103–114.

Hamacher DW, Nunn PD, Gantevoort M, Taylor R, Lehman G *et al.* (2021 in press) Dating Aboriginal Tasmanian oral traditions to the Late-Pleistocene. Preprint. https://www.academia.edu/50962664/Dating_Aboriginal_Tasmanian_oral_traditions_to_the_Late_Pleistocene?from_navbar=true

Hamby L (Ed.) (2005) *Twined Together: Kunmadj Njalehnjaleken*. Injalak Arts & Crafts, Gunbalanya, Northern Territory.

Hamby L, Young D (2001) *Art on a String: Aboriginal Threaded Objects from the Central Desert and Arnhem Land*. Australian Centre for Craft & Design and the Centre for Cross Cultural Research, Canberra.

Hamlyn-Harris R, Smith F (1916) On fish poisoning and poisons employed among the Aborigines of Queensland. *Memoirs of the Queensland Museum* **5**, 1–9.

Hancock P (2014) Ancient tales of Perth's fascinating birds. *Sydney Morning Herald*, New South Wales. 22 February. http://www.smh.com.au/entertainment/about-town/wrens-splendid-theft-sparks-blue-with-eagle-20140222-3384z.html

Hanspach J, Haider LJ, Oteros-Rozas E, Olafsson AS, Gulsrud NM *et al.* (2020) Biocultural approaches to sustainability: a systematic review of the scientific literature. *People and Nature* **2**, 643–659. doi:10.1002/pan3.10120

Hardwick J (2019a) *Wadeye Kardu Murntak Warra: Old People Before Weapons – Traditional Knowledge, Language and Skills of the Aboriginal People from the Wadeye Region NT Australia*. Book 1. The author, Batchelor, Northern Territory.

Hardwick J (2019b) *Wadeye Kardu Murntak Warra: Old People Before Hunting and Fishing – Traditional Knowledge, Language and Skills of the Aboriginal People from the Wadeye Region NT Australia*. Book 2. The author, Batchelor, Northern Territory.

Hardwick J (2019c) *Wadeye Kardu Murntak Warra: Old People Before Plant Food Collecting and Processing – Traditional Knowledge, Language and Skills of the Aboriginal People from the Wadeye Region NT Australia*. Book 3. The author, Batchelor, Northern Territory.

Hardwick J (2019d) *Wadeye Kardu Murntak Warra: Old People Before Bark and Fibres – Traditional Knowledge, Language and Skills of the Aboriginal People from the Wadeye Region NT Australia*. Book 4. The author, Batchelor, Northern Territory.

Hardwick J (2019e) *Wadeye Kardu Murntak Warra: Old People Before Ceremonial Art and Craft – Traditional Knowledge, Language and Skills of the Aboriginal People from the Wadeye Region NT Australia.* Book 5. The author, Batchelor, Northern Territory.

Harney WE (1957) *Life Among the Aborigines.* Robert Hale, London.

Harney WE (1959) *Tales from the Aborigines.* Robert Hale, London.

Harney WE, Thompson P (1960) *Bill Harney's Cook Book.* Lansdowne Press, Melbourne.

Harris JK (1894–95) Letters. D6510(L)13–17. State Library of South Australia, Adelaide.

Harvey WJ (1932) Coombe district notes. *South Australian Ornithologist* **11**(8), 225–226.

Harvey A (1939) *Field Notebook.* AA105 Fry Collection, South Australian Museum Archives, Adelaide.

Harvey A (1943) A fishing legend of the Jaralde tribe of Lake Alexandrina, South Australia. *Mankind* **4**(4), 108–112.

Harvey A (1945) Food preservation in Australian tribes. *Mankind* **3**(7), 191–192.

Hassell E (1934a) Myths and folk-tales of the Wheelman tribe of south-Western Australia. *Folklore* **45**(3), 232–248. doi:10.1080/0015587X.1934.9718560

Hassell E (1934b) Myths and folk-tales of the Wheelman tribe of south-Western Australia: II. *Folklore* **45**(4), 317–341. doi:10.1080/0015587X.1934.9718572

Hassell E (1936) Notes on the ethnology of the Wheelman tribe of south-Western Australia. *Anthropos* **31**, 679–711.

Hassell E (1975) *My Dusky Friends: Aboriginal Life Customs and Legends and Glimpses of Station Life at Jarramungup in the 1880s.* CW Hassell, Perth.

Hawker JC (1841–45) *Journal of an Expedition to the River Murray Against the Natives, in Order to Recover Sheep Taken by Them from Messrs Field and Inman on Their Overland Journey from New South Wales to South Australia; Also to Protect Another Overland Party Expected Almost Immediately.* (Ed. I Palios, 1981). State Library of South Australia, Adelaide.

Hays T (1982) Utilitarian/adaptationist explanations of folk biological classification: some cautionary notes. *Journal of Ethnobiology* **2**, 89–94.

Heagney J (1886) Vocabulary of the dialect of the Kungarditchi tribe: vocabulary of the Koongerri language. In *The Australian Race: Its Origins, Languages, Customs, Place of Landing in Australia and the Routes by which it Spread Itself over the Continent.* Vol. 2. (Ed. EM Curr) pp. 380–383. Trubner, London.

Healey C (2007) Book review: *Some Indigenous Names for Australian Birds.* John M. Peter (2006). *Journal of Ethnobiology* **27**(1), 133–135. doi:10.2993/0278-0771(2007)27[133:SINFAB]2.0.CO;2

Heath J (1978) Linguistic approaches to Nunggubuyu ethnozoology and ethnobotany. In *Australian Aboriginal Concepts.* (Ed. LR Hiatt) pp. 40–55. Australian Institute of Aboriginal Studies, Canberra.

Helms R (1896) Anthropology. *Transactions and Proceedings of the Royal Society of South Australia* **16**, 237–332.

Hemming SJ, Clarke PA (1989) *Aboriginal Culture in South Australia.* Department of Aboriginal Affairs, Canberra.

Hemming SJ, Jones PG, Clarke PA (2000) *Ngurunderi: An Aboriginal Dreaming.* South Australian Museum, Adelaide.

Henderson J, Dobson V (1994) *Eastern and Central Arrernte to English Dictionary.* IAD Press, Alice Springs.

Henderson J, Nash D (2002) *Language in Native Title.* Aboriginal Studies Press, Canberra.

Henshall T, Jambijinpa D, Spencer JN, Kelly FJ, Bartlett P *et al.* (1980) *Ngurrju Maninja Kurlangu. Yapa Nyurnu Kurlangu: Bush Medicine.* Rev. edn. Warlpiri Literature Production Centre, Yuendumu, Northern Territory.

Hercus LA (1966) Notes on some Victorian names for plants and animals. *Victorian Naturalist* **83**, 189–192.

Hercus LA (1971) Eaglehawk and crow: a Madimadi version. *Mankind* **8**(2), 137–140.

Hercus LA (1982) *The Bagandji Language.* Pacific Linguistics Series B, No. 67. Department of Linguistics, Research School of Pacific Studies, Australian National University, Canberra.

Hercus LA (1992) *A Nukunu Dictionary*. The author, Canberra.

Hercus LA, Potezny V (1999) 'Finch' versus 'Finch-Water': a study of Aboriginal place-names in South Australia. *Records of the South Australian Museum* **31**(2), 165–180.

Heuzé V, Thiollet H, Tran G, Hassoun P, Lebas F (2018) *Ahuhu* (*Tephrosia purpurea*). Feedipedia A programme by INRA CIRAD AFZ and FAO. https://www.feedipedia.org/node/654 (last updated on 25 January 2018, 14, 36).

Hiatt B (1967) The food quest and the economy of the Tasmanian Aborigines. *Oceania* **38**(2), 99–133. doi:10.1002/j.1834-4461.1967.tb00947.x

Hiatt LR (Ed.) (1975a) *Australian Aboriginal Mythology: Essays in Honour of W.E.H. Stanner*. Australian Institute of Aboriginal Studies, Canberra.

Hiatt LR (1975b) Introduction. In *Australian Aboriginal Concepts*. (Ed. LR Hiatt) pp. 1–23. Australian Institute of Aboriginal Studies, Canberra.

Hicks D, Beaudry M (Eds) (2010) *The Oxford Handbook of Material Culture Studies*. Oxford University Press, Oxford.

Hill R (1968) *Bush Quest*. Lansdowne Press, Melbourne.

Hill R, Pert PL, Davies J, Robinson CJ, Walsh F *et al.* (2013) *Indigenous Land Management in Australia: Extent, Scope, Diversity Barriers and Success Factors*. Australian Landcare Council Secretariat, Landcare and Regional Delivery Improvement Branch, Department of Agriculture Fisheries and Forestry, Canberra.

Hilliard W (1968) *The People in Between: The Pitjantjatjara People of Ernabella*. Hodder & Stoughton, London.

HJL (1927) Sacred brolgas. *The Mail*, South Australia. 31 December, p. 10.

Hocknull SA, Lewis R, Arnold LJ, Pietsch T, Joannes-Boyau R *et al.* (2020) Extinction of eastern Sahul megafauna coincides with sustained environmental deterioration. *Nature Communications* **11**(1), 2250-1–2250-14. doi:10.1038/s41467-020-15785-w

Holden R, Holden N (2001) *Bunyips: Australia's Folklore of Fear*. National Library of Australia, Canberra.

Holmer NM, Holmer VE (1969) *Stories from Two Native Tribes of Eastern Australia: Australian Essays and Studies*. (Ed. SB Liljegran). A.-B. Lundequistska Bokhandeln, Uppsala.

Hope GS, Coutts PJF (1971) Past and present Aboriginal food resources at Wilsons Promontory, Victoria. *Mankind* **8**(2), 104–114.

Hope JH, Lampert RJ, Edmondson E, Smith MJ, van Tets GF (1977) Late Pleistocene faunal remains from Seton Rock Shelter, Kangaroo Island, South Australia. *Journal of Biogeography* **4**(4), 363–385. doi:10.2307/3038194

Horne G, Aiston G (1924) *Savage Life in Central Australia*. Macmillan, London.

Horton DR (1978) Preliminary notes on the analysis of Australian coastal middens. *Newsletter – Australian Institute of Aboriginal Studies* **10**, 30–33.

Horton DR (Ed.) (1994) *The Encyclopaedia of Aboriginal Australia*. 2 volumes. Australian Institute of Aboriginal and Torres Strait Islander Studies, Canberra.

Horton P, Blaylock B, Black A (2013) Section 3: Birds (September 2013 update). In *Census of South Australian Vertebrates*. (Eds H Owens and A Graham) pp. 1–38. South Australian Department of Environment and Natural Resources and South Australian Museum, Adelaide. http://www.birdssa.asn.au/

Howchin W (1934) *The Stone Implements of the Adelaide Tribe of Aborigines Now Extinct*. Gillingham & Co., Adelaide.

Howitt AW (1884) On some Australian beliefs. *Journal of the Anthropological Institute of Great Britain and Ireland* **13**(2), 185–198. doi:10.2307/2841724

Howitt AW (1904) *Native Tribes of South-east Australia*. Macmillan, London.

Humphries P (2007) Historical Indigenous use of aquatic resources in Australia's Murray-Darling Basin and its implications for river management. *Ecological Management & Restoration* **8**(2), 106–113. doi:10.1111/j.1442-8903.2007.00347.x

Hunn ES (1982) The utilitarian factor in folk biological classification. *American Anthropologist* **84**(4), 830–847. doi:10.1525/aa.1982.84.4.02a00070

Hunn ES (2010) Foreword. In *Ethno-ornithology: Birds, Indigenous Peoples, Culture and Society.* (Eds S Tidemann and A Gosler) pp. xi–xii. Earthscan, London.

Hunn ES, Thornton TF (2010) Tlingit birds: an annotated list with a statistical comparative analysis. In *Ethno-ornithology: Birds, Indigenous Peoples, Culture and Society.* (Eds S Tidemann and A Gosler) pp. 181–209. Earthscan, London.

Huntington HP (2000) Using traditional ecological knowledge in science: methods and applications. *Ecological Applications* **10**(5), 1270–1274. doi:10.1890/1051-0761(2000)010[1270:UTEKIS]2.0.CO;2

Hyam GN (1943) Living off the land in Victoria. *Victorian Naturalist* **59**, 171–173.

Ibarra JT, Caviedes J, Benavides P (2020) Winged voices: Mapuche ornithology from South American temperate forests. *Journal of Ethnobiology* **40**(1), 89–100. doi:10.2993/0278-0771-40.1.89

Ichikawa M (1998) The birds as indicators of the invisible world: ethno-ornithology of the Mbuti hunter-gatherers. *African Study Monographs. Supplementary Issue* **25**, 105–121.

Ingram C (1906) Wild turkeys in South Australia. *Badminton Magazine of Sports and Pastimes* **22**, 334–337.

Isaacs J (1980) *Australian Dreaming: 40,000 Years of Aboriginal History.* Lansdowne Press, Sydney.

Iwaniszewski S (2014) Australian Aboriginal astronomy and cosmology. In *Handbook of Archaeoastronomy and Ethnoastronomy.* (Ed. CLN Ruggles) pp. 3–14. Springer, New York.

Jackson S, Finn M, Featherston P (2012) Indigenous aquatic resource use in two tropical Australian river catchments: the Fitzroy River (W.A.) and Daly River (N.T.). *Human Ecology* **40**(6), 893–908. doi:10.1007/s10745-012-9518-z

Jacob T (1991) *In the Beginning: A Perspective on Traditional Aboriginal Societies.* Ministry of Education Western Australia, Perth.

Janca A, Bullen C (2003) The Aboriginal concept of time and its mental health implications. *Australasian Psychiatry* **11**(s1), S40–S44. doi:10.1046/j.1038-5282.2003.02009.x

Jepson P (2010) Towards an Indonesian bird conservation ethos: reflections from a study of bird-keeping in the cities of Java and Bali. In *Ethno-ornithology: Birds, Indigenous Peoples, Culture and Society.* (Eds S Tidemann and A Gosler) pp. 313–330. Earthscan, London.

JM (1856) Sketches of Australian zoology. No. VII. *Adelaide Observer*, South Australia. 8 November, p. 7.

Johnson D (1998) *Night Skies of Aboriginal Australia.* Oceania Monograph No. 47. University of Sydney, Sydney.

Johnston TH (1943) Aboriginal names and utilization of the fauna in the Eyrean region. *Transactions of the Royal Society of South Australia* **67**(2), 244–311.

Johnston TH, Cleland JB (1933–34) The history of the Aboriginal narcotic, pituri. *Oceania* **4**(2), 201–223; **4**(3), 268–289.

Johnston TH, Cleland JB (1943) Native names and uses of plants in the north-eastern corner of South Australia. *Transactions of the Royal Society of South Australia* **67**(1), 149–173.

Jones R (1969) Fire-stick farming. *Australian Natural History* **16**, 224–228.

Jones R (1980) Cleaning the country: the Gidjingali and their Arnhem Land environment. *BHP Journal* **1**, 10–15.

Jones R (1985) Ordering the landscape. In *Seeing the First Australians.* (Eds I Donaldson and T Donaldson) pp. 181–209. George Allen Unwin Australia, Sydney.

Jones PG (1990) Unaipon, David (1872–1967). In *Australian Dictionary of Biography.* National Centre of Biography, Australian National University, Canberra. http://adb.anu.edu.au/biography/unaipon-david-8898

Jones PG (1996a) *Boomerang: Behind an Australian Icon.* Wakefield Press, Adelaide.

Jones PG (1996b) 'A box of native things': ethnographic collectors and the South Australian Museum, 1830s–1930s. PhD thesis. Department of History, University of Adelaide, Adelaide.

Jones R (1998) The fifth continent: problems concerning the human colonisation of Australia. In *Archaeology of Aboriginal Australia: A Reader*. (Ed. T Murray) pp. 102–118. Allen & Unwin, Sydney.

Jones PG (2007) *Ochre and Rust: Artefacts and Encounters on Australian Frontiers*. Wakefield Press, Adelaide.

Jones PG (2017) Beyond songlines. *Australian Book Review*, Victoria. September, pp. 21–30.

Jones R, Allen J (1978) Caveat excavator: a sea bird midden on Steep Head Island, north west Tasmania. *Australian Archaeology* **8**, 142–145. doi:10.1080/03122417.1978.12093346

Jones DS, Clarke PA (2018) Australian Aboriginal culture and food landscape relationships: possibilities of Indigenous knowledge for the future Australian landscape. In *Routledge Companion to Landscape and Food*. (Eds J Zeunert and T Waterman) pp. 41–60. Routledge, London.

Jones PG, Sutton P (1986) *Art and Land: Aboriginal Sculptures of the Lake Eyre Basin*. South Australian Museum and Wakefield Press, Adelaide.

Jones DS, Low Choy D, Clarke PA, Hale R (2013) Watching clouds over Country: reconsidering Australian Indigenous perspectives about environmental change and climate change. In *UPE10 2012: NEXT CITY: Planning for a New Energy and Climate Future. 10th International Urban Planning and Environment Association Symposium*. (Eds N Gurran, P Phibbs and S Thompson) pp. 148–163. ICMS, Sydney.

Jones DS, Roös P, Dearnaley J, Threadgold H, Nicholson M *et al.* (2018) Recrafting urban climate change resilience understandings: learning from Australian Indigenous cultures. In *IFLA 2018: Biophilic City, Smart Nation and Future Resilience: Proceedings of the 55th International Federation of Landscape Architects World Congress 2018*, pp. 401–416. International Federation of Landscape Architects, Singapore.

Joralemon D (2015) *Exploring Medical Anthropology*. Pearson, Boston.

Kaberry PM (1935) Death and deferred mourning ceremonies in the Forrest River tribes, north-west Australia. *Oceania* **6**(1), 33–47. doi:10.1002/j.1834-4461.1935.tb01684.x

Kaplan G (2019) *Bird Bonds: Sex, Mate-choice and Cognition in Australian Native Birds*. Macmillan, Sydney.

Karadada J, Karadada L, Goonack W, Mangolamara G, Bunjuck W *et al.* (2011) *Plants and Animals: Aboriginal Biological Knowledge from Wunambal Gaambera Country in the North-west Kimberley Australia*. Northern Territory Botanical Bulletin No. 35. Department of Natural Resources, Environment; the Arts and Sport, Kalumburu, Western Australia.

Karskens GE (2020) *People of the River*. Allen & Unwin, Sydney.

Keen I (1994) *Knowledge and Secrecy in an Aboriginal Religion: Yolngu of North-east Arnhem Land*. Oxford University Press, Oxford.

Kelly D (2014) Foundational sources and purposes of authority in Madayin. *Victoria University Law and Justice Journal* **4**(1), 33–45.

Kendon A (1988) *Sign Languages of Aboriginal Australia: Cultural, Semiotic and Communicative Perspectives*. Cambridge University Press, Cambridge.

Kenyon AS (1912) Camping places of the Aborigines of south-east Australia. *Victorian Historical Magazine* **2**(3), 97–110.

Kilham C, Pamulkan M, Pootchemunka J, Wolmby T (1986) *Dictionary and Source Book of the Wik-Mungkan Language*. Summer Institute of Linguistics, Australian Aborigines Branch, Darwin.

Kimber R (1983) Black lightning: Aborigines and fire in Central Australia and the Western Desert. *Archaeology in Oceania* **18**, 38–45. doi:10.1002/arco.1983.18.1.38

Kimber RG (1997) Cry of the plover, song of the desert rain. In *Windows on Meteorology: Australian Perspective*. (Ed. EK Webb) pp. 7–13. CSIRO Publishing, Melbourne.

Kizungu B, Ntabaza M, Mburunge M (1998) Ethnoornithology of the Tembo in eastern DRC (former Zaire): Part 1, Kalehe zone. *African Study Monographs* **19**(2), 103–113.

Kloot T, McCulloch EM (1980) *Birds of Australian Gardens. Paintings by Peter Trusler*. Rigby, Adelaide.

Koch H, Hercus LA, Kelly P (2018) Moiety names in south-eastern Australia: distribution and reconstructed history. In *Skin, Kin and Clan: The Dynamics of Social Categories in Indigenous Australia*. (Eds P McConvell, P Kelly and S Lacrampe) pp. 139–178. ANU Press, Canberra.

Kofron CP (1999) Attacks to humans and domestic animals by the southern cassowary (*Casuarius casuarius johnsonii*) in Queensland, Australia. *Journal of Zoology* **249**, 375–381. doi:10.1111/j.1469-7998.1999.tb01206.x

Krefft G (1865) On the manners and customs of the Aborigines of the Lower Murray and Darling. *Transactions of the Philosophical Society of New South Wales 1862–1865* **1**, 357–374.

Kyriazis S (1995) *Bush Medicine of the Northern Peninsula Area of Cape York*. Nai Beguta Agama Aboriginal Corporation, Bamaga, Queensland.

Lakoff G (1987) *Women, Fire and Dangerous Things: What Categories Reveal About the Mind*. University of Chicago Press, Chicago.

Lang A (1902) The origin of totem names and beliefs. *Folklore* **13**(4), 347–397. doi:10.1080/001558 7X.1902.9719321

Langloh Parker K (1905) *The Euahlayi Tribe: A Study of Aboriginal Life in Australia*. Archibald Constable, London.

Langloh Parker K (1953) *Australian Legendary Tales*. Angus & Robertson, Sydney.

Langloh Parker K (1991) *Tales of the Dreamtime*. Cornstalk Publishing, Sydney.

Langton M, Rhea ZM (2005) Traditional Indigenous biodiversity-related knowledge. *Australian Academic and Research Libraries* **36**(2), 45–69. doi:10.1080/00048623.2005.10721248

Lassak EV, McCarthy T (1983) *Australian Medicinal Plants*. Methuen, Melbourne.

Latz P (1995) *Bushfires and Bushtucker: Aboriginal Plant Use in Central Australia*. IAD Press, Alice Springs.

Laughren M, Curran G, Turpin M, Peterson N (2016) Women's *yawulyu* songs as evidence of connections to and knowledge of land: the *Jardiwanpa*. In *Language, Land and Song: Studies in Honour of Luise Hercus*. (Eds PK Austin, H Koch and J Simpson) pp. 419–449. EL Publishing, London.

Lawrence R (1968) *Aboriginal Habitat and Economy*. Geography Occasional Paper No. 6. Australian National University, Canberra.

Lee RB (1968) What hunters do for a living, or, how to make out on scarce resources. In *Man the Hunter*. (Eds RB Lee, I DeVore and J Nash-Mitchell) pp. 30–48. Aldine Publishing, New York.

Lee M, Wei K (2000) Australasian microtektites in the South China Sea and the West Philippine Sea: implications for age, size and location of the impact crater. *Meteoritics & Planetary Science* **35**, 1151–1155. doi:10.1111/j.1945-5100.2000.tb01504.x

Lefebvre L, Nicolakakis N, Boire D (2002) Tools and brains in birds. *Behaviour* **139**(7), 939–973. doi:10.1163/156853902320387918

Leichhardt FWL (1847) *Journal of an Overland Expedition in Australia: From Moreton Bay to Port Essington A Distance of Upwards of 3000 miles, During the Years 1844–1845*. Republished in 2000. Corkwood Press, Adelaide.

Leonard S, Parsons M, Olawsky K, Kofod K (2013) The role of culture and traditional knowledge in climate change adaptation: insights from East Kimberley, Australia. *Global Environmental Change* **23**, 623–632. doi:10.1016/j.gloenvcha.2013.02.012

Lepofsky D (2009) The past, present and future of traditional resource and environmental management. *Journal of Ethnobiology* **29**, 161–166. doi:10.2993/0278-0771-29.2.161

Lertzman K (2009) The paradigm of management, management systems and resource stewardship. *Journal of Ethnobiology* **29**, 339–358. doi:10.2993/0278-0771-29.2.339

Letch D (2010) *Chain of Bays: Preserving the West Coast of South Australia*. Friends of Sceale Bay, Sceale Bay, South Australia.

Lewis D (1988) Hawk hunting hides in the Victoria River district. *Australian Aboriginal Studies* **2**, 74–78.

Lockwood D (1963) *I, The Aboriginal*. Cassell, London.

Long J (2002) *The Dinosaur Dealers*. Allen & Unwin, Sydney.

Loss ATG, Neto EMC, Machado CG, Flores FM (2014) Ethnotaxonomy of birds by the inhabitants of Pedra Branca village, Santa Teresinha municipality, Bahia state, Brazil. *Journal of Ethnobiology and Ethnomedicine* **10**(1), 55. doi:10.1186/1746-4269-10-55

Love JRB (1936) *Stone Age Bushmen of Today*. Republished in 2009. David M Welch, Virginia, Northern Territory.

Love JRB (1946) The tale of the winking owl: a Worora bird legend. *Mankind* **3**(9), 258–261.

Lover N (1918) Women's column. *Sydney Morning Herald*, New South Wales. 7 August, p. 8.

Low T (2014) *Where Song Began: Australia's Birds and How They Changed the World*. Viking, Melbourne.

Low Choy D, Clarke PA, Jones DS, Serrao-Neumann S, Hales R *et al.* (2013) *Indigenous Climate Change Adaptation: Understanding Coastal Urban and Peri-urban Indigenous People's Vulnerability and Adaptive Capacity to Climate Change*. National Climate Change Adaptation Research Facility, Griffith University, Brisbane.

Lowe P (2002) *Hunters and Trackers of the Australian Desert*. Rosenberg Publishing, Sydney.

Lumholtz CS (1889) *Among Cannibals: An Account of Four Years, Travels in Australia and of Camp Life with the Aborigines of Queensland*. Translated by R.B. Anderson. Charles Scribner's Sons, New York.

Lycett J (1820–22) *The Lycett Album: Drawings of Aborigines and Australian Scenery*. Published in 1990, National Library of Australia, Canberra.

Lynch AJJ, Thackway R, Specht A, Beggs PJ, Brisbane S *et al.* (2015) Transdisciplinary synthesis for ecosystem science, policy and management: the Australian experience. *Science of the Total Environment* doi:10.1016/j.scitotenv.2015.04.100.

MacGlashan J (1887) No.142: Main range between the Belyando and Cape rivers waters. In *The Australian Race*. Vol. 3. (Ed. EM Curr) pp. 18–25. Trubner, London.

Macindoe A (2012) Avian nomenclature. importance of research into names of birds in various languages, dialects and cultures. *Civilization* **21**(29), 147–161.

Mackinolty C, Wainburranga P (1988) Too many Captain Cooks. In *Aboriginal Australians and Christian Missions*. (Eds T Swain and D Bird Rose) pp. 355–360. Australian Association for the Study of Religions, Adelaide.

MacPherson J (1931) Some Aboriginal animal names. *Australian Zoologist* **6**(4), 368–371.

Maddock K (1970) Myths of the acquisition of fire in northern and eastern Australia. In *Australian Aboriginal Anthropology: Modern Studies in the Social Anthropology of the Australian Aborigines*. (Ed. RM Berndt) pp. 174–199. Australian Institute of Aboriginal Studies and University of Western Australia, Canberra.

Maddock K (1975) The emu anomaly. In *Australian Aboriginal Mythology*. (Ed. LR Hiatt) pp. 102–122. Australian Institute of Aboriginal Studies, Canberra.

Maddock K (1978a) Taxonomy and Kakadu totemic relationships. In *Australian Aboriginal Concepts*. (Ed. LR Hiatt) pp. 121–133. Australian Institute of Aboriginal Studies, Canberra.

Maddock K (1978b) Metaphysics in a mythical view of the world. In *The Rainbow Serpent: A Chromatic Piece*. (Eds IR Buchler and K Maddock) pp. 99–118. Mouton Publishers, The Hague.

Maddock K (1982) *The Australian Aborigines. A Portrait of their Society*, 2nd edn. Penguin Books, Melbourne.

Maddock K, Cawte JE (1970) Aboriginal law and medicine. *Proceedings of the Medico-Legal Society of New South Wales* **4**, 170–190.

Magarey AT (1895) Aboriginal water quest. *Proceedings of the Royal Geographical Society of Australasia: South Australia Branch* **3**, 67–82.

Magarey AT (1899) Tracking by the Australian Aborigine. *Proceedings of the Royal Geographical Society of Australasia: South Australia Branch* **3**, 119–126.

Maher P (1999) A review of traditional Aboriginal health beliefs. *Australian Journal of Rural Health* **7**, 229–236. doi:10.1046/j.1440-1584.1999.00264.x

Mann JF (1883) Notes on the Aborigines of Australia. III. *Sydney Mail and New South Wales Advertiser*, New South Wales. 6 October, p. 640.

Mansergh I, Hercus LA (1981) An Aboriginal vocabulary of the fauna of Gippsland. *Memoirs of the National Museum of Victoria* **42**, 107–122. doi:10.24199/j.mmv.1981.42.06

Marshallsea T (1991) North Americans are breeding emus for the table. *Canberra Times*, Australian Capital Territory. 14 December, p. 14.

Mashtoub S, Lampton LS, Eden GL, Cheah KY, Lymn KA *et al.* (2016) Emu oil combined with Lyprinol™ reduces small intestinal damage in a rat model of chemotherapy-induced mucositis. *Nutrition and Cancer* **68**(7), 1171–1180. doi:10.1080/01635581.2016.1208829

Massola AS (1957) The Challicum bun-yip. *Victorian Naturalist* **74**, 76–83.

Massola AS (1968) *Bunjil's Cave: Myths, Legends and Superstitions of the Aborigines of South-east Australia*. Lansdowne Press, Melbourne.

Massola AS (1969) *Journey to Aboriginal Victoria*. Rigby, Adelaide.

Mathew J (1899) *Eaglehawk and Crow: A Study of the Australian Aborigines, Including an Inquiry into Their Origin and a Survey of Australian Languages*. Melville, Mullen & Slade, Melbourne.

Mathew J (1910) The origin of the Australian phratries and explanations of some of the phratry names. *Journal of the Royal Anthropological Institute of Great Britain and Ireland* **40**, 164–170. doi:10.2307/2843147

Mathews RH (1893) Rock paintings by the Aborigines in caves on Bulgar Creek, near Singleton. *Journal and Proceedings of the Royal Society of New South Wales* **27**, 353–358.

Mathews RH (1897a) The Burbung or initiation ceremonies of the Murrumbidgee tribes. *Journal and Proceedings of the Royal Society of New South Wales* **31**, 111–153.

Mathews RH (1897b) The totemic divisions of Australian tribes. *Journal and Proceedings of the Royal Society of New South Wales* **31**, 154–176.

Mathews RH (1898a) Victorian Aborigines: their initiation ceremonies and divisional systems. *American Anthropologist* **11**, 325–343. doi:10.1525/aa.1898.11.11.02a00000

Mathews RH (1898b) Folk-lore of the Australian Aborigines. *Science of Man* **1**(4), 91–93.

Mathews RH (1898c) Folk-lore of the Australian blacks. *Science of Man* **1**(5), 117–119.

Mathews RH (1898d) Folk-lore of the Australian blacks. *Science of Man* **1**(6), 142–143.

Mathews RH (1900) Phallic rites and initiation ceremonies of the South Australian Aborigines. *Proceedings of the American Philosophical Society* **39**(64), 622–638.

Mathews RH (1901) Ethnological notes on the Aboriginal tribes of the Northern Territory. *Queensland Geographical Journal* **16**, 69–90.

Mathews RH (1904) Ethnological notes on the Aboriginal tribes of New South Wales and Victoria. *Journal and Proceedings of the Royal Society of New South Wales* **38**, 203–381.

Mathews RH (1909) The wallaroo and the willy-wagtail: a Queensland folk-tale. *Folklore* **20**(2), 214–216. doi:10.1080/0015587X.1909.9719878

Mattingley AHE (1905) Bird-names of Aborigines. *Emu* **4**(4), 168. doi:10.1071/MU904168b

Mayor A (2011) *The First Fossil Hunters: Dinosaurs, Mammoths and Myth in Greek and Roman Times*, 2nd edn. Princeton University Press, Princeton.

Mayor A, Sarjeant WAS (2001) The folklore of footprints in stone: from classical antiquity to the present. *Ichnos* **8**(2), 143–163. doi:10.1080/10420940109380182

McCarthy FD (1940) Aboriginal Australian material culture: causative factors in its composition. *Mankind* **2**(8), 241–269; **2**(9), 94–320.

McCarthy FD (1952) A werpoo, or bone dagger, from South Australia. *Australian Museum Magazine* **10**(9), 290–292.

McCarthy FD (1953) Aboriginal rain-makers. *Weather* **8**, 72–77. doi:10.1002/j.1477-8696.1953.tb01603.x

McCarthy FD (1961) The boomerang. *Australian Museum Magazine* **13**(11), 343–349.

McCarthy FD (1965) The emu and the Aborigines. *Australian Natural History* **15**(1), 16–21.

McCarthy FD (1974) *Australian Aboriginal Decorative Art*, 8th edn. Australian Museum, Sydney.

McCarthy FD (1976) *Australian Aboriginal Stone Implements*, 2nd edn. Australian Museum Trust, Sydney.

McCarthy FD (1978) Australian Aboriginal material culture. In *Aboriginal Art in Australia*. (Ed. R Edwards) pp. 21–28. Ure Smith, Sydney.

McConnel UH (1930) The Wik-Mungkan tribe. Part II: totemism. *Oceania* **1**(2), 181–205. doi:10.1002/j.1834-4461.1930.tb01644.x

McConnel UH (1931) A moon legend from the Bloomfield River, north Queensland. *Oceania* **2**(1), 9–25. doi:10.1002/j.1834-4461.1931.tb00020.x

McConnel UH (1936) Illustration of the myth of Shiveri and Nyunggu. *Oceania* **7**(2), 217–219. doi:10.1002/j.1834-4461.1936.tb00452.x

McConnel UH (1939) Social organisation of the tribes of Cape York Peninsula, north Queensland. *Oceania* **10**(1), 54–72. doi:10.1002/j.1834-4461.1939.tb00256.x

McConnel UH (1953) Native arts and industries on the Archer, Kendall and Holroyd rivers, Cape York Peninsula, north Queensland. *Records of the South Australian Museum* **11**(1), 1–42.

McConnel UH (1957) *Myths of the Mungkan.* Melbourne University Press, Melbourne.

McConvell P, Ponsonnet M (2018) Generic terms for subsections ('skins') in Australia: sources and semantic networks. In *Skin, Kin and Clan: The Dynamics of Social Categories in Indigenous Australia.* (Eds P McConvell, P Kelly and S Lacrampe) pp. 271–315. ANU Press, Canberra.

McConvell P, Thieberger N (2006) Keeping track of language endangerment in Australia. In *Language Diversity in the Pacific.* (Eds D Cunningham, D Ingram and K Sumbuk) pp. 54–84. Multilingual Matters, Clevedon, UK.

McCourt T (1975) *Aboriginal Artefacts.* Rigby, Adelaide.

McEntee J, McKenzie P (1992) *Adna-mat-na English Dictionary.* The authors, Adelaide.

McEntee J, McKenzie P, McKenzie J (1986) *Witi-ita-nanalpila: Plants and Birds of the Northern Flinders Ranges and Adjacent Plains with Aboriginal Names.* The authors, Adelaide.

McGregor D, Restoule JP, Johnston R (Eds) (2018) *Indigenous Research: Theories, Practices and Relationships.* Canadian Scholars Press, Toronto.

McKemey M, Ens E, Yugul Mangi Rangers, Costello O, Reid N (2020) Indigenous knowledge and seasonal calendar inform adaptive savanna burning in northern Australia. *Sustainability* **12**(3), 995. doi:10.3390/su12030995

McKnight D (1973) Sexual symbolism of food among the Wik-mungkan. *Man* **8**(2), 194–209. doi:10.2307/2800846

McKnight D (1999) *People, Countries and the Rainbow Serpent. Systems of Classification Among the Lardil of Mornington Island.* Oxford University Press, New York.

Meagher SJ (1974) The food resources of the Aborigines of the south-west of Western Australia. *Records of the Western Australian Museum* **3**(10), 14–65.

Medin DL, Atran S (1999) *Folkbiology.* MIT Press, Cambridge.

Meehan B (1982a) *Shell Bed to Shell Midden.* Australian Institute of Aboriginal Studies, Canberra.

Meehan B (1982b) Ten fish for one man: some Anbarra attitudes towards food and health. In *Body, Land and Spirit: Health and Healing in Aboriginal Society.* (Ed. JS Reid) pp. 96–120. University of Queensland Press, Brisbane.

Meggitt MJ (1962) *Desert People. A Study of the Walbiri Aborigines of Central Australia.* Angus & Robertson, Sydney.

Memmott P (2010) *Material Culture of the North Wellesley Islands.* Aboriginal and Torres Strait Islander Studies Unit, University of Queensland, Brisbane.

Menkhorst P, Ryan E (2015) C.H. McLennan ('Mallee Bird') and his Aboriginal informant Jowley: the source of early records of the night parrot *Pezoporus occidentalis* in Victoria? *Memoirs of the Museum of Victoria* **73**, 107–115. doi:10.24199/j.mmv.2015.73.09

Meston A (1921) Aboriginal names. *Sydney Morning Herald*, New South Wales. 19 October, p. 11.

Meyer HAE (1843) *Vocabulary of the Language Spoken by the Aborigines of South Australia.* Allen, Adelaide.

Meyer HAE (1846) Manners and customs of the Aborigines of the Encounter Bay tribe, South Australia. In *The Native Tribes of South Australia.* (Ed. JD Woods, 1879) pp. 183–206. ES Wiggs, Adelaide.

Miller G, Hercus LA, Monaghan P, Naessan P (2010) *A Dictionary of the Wirangu Language of the Far West Coast of South Australia.* Tjutjunaku Worka Tjuta and University of Adelaide, Adelaide.

Miller G, Magee J, Smith M, Spooner N, Baynes A *et al.* (2016) Human predation contributed to the extinction of the Australian megafaunal bird *Genyornis newtoni* ~47 ka. *Nature Communications* **7**(1), 1–7. doi:10.1038/ncomms10496

Mills K (2019) *Jungung Jack McGinness: Plaiting the Grass for Family Community and the Future, 1902–1973.* Uniprint Charles Darwin University, Darwin.

Mintz SW, Du Bois CM (2002) The anthropology of food and eating. *Annual Review of Anthropology* **31**(1), 99–119. doi:10.1146/annurev.anthro.32.032702.131011

Mitchell TL (1839) *Three Expeditions into the Interior of Eastern Australia: with Descriptions of the Recently Explored Region of Australia Felix and of the Present Colony of New South Wales.* T & W Boone, London.

Mitchell SR (1949) *Stone-age Craftsmen. Stone Tools and Camping Places of the Australian Aborigines.* Tait Book Co., Melbourne.

Mjöberg E (1915) *Among Wild Animals and People in Australia [Bland Vilda Djur och Folk i Australien].* (Ed. K Akerman). Republished in 2012. Hesperian Press, Perth.

Mjöberg E (1918) *Amongst Stone Age People in the Queensland Wilderness [Bland Stenåldersmänniskor I Queenslands Wildmarker].* Republished in 2015. Hesperian Press, Perth.

Molineux A (1905) The Australian pelican. *Evening Journal,* South Australia. 12 August, p. 6.

Monaghan P, Mühlhäusler P (2015) State versus community approaches to language revival. *Education in Languages of Lesser Power: Asia-Pacific Perspectives* **35**, 185–203. doi:10.1075/impact.35.11mon

Moore GF (1884) *Diary of Ten Years' Eventful Life of an Early Settler in Western Australia: And a Descriptive Vocabulary of the Language in Common Use Amongst the Aborigines of Western Australia.* 2 parts. M Walbrook, London.

Moreman CM (2014) On the relationship between birds and spirits of the dead. *Society & Animals* **22**(5), 481–502. doi:10.1163/15685306-12341328

Morphy H (1984) *Journey to the Crocodile's Nest.* Australian Institute of Aboriginal Studies, Canberra.

Morphy H (1998) *Aboriginal Art.* Phaidon Press, London.

Morris B (1987) *Anthropological Studies of Religion: An Introductory Text.* Cambridge University Press, Cambridge.

Morrison RGB (1981) *A Field Guide to the Tracks and Traces of Australian Animals.* Rigby, Adelaide.

Morwood MJ (2002) *Visions from the Past: The Archaeology of Australian Aboriginal Art.* Allen & Unwin, Sydney.

Mountford CP (1929) A unique example of Aboriginal rock carving at Panaramitee North. *Transactions of the Royal Society of South Australia* **53**, 245–248.

Mountford CP (1937) Why the crow is black: legends and customs of Flinders Ranges natives. *The Advertiser,* South Australia. 29 June, p. 18.

Mountford CP (1939) Aboriginal methods of fishing and cooking as used on the southern coast of Eyre's Peninsula, South Australia. *Mankind* **2**(7), 196–200.

Mountford CP (1941a) 'News' to tell Aboriginal fairy tales. *News,* South Australia. 25 September, p. 4.

Mountford CP (1941b) How the crow got his black feathers. *News,* South Australia. 25 October, p. 2.

Mountford CP (1941c) Wants more native legends gathered. *News,* South Australia. 20 December, p. 2.

Mountford CP (1943) Native fairy tales. *News,* South Australia. 25 June, p. 2.

Mountford CP (1948) *Brown Men and Red Sand.* Revised edn, 1981. Angus & Robertson, Sydney.

Mountford CP (1949) Gesture language of the Warlpari tribe, Central Australia. *Transactions of the Royal Society of South Australia* **73**(1), 100–101.

Mountford CP (Ed.) (1956) *Records of the American-Australian Scientific Expedition to Arnhem Land. Vol. 1: Art, Myth and Symbolism.* Melbourne University Press, Melbourne.

Mountford CP (1958) *The Tiwi: Their Art, Myth and Ceremony.* Phoenix House, London.

Mountford CP (1965) *Ayers Rock: Its People, Their Beliefs and Their Art.* Angus & Robertson, Sydney.

Mountford CP (1976a) *Nomads of the Australian Desert.* Rigby, Adelaide.

Mountford CP (1976b) *Before Time Began.* Nelson, Sydney.

Mountford CP (1976c) The cave paintings of Arnhem Land. *Journal of the Anthropological Society of South Australia* **14**(3), 5–10.

Mountford CP (1978) The Rainbow-serpent myths of Australia. In *The Rainbow Serpent: A Chromatic Piece*. (Eds IR Buchler and K Maddock) pp. 23–98. Mouton Publishers, The Hague.

Mountford CP, Edwards R (1962) Aboriginal rock engravings of extinct creatures in South Australia. *Man* **62**, 97–99. doi:10.2307/2796665

Mountford CP, Edwards R (1963) Rock engravings of Panaramitee Station, north-eastern South Australia. *Transactions of the Royal Society of South Australia* **86**, 131–146.

Mountford CP, Roberts A (1965) *The Dreamtime*. Rigby, Adelaide.

Mountford CP, Roberts A (1969) *The Dawn of Time*. Rigby, Adelaide.

Mountford CP, Roberts A (1971) *The First Sunrise*. Rigby, Adelaide.

Mühlhäusler P (2003) *Language of Environment, Environment of Language: A Course in Ecolinguistics*. Battlebridge, London.

Mühlhäusler P, Fill A (Eds) (2001) *The Ecolinguistics Reader: Language, Ecology and Environment*. Continuum, London.

Muir M (1990) Stow, Catherine Eliza (Katie) (1856–1940). In *Australian Dictionary of Biography*. https://adb.anu.edu.au/biography/stow-catherine-eliza-katie-8691

Muiruri MN, Maundu P (2010) Birds, people and conservation in Kenya. In *Ethno-ornithology: Birds, Indigenous Peoples, Culture and Society*. (Eds S Tidemann and A Gosler) pp. 279–289. Earthscan, London.

Muloin S, Zeppel H, Higginbottom K (2001) *Indigenous Wildlife Tourism in Australia*. Wildlife Tourism Research Report Series No. 15. Cooperative Research Centre for Sustainable Tourism, Gold Coast, Queensland.

Mulvaney DJ (1959) What valley hunters ate 5,000 years ago. *Riverlander*, South Australia. September, pp. 8, 36.

Mulvaney DJ (1960) Archaeological excavations at Fromms Landing on the lower Murray River, South Australia. *Proceedings of the Royal Society of Victoria* **72**, 53–85.

Mulvaney DJ (1989) *Encounters in Place: Outsiders and Aboriginal Australians 1606–1985*. University of Queensland Press, Brisbane.

Mulvaney DJ (Ed.) 1991. *The Humanities and the Australian Environment*. Occasional Paper No. 11. Papers from the Australian Academy of the Humanities Symposium 1990. Australian Academy of the Humanities, Canberra.

Mulvaney KJ (1993) Hunting with hides: ethno-historical reflections on Victoria River stone structures. *Records of the South Australian Museum* **26**(2), 111–120.

Mulvaney DJ (1994) The Namoi bunyip. *Australian Aboriginal Studies* **1**, 36–38.

Mulvaney DJ, Kamminga J (1999) *The Prehistory of Australia*. Allen & Unwin, Sydney.

Munn N (1973) *Walbiri Iconography: Graphic Representation and Cultural Symbolism in a Central Australian Society*. Cornell University Press, Ithaca.

Murphy GR (2002) *The Owl, the Raven and the Dove: The Religious Meanings of the Grimms' Magic Fairy Tales*. Oxford University Press, New York.

Murray PF, Vickers-Rich P (2004) *Magnificent Mihirungs: The Colossal Flightless Birds of the Australian Dreamtime*. Indiana University Press, Bloomington.

Musharbash Y (2014) Monstrous transformations: a case study from Central Australia. In *Monster Anthropology in Australasia and Beyond*. (Eds Y Musharbash and GH Presterudstuen) pp. 39–55. Springer, The Hague.

Musharbash Y, Presterudstuen GH (Eds) (2014) *Monster Anthropology in Australasia and Beyond*. Springer, The Hague.

Myers FR (1986) *Pintupi Country, Pintupi Self*. Smithsonian Institution Press, Washington.

Nabhan G (2009) Ethnoecology: bridging disciplines, cultures and species. *Journal of Ethnobiology* **29**, 3–7. doi:10.2993/0278-0771-29.1.3

Naessan PA (2017) A preliminary outline of Antikirrinya bird classification: a comparative approach. *Journal of the Anthropological Society of Oxford* **9**(3), 344–367.

Naturalist A (1886) Notes upon additions from the Museum by an 'Amateur Naturalist'. *South Australian Register*, South Australia. 16 August, p. 6.

Naturalist A (1887) *Notes upon Additions to the Museum of the South Australian Public Library, Museum and Art Gallery by 'An Amateur Naturalist'*. Reprinted from *South Australian Register* and *Adelaide Observer*. WK Thomas & Co. Printers, Adelaide.

Nettle D, Romaine S (2000) *Vanishing Voices: The Extinction of the World's Languages*. Oxford University Press, Oxford.

Newland S (1890) *Parkengees or Aboriginal Tribes on the Darling River*. Royal Geographical Society of Australasia, South Australian Branch. HF Leader, Government Printer, Adelaide.

Newsome AE (1980) The eco-mythology of the red kangaroo in Central Australia. *Mankind* **12**(4), 327–333.

Ng'weno F (2010) Sound, sight, stories and science: avoiding methodological pitfalls in ethno-ornithological research, with examples from Kenya. In *Ethno-ornithology" Birds, Indigenous Peoples, Culture and Society*. (Eds S Tidemann and A Gosler) pp. 103–113. Earthscan, London.

Ngaanyatjarra Pitjantjatjara Yankunytjatjara Women's Council Aboriginal Corporation (Ed.) (2003) *Ngangkari Work – Anangu Way: Traditional Healers of Central Australia*. Ngaanyatjarra Pitjantjatjara Yankunytjatjara Women's Council Aboriginal Corporation, Alice Springs.

Ngakulmungan Kangka Leman (1997) *Lardil Dictionary: A Vocabulary of the Language of the Lardil People, Mornington Island, Gulf of Carpentaria, Queensland*. Mornington Shire Council, Mornington Island, Queensland.

Nicholls CJ (2008) The hitchhiker's guide to Australian Aboriginal bodily adornment, object making and jewellery BC–AC (Before Cook–After Colonisation). Presented at *Inside/Out: 13th Biennial Conference of the Jewellers and Metalsmiths Group of Australia*. January 2008, Adelaide. https://core.ac.uk/download/pdf/14947082.pdf

Nicholls CJ (2014) 'Dreamings' and place: Aboriginal monsters and their meanings. *The Conversation* http://theconversation.com/dreamings-and-place-aboriginal-monsters-and-their-meanings-25606

Nicholls CJ (2019) A wild roguery: Bruce Chatwin's *The Songlines* reconsidered. *Text Matters* **9**(9), 22–49. doi:10.18778/2083-2931.09.02

Nobbs C (1989) A pillar of the sky. the significance of the fossilised tree which stands outside the South Australian Museum. *Journal of the Anthropological Society of South Australia* **27**(6), 37–54.

Noetling F (1910) The food of the Tasmanian Aborigines. *Papers and Proceedings of the Royal Society of Tasmania* **69**, 279–305.

Norman J (1951) Aboriginal legends and the bunyip. *The Advertiser*, South Australia. 24 February, p. 6.

Norris R, Norris C (2009) *Emu Dreaming. An Introduction to Australian Aboriginal Astronomy*. Emu Dreaming, Sydney.

North AJ, Keartland GA (1898) List of birds collected by the Calvert Exploring Expedition in Western Australia. *Transactions and Proceedings of the Royal Society of South Australia* **22**, 125–192.

Nott WJ (1856–c.1935) *When I Was a Boy. Memoirs of William John Nott 23/3/1856–c.1935*. South Australian Museum Archives, Adelaide.

Nunn PD (2020) In anticipation of extirpation: how ancient peoples rationalized and responded to postglacial sea level rise. *Environmental Humanities* **12**(1), 113–131. doi:10.1215/22011919-8142231

Nunn PD, Reid NJ (2015) Aboriginal memories of inundation of the Australian coast dating from more than 7000 years ago. *Australian Geographer* **47**(1), 11–47.

O'Bader M (1945) Of northern visitors, legends, birds, mines and Aborigines. *Chronicle*, South Australia. 27 September, p. 35.

O'Connell JF, Allen J, Williams MA, Williams AN, Turney CS *et al.* (2018) When did *Homo sapiens* first reach southeast Asia and Sahul? *Proceedings of the National Academy of Sciences of the United States of America* **115**(34), 8482–8490. doi:10.1073/pnas.1808385115

O'Dea K (1984) Marked improvement in carbohydrate and lipid metabolism in diabetic Australian Aborigines after temporary reversion to traditional lifestyle. *Diabetes* **33**(6), 596–603. doi:10.2337/diab.33.6.596

O'Dea K, Spargo RM, Akerman K (1980) The effect of transition from traditional to urban life-style on the insulin secretory response in Australian Aborigines. *Diabetes Care* **3**, 31–37. doi:10.2337/diacare.3.1.31

O'Dea K, Spargo RM, Nestel PJ (1982) Impact of westernization on carbohydrate and lipid metabolism in Australian Aborigines. *Diabetologia* **22**(3), 148–153. doi:10.1007/BF00283742

O'Malley P (1998) Emu farming. In *The New Rural Industries: A Handbook for Farmers and Investors.* (Ed. K Hyde) pp. 40-49. Rural Industries Research and Development Corporation, Canberra.

Olsen P (2018) *Night Parrot: Australia's Most Elusive Bird.* CSIRO Publishing, Melbourne.

Olsen P, Russell L (2019) *Australia's First Naturalists: Indigenous Peoples' Contribution to Early Zoology.* NLA Publishing, Canberra.

Padman EL (1987) *The Story of Narrung: The Place of Large She-oaks.* Lutheran Publishing House, Adelaide.

Palci A, Hutchinson MN, Caldwell MW, Scanlon JD, Lee MS (2018) Palaeoecological inferences for the fossil Australian snakes Yurlunggur and Wonambi (Serpentes, Madtsoiidae). *Royal Society Open Science* **5**(3), 172012. doi:10.1098/rsos.172012

Palmer K (1998) *Swinging the Billy: Indigenous and Other Styles of Australian Bush Cooking.* Aboriginal Studies Press, Canberra.

Parker SA, Reid NCH (1983) Birds. In *Natural History of the South East.* (Eds MJ Tyler, CR Twidale, JK Ling and JW Holmes) pp. 135–150. Royal Society of South Australia, Adelaide.

Parliament of the Commonwealth of Australia (1998) *Commercial Utilisation of Australian Native Wildlife.* Report of the Senate Rural and Regional Affairs and Transport References Committee. Parliament House, Canberra.

Pascoe B (2014) *Dark Emu: Black Seeds. Agriculture or Accident?* Magabala Books, Broome.

Patak AE, Baldwin J (1998) Pelvic limb musculature in the emu *Dromaius novaehollandiae* (Aves: Struthioniformes: Dromaiidae): adaptations to high-speed running. *Journal of Morphology* **238**(1), 23–37. doi:10.1002/(SICI)1097-4687(199810)238:1<23::AID-JMOR2>3.0.CO;2-O

Patterson C, Rich PV (1987) The fossil history of the emus, *Dromaius* (Aves: Dromaiinae). *Records of the South Australian Museum* **21**(2), 85–117.

Peckover WS, Filewood LWC (1976) *Birds of New Guinea and Tropical Australia: The Birds of Papua New Guinea, Irian Jaya, the Solomon Islands and Tropical North Australia.* AH & AW Reed, Sydney.

Peile AR (1980) Preliminary notes on the ethno-botany of the Gugadja Aborigines at Balgo, Western Australia. *Western Australian Herbarium Research Notes* **3**, 59–64.

Peile AR (1997) *Body and Soul: An Aboriginal View.* (Ed. P Bindon). Hesperian Press, Perth.

Pellis SM (1981) A description of social play by the Australian magpie *Gymnorhina tibicen* based on Eshkol-Wachman notation. *Bird Behaviour* **3**(3), 61–79. doi:10.3727/015613881791560685

Penney R [Cuique] (1842–43) The spirit of the Murray. Republished in 1991. *Journal of the Anthropological Society of South Australia* **29**(1), 24–87.

Pert PL, Ens EJ, Locke J, Clarke PA, Packer JM *et al.* (2015) An online spatial database of Australian Indigenous biocultural knowledge for contemporary natural and cultural resource management. *Science of the Total Environment* doi:10.1016/j.scitotenv.2015.01.073.

Peter JM (2006) *Some Indigenous Names for Australian Birds.* Birds Australia Report No.20. Birds Australia, Melbourne.

Peterson AM (1969) The plan of salvation as revealed in Aboriginal legends. *Signs of the Times* **84**(2), 18–21.

Peterson N (1977) Aboriginal uses of Australian Solanaceae. In *The Biology and Taxonomy of Solanaceae.* (Eds JG Hawkes, RN Lester and AD Skelding) pp. 171–188. Academic Press, London.

Petri H (1952) *Der Australische Medizenmann.* Republished 2014 as *The Australian Medicine Man.* (Ed. K Akerman). Hesperian Press, Perth.

Petty AM, deKoninck V, Orlove B (2015) Cleaning, protecting, or abating? Making Indigenous fire management 'work' in northern Australia. *Journal of Ethnobiology* **35**(1), 140–162. doi:10.2993/0278-0771-35.1.140

Piddington R (1932) Totemic system of the Karadjeri tribe. *Oceania* **2**(4), 373–400. doi:10.1002/j.1834-4461.1932.tb00041.x

Pine T (1897) A legend of 'Ti-ya-tinity', the screech owl of Australia. *Journal of the Polynesian Society (N. Z.)* **64**(4), 169–173.

Plomley NJB (1966) *Friendly Mission: The Tasmanian Journals and Papers of George Augustus Robinson. 1829–1834*. Tasmanian Historical Research Association, Hobart.

Pretty GL (1977) The cultural chronology of the Roonka Flat. In *Stone Tools as Cultural Markers*. (Ed. RVS Wright) pp. 288–331. Prehistory and Material Cultural. Series No.12. Australian Institute of Aboriginal Studies, Canberra.

Prober SM, O'Connor MH, Walsh FJ (2011) Australian Aboriginal peoples' seasonal knowledge: a potential basis for shared understanding in environmental management. *Ecology and Society* **16**(2), 12. doi:10.5751/ES-04023-160212

Purnomo GA, Mitchell KJ, O'Connor S, Kealy S, Taufik L *et al.* (2021) Mitogenomes reveal two major influxes of Papuan ancestry across Wallacea following the last glacial maximum and Austronesian contact. *Genes* **12**(7), 965. doi:10.3390/genes12070965

Pursche K (2004) *Aboriginal Management and Planning for Country: Respecting and Sharing Traditional Knowledge. Full Report on Subprogram 5 of the Ord-Bonaparte Program*. Australian Government, Land and Water Australia; Kimberley Land Council, Canberra.

Puruntatameri J, Puruntatameri R, Pangiraminni A, Burak L, Tipuamantymirri C *et al.* (2001) *Tiwi Plants and Animals: Aboriginal Flora and Fauna Knowledge from Bathurst and Melville Islands, Northern Australia*. Parks and Wildlife Commission of the Northern Territory and Tiwi Land Council, Darwin.

Pyke GH, Ehrlich PR (2014) Conservation and the holy grail: the story of the night parrot. *Pacific Conservation Biology* **20**(2), 221–226. doi:10.1071/PC140221

Radcliffe-Brown AR (1923) Notes on the social organization of Australian tribes. Part II. The Wongaibon tribe. *Journal of the Royal Anthropological Institute of Great Britain and Ireland* **53**, 424–447. doi:10.2307/2843580

Radcliffe-Brown AR (1929) Notes on totemism in eastern Australia. *Journal of the Royal Anthropological Institute of Great Britain and Ireland* **59**, 399–415. doi:10.2307/2843892

Radcliffe-Brown AR (1958) The comparative method in social anthropology. In *Method in Social Anthropology: Selected Essays by A.R. Radcliffe-Brown*. (Ed. MN Srinivas) pp. 108–129. University of Chicago Press, Chicago.

Ramsay Smith W (1909) The Aborigines of Australia. Reprinted from *Official Year Book of the Commonwealth No. 3*. McCarron Bird & Co., Melbourne.

Ramsay Smith W (1930) *Myths and Legends of the Australian Aboriginals*. Harrap, Sydney.

Ramson WS (1966) *Australian English: An Historical Study of the Vocabulary 1788–1898*. ANU Press, Canberra.

Ramson WS (Ed.) (1988) *The Australian National Dictionary: A Dictionary of Australianisms on Historical Principles*. Oxford University Press, Oxford.

Rankine HJ (1991) A talk by Henry Rankine. *Journal of the Anthropological Society of South Australia* **29**(2), 108–127.

Ratsch A, Steadman KJ, Bogossian F (2010) The pituri story: a review of the historical literature surrounding traditional Australian Aboriginal use of nicotine in Central Australia. *Journal of Ethnobiology and Ethnomedicine* **6**, 26. doi:10.1186/1746-4269-6-26

Raven M, Robinson D, Hunter J (2021) The emu: more-than-human and more-than-animal geographies. *Antipode* **53**(5), 1526–1545. https://onlinelibrary.wiley.com/doi/full/10.1111/anti.12736.

Raymond E, Blutja J, Gingina L, Raymond M, Raymond O *et al.* (1999) *Wardaman Ethnobiology*. Northern Territory Botanical Bulletin No. 25. Parks and Wildlife Commission of the Northern Territory and the Northern Territory University, Darwin.

Reed AW (1978) *Aboriginal Legends: Animal Tales*. Reed Books, Sydney.

Reed AW (1980) *Aboriginal Stories of Australia*. Reed Books, Sydney.

Reid E (1977) *The Records of Western Australian Plants Used by Aboriginals as Medicinal Agents*. School of Pharmacy, Western Australian Institute of Technology, Perth.

Reid JC (1978a) The role of the Marrnggitji in contemporary health care. *Oceania* **49**(2), 96–109. doi:10.1002/j.1834-4461.1978.tb01381.x

Reid JC (1978b) Change in the Indigenous medical system of an Aboriginal community. *Newsletter – Australian Institute of Aboriginal Studies* **9**, 61–72.

Reid JC (1979) Health as harmony, sickness as conflict. *Hemisphere* **23**(4), 194–199.

Reid JC (Ed.) (1982a) *Body, Land and Spirit: Health and Healing in Aboriginal Society.* University of Queensland Press, Brisbane.

Reid JC (1982b) Introduction: 'The Australian Problem'. In *Body, Land and Spirit: Health and Healing in Aboriginal Society.* (Ed. JS Reid) pp. ix–xvi. University of Queensland Press, Brisbane.

Reid JC (1983) *Sorcerers and Healing Spirits: Continuity and Change in an Aboriginal Medical System.* ANU Press, Canberra.

Reid A (1995) *Banksias and Bilbies: Seasons of Australia.* Gould League, Melbourne.

Rhodes J (2018) *Cage of Ghosts.* Darkwood, New South Wales.

Rich PV (1979) *The Dromornithidae: An Extinct Family of Large Ground Birds Endemic to Australia.* BMR Bulletin No. 184. Australian Government Publishing Service, Canberra.

Rich PV (1985) *Genyornis newtoni* A mihirung. In *Kadimakara: Extinct Vertebrates of Australia.* (Ed. PV Rich, GF van-Tets and F Knight) pp. 188–194. Pioneer Design Studio, Melbourne.

Richards C (1903) Marraa' Warree', or Marrae' Arree'. *Science of Man* **31**(December), 163–169.

Roberts MJ, Roberts A (1975) *Dreamtime Heritage.* Rigby, Adelaide.

Roberts RG, Flannery TF, Ayliffe LK, Yoshida H, Olley JM (2001) New ages for the last Australian megafauna: continent-wide extinction about 46,000 years ago. *Science* **292**(5523), 1888–1892. doi:10.1126/science.1060264

Roberts A, van Duivenvoorde W, Morrison M, Moffat I, Burke H *et al.* (2017) 'They call 'im Crowie': an investigation of the Aboriginal significance attributed to a wrecked River Murray barge in South Australia. *International Journal of Nautical Archaeology* **46**(1), 132–148. doi:10.1111/1095-9270.12208

Roheim G (1971) *Australian Totemism: A Psychoanalytic Study in Anthropology.* Frank Cass & Co., London.

Roheim G (1974) *Children of the Desert: The Western Tribes of Central Australia.* Basic Books, New York.

Romance M (1994) Emu oil has arthritis on the run. *Times*, South Australia. 14 October, p. 17.

Rose FGG (1947) Malay influence on Aboriginal totemism in northern Australia. *Man* **47**, 129. doi:10.2307/2791468

Rose FGG (1987) *The Traditional Mode of Production of the Australian Aborigines.* Angus & Robertson, Sydney.

Ross MC (1986) Australian Aboriginal oral traditions. *Oral Tradition* **1**(2), 231–271.

Roth WE (1897) *Ethnological Studies Among the North-West-Central Queensland Aborigines.* Government Printer, Brisbane.

Roth WE (1901a) *String and Other Forms of Strand: Basketry, Woven Bag and Net-work.* North Queensland Ethnography Bulletin No. 1. Government Printer, Brisbane.

Roth WE (1901b) *The Structure of the Koko-Yimidir Language.* North Queensland Ethnography Bulletin No. 2. Government Printer, Brisbane.

Roth WE (1901c) *Food: Its Search, Capture and Preparation.* North Queensland Ethnography Bulletin No. 3. Government Printer, Brisbane.

Roth WE (1902) *Games, Sports and Amusements.* North Queensland Ethnography Bulletin No. 4. Government Printer, Brisbane.

Roth WE (1903) *Superstition, Magic and Medicine.* North Queensland Ethnography Bulletin No. 5. Government Printer, Brisbane.

Roth WE (1904) *Domestic Implements, Arts and Manufactures.* North Queensland Ethnography Bulletin No. 6. Government Printer, Brisbane.

Russell-Smith J, Whitehead PJ, Cooke P (2009) *Culture, Ecology and Economy of Fire Management in Northern Australian Savannas: Rekindling the Wurrk Tradition.* CSIRO Publishing, Melbourne.

Sackett L (1979) The pursuit of prominence: hunting in an Australian Aboriginal community. *Anthropologica* **21**(2), 223–246. doi:10.2307/25605025

Saethre E (2007) Close encounters: UFO beliefs in a remote Australian Aboriginal community. *Journal of the Royal Anthropological Institute* **13**, 901–915. doi:10.1111/j.1467-9655.2007.00463.x

Saitoh F, Noma M, Kawashima N (1985) The alkaloid contents of sixty *Nicotiana* species. *Phytochemistry* **24**(3), 477–480. doi:10.1016/S0031-9422(00)80751-7

Sales J (2007) The emu (*Dromaius novaehollandiae*): a review of its biology and commercial products. *Avian and Poultry Biology Reviews* **18**(1), 1–20. doi:10.3184/147020607X245048

Sanger EB (1883) Notes on the Aborigines of Cooper's Creek, Australia. *American Naturalist* **17**, 1220–1225. doi:10.1086/273533

Satterthwait L (1986) Aboriginal Australian net hunting. *Mankind* **16**(1), 31–48.

Sault N (2010) Bird messengers for all seasons: landscapes of knowledge among the Bribri of Costa Rica. In *Ethno-ornithology: Birds, Indigenous Peoples, Culture and Society*. (Eds S Tidemann and A Gosler) pp. 291–300. Earthscan, London.

Sayers A (1994) *Aboriginal Artists of the Nineteenth Century*. Oxford University Press, Oxford.

Scanlon JD (1992) A new large madtsoiid snake from the Miocene of the Northern Territory. *The Beagle: Records of the Northern Territory Museum of Arts amd Sciences* **9**(1), 49–60. doi:10.5962/p.263117

Scarlett N, White N, Reid JC (1982) 'Bush medicines': the pharmacopoeia of the Yolngu of Arnhem Land. In *Body, Land and Spirit: Health and Healing in Aboriginal Society*. (Ed. JC Reid) pp. 154–191. University of Queensland Press, Brisbane.

Schell (1914) The river blacks: a vanished race – interesting native customs. *Murray Pioneer and Australian River Record*, South Australia. 30 April, p. 8.

Schmidt A (1993) *The Loss of Australia's Aboriginal Language Heritage*. Aboriginal Studies Press, Canberra.

Schodde R, Glover B, Kinsky FC, Marchant S, McGill AR *et al.* (1978) Recommended English names for Australian birds. *Emu* **77**(Suppl), 245–307. doi:10.1071/MU9770245s

Schürmann CW (1844) *Vocabulary of the Parnkalla Language Spoken by the Natives Inhabiting the Western Shores of Spencer's Gulf*. Dehane, Adelaide.

Schürmann CW (1846) The Aboriginal tribes of Port Lincoln in South Australia: their mode of life, manners, customs … In *The Native Tribes of South Australia*. (Ed. JD Woods, 1879) pp. 207–252. Dehane, Adelaide.

Scott J, Laurie R (2007) Colonialism on display: Indigenous people and artefacts at an Australian agricultural show. *Aboriginal History* **31**, 45–62.

Scott HH, Scott EOG (1934) Nature and science: curator's notes – blackfellow buttons. *Examiner*, Tasmania. 23 January, p. 10.

Sculthorpe G, La Combe M, Lakic M (1990) *Guide to Victorian Aboriginal Collections in the Museum of Victoria*. Museum of Victoria Council, Melbourne.

Seddon PJ, Launay F (2008) Arab falconry: changes, challenges and conservation opportunities of an ancient art. In *Tourism and the Consumption of Wildlife: Hunting, Shooting and Sport Fishing*. (Ed. B Lovelock) pp. 196–210. Routledge, London.

SENEX (1923) Life and lore of the bush: the bunyip myth. *Sunday Times*, Western Australia. 24 June, p. 5.

SENEX (1927) Life and lore of the bush: some Aboriginal names and myths. *Sunday Times*, Western Australia. 27 February, p. 44.

Shackeroff JM, Campbell LM (2007) Traditional ecological knowledge in conservation research: problems and prospects for their constructive engagement. *Conservation & Society* **5**, 343–360.

Shakespeare W (1597) *The Tragedy of Romeo and Juliet*. https://gutenberg.org/ebooks/1513

Sharp RL (1939) Tribes and totemism in north-east Australia. *Oceania* **9**(3), 254–275; **9**(4), 439–461.

Sherrie WM (1918) On the land: bird names. *Sydney Morning Herald*, New South Wales. 26 January, p. 11.

Sherwood J, McNiven IJ, Laurenson L, Richards T, Bowler J (2016) Prey selection, size and breakage differences in *Turbo undulatus opercula* found within Pacific Gull (*Larus pacificus*) middens compared to Aboriginal middens and natural beach deposits, southeast Australia. *Journal of Archaeological Science, Reports* **6**, 14–23. doi:10.1016/j.jasrep.2016.01.018

Shoemaker A (2000) The headless state. *Griffith Law Review* **9**(2), 358.

Shute E, Prideaux GJ, Worthy TH (2017) Taxonomic review of the late Cenozoic megapodes (Galliformes: Megapodiidae) of Australia. *Royal Society Open Science* **4**(6), 170233. doi:10.1098/rsos.170233

Si A (2019) Flora–fauna loanwords in Arnhem Land and beyond: an ethnobiological approach. *Australian Journal of Linguistics* **39**(2), 202–256. doi:10.1080/07268602.2019.1566888

Silberbauer GB (1981) *Hunter and Habitat in the Central Kalahari Desert*. Cambridge University Press, Cambridge.

Silcock JL, Tischler M, Smith MA (2012) Quantifying the Mulligan River pituri, *Duboisia hopwoodii* ((F. Muell.) F.Muell.) (Solanaceae), trade of Central Australia. *Ethnobotany Research and Applications* **10**, 37–44. doi:10.17348/era.10.0.037-044

Silverleaf (1883) The Aborigines of New South Wales: their habits, laws and customs. *Illustrated Sydney News*, New South Wales. 14 April, p. 14.

Simpson J (1998) Aboriginal personal names on the Fleurieu Peninsula at the time of the invasion. In *History in Portraits: Biographies of Nineteenth Century South Australian Aboriginal People*. Monograph 6. (Eds J Simpson and LA Hercus) pp. 1–13. Aboriginal History, Canberra.

Sinclair JR, Tuke L, Opiang M (2010) What the locals know: comparing traditional and scientific knowledge of megapodes in Melanesia. In *Ethno-ornithology: Birds, Indigenous Peoples, Culture and Society*. (Eds S Tidemann and A Gosler) pp. 115–137. Earthscan, London.

Singer R, Garidjalalug N, Urabadi R, Hewett H, Mirwuma P *et al.* (2021) *Mawng to English Dictionary*. Aboriginal Studies Press, Canberra.

Skira I (1996) Aboriginal people and muttonbirding in Tasmania. In *Sustainable Use of Wildlife by Aboriginal Peoples and Torres Strait Islanders*. (Eds M Bomford and J Caughley) pp. 167–175. Australian Government Publishing Service, Canberra.

Slater P (1970) *A Field Guide to Australian Birds: Non-passerines*. Rigby, Adelaide.

Slater P (1974) *A Field Guide to Australian Birds: Passerines*. Rigby, Adelaide.

Slater P (1978) *Rare and Vanishing Australian Birds*. Rigby, Adelaide

Smith C (1880) *The Boandik Tribe of South Australian Aborigines: A Sketch of Their Habit,s Customs, Legends and Language*. Government Printer, Adelaide.

Smith H (1990) *Tiwi: The Life and Art of Australia's Tiwi People*. Angus & Robertson, Sydney.

Smith MA (2013) *The Archaeology of Australia's Deserts*. Cambridge University Press, Cambridge.

Smith MA (2021) Historizing the 'Dreaming': an archaeological perspective from arid Australia. In *The Oxford Handbook of the Archaeology of Indigenous Australia and New Guinea*. (Eds IJ McNiven and B David). Oxford Handbooks Online, Oxford. https://www.oxfordhandbooks.com/view/10.1093/oxfordhb/9780190095611.001.0001/oxfordhb-9780190095611

Smith MJ (1976) Small fossil vertebrates from Victoria Cave, Naracoorte, South Australia. IV. Reptiles. *Transactions of the Royal Society of South Australia* **100**(1), 39–51.

Smith NM (1991) Ethnographic field notes from the Northern Territory Australia. *Journal of the Adelaide Botanic Gardens* **14**(1), 1–65.

Smith NM (2013) Contested discourses: Aboriginal attitudes towards non-native plants and engagement in weed management in Cape York, Northern Australia. PhD thesis. Research Institute for the Environment and Livelihoods, Faculty of Education, Health, Science and Environment, Charles Darwin University, Brisbane.

Smith NM, Wididburu B, Harrington RN, Wightman GM (1993) *Ngarinyman Ethnobotany: Aboriginal Plant Use from the Victoria River Area, Northern Australia*. Northern Territory Botanical Bulletin No. 16. Conservation Commission of the Northern Territory, Darwin.

Smith PA (1965) *Moonbird People*. Rigby, Adelaide.

Smyth RB (1878) *The Aborigines of Victoria*. 2 volumes. Government Printer, Melbourne.

Smyth D, Szabo S, George M (2004) *Case Studies in Indigenous Engagement in Natural Resource Management in Australia*. Department of Environment and Heritage, Canberra.

Soong FS (1983) Role of the margidbu (traditional healer) in western Arnhem Land. *Medical Journal of Australia* 1, 474–477. doi:10.5694/j.1326-5377.1983.tb136170.x

Sorenson ES (1909) The emu. *Queensland Times*, Queensland. 29 March, p. 2.

Sorenson ES (1920) Aboriginal names of birds. *Emu* 20, 32–33. doi:10.1071/MU920032

Spencer WB (1896) *Report on the Work of the Horn Scientific Expedition to Central Australia. Part 1.* Melville, Mullin & Slade, Melbourne.

Spencer WB (1913) *Preliminary Report on the Aboriginals of the Northern Territory.* Bulletin of the Northern Territory No. 7. Commonwealth of Australia, Melbourne.

Spencer WB (1918) Kitchen middens and native ovens. *Victorian Naturalist* 35, 113–118.

Spencer WB (1922) *Guide to the Australian Ethnological Collection Exhibited in the National Museum of Victoria.* Trustees of the Public Library, Museums and National Gallery of Victoria, Melbourne.

Spencer WB (1928) *Wanderings in Wild Australia.* Macmillan, London.

Spencer WB, Gillen FJ (1899) *The Native Tribes of Central Australia.* Macmillan, London.

Spencer WB, Gillen FJ (1904) *The Northern Tribes of Central Australia.* Macmillan, London.

Spencer WB, Gillen FJ (1927) *The Arunta: A Study of a Stone Age People.* Macmillan, London.

Stanbridge WE (1857) On the astronomy and mythology of the Aborigines of Victoria. *Transactions of the Philosophical Institute of Victoria* 2, 137–140.

Stanbridge WE (1861) Some particulars of the general characteristics, astronomy and mythology of the tribes in the central part of Victoria, southern Australia. *Transactions of the Ethnological Society of London* 1, 286–304. doi:10.2307/3014201

Stanner WEH (1937) Aboriginal modes of address and reference in the north-west of the Northern Territory. *Oceania* 7, 300–315. doi:10.1002/j.1834-4461.1937.tb00385.x

Stanner WEH (1953) The Dreaming. In *White Man Got No Dreaming: Essays 1938–1973.* (Ed. WEH Stanner, 1979) pp. 23–40. ANU Press, Canberra.

Stanner WEH (1963) *On Aboriginal Religion.* Republished in 2014. Sydney University Press, Sydney.

Stanner WEH (1965) Religion, totemism and symbolism. In *Aboriginal Man in Australia.* (Eds RM Berndt and CH Berndt) pp. 207–237. Angus & Robertson, Sydney.

Stephens E (1889) The Aborigines of Australia. *Journal and Proceedings of the Royal Society of New South Wales* 23, 476–503.

Stirling EC (1911) Preliminary report on the discovery of native remains at Swanport, River Murray, with an enquiry into the alleged occurrence of a pandemic among the Australian Aborigines. *Transactions of the Royal Society of South Australia* 35, 4–46.

Stirling EC, Zeitz AHC (1896a) Preliminary notes on *Genyornis newtoni*; a new genus and species of fossil struthious bird found at Lake Callabonna, South Australia. *Transactions and Proceedings of the Royal Society of South Australia* 20, 171–190.

Stirling EC, Zeitz AHC (1896b) *Genyornis newtoni*: a fossil struthious bird from Lake Callabonna, South Australia. Description of the bones of the leg and foot. *Transactions and Proceedings of the Royal Society of South Australia* 20, 191–211.

Stone AC (1911) Aborigines of Lake Boga. *Proceedings of the Royal Society of Victoria* 23, 433–468.

Strehlow C (1909) *Comparative Heritage Dictionary: An Aranda, German, Loritja and Dieri to English Dictionary, with Introductory Essays.* Republished in 2018. ANU Press, Canberra.

Strehlow TGH (1933) Ankotarinja: an Aranda myth. *Oceania* 4(2), 187–200. doi:10.1002/j.1834-4461.1933.tb00100.x

Strehlow TGH (1947) *Aranda Traditions.* Melbourne University Press, Melbourne.

Strehlow TGH (1970) Geography and the totemic landscape in Central Australian: a functional study. In *Australian Aboriginal Anthropology: Modern Studies in the Social Anthropology of the Australian Aborigines.* (Ed. RM Berndt) pp. 92–140. UWA Publishing, Perth.

Striedter GF (2013) Bird brains and tool use: beyond instrumental conditioning. *Brain Behavior and Evolution* 82(1), 55–67. doi:10.1159/000352003

Stuart JM (1865) *Explorations in Australia: The Journals of John McDouall Stuart During the Years 1858, 1859, 1860, 1861, and 1862, When he Fixed the Centre of the Continent and Successfully Crossed it from Sea to Sea.* Saunders & Otley, London.

Sturtevant WC (1964) Studies in ethnoscience. *American Anthropologist* **66**(3), 99–131. doi:10.1525/aa.1964.66.3.02a00850

Sullivan C (1928) Bird notes from the west coast. *South Australian Ornithologist* **9**(5), 164–169.

Sutton P (1978a) WIK: Aboriginal society, territory and language at Cape Keerweer, Cape York Peninsula, Australia. PhD thesis. Department of Anthropology and Sociology, University of Queensland, Brisbane.

Sutton P (1978b) Bird names in Wik-ngathan. Manuscript. South Australian Museum, Adelaide.

Sutton P (1982) Personal power, kin classification and speech etiquette in Aboriginal Australia. In *Languages of Kinship in Aboriginal Australia*. (Eds J Heath, F Merlan and A Rumsey) pp. 182–200. Oceania Linguistic Monographs No. 24. Oceania, Sydney.

Sutton P (Ed.) (1988a) *Dreamings: The Art of Aboriginal Australia*. Penguin Books, Melbourne.

Sutton P (1988b) Dreamings. In *Dreamings: The Art of Aboriginal Australia*. (Ed. P Sutton) pp. 13–32. Penguin Books, Melbourne.

Sutton P (1988c) Myth as history, history as myth. In *Being Black. Aboriginal Cultures in 'Settled' Australia*. (Ed. I Keen) pp. 251–268. Aboriginal Studies Press, Canberra.

Sutton P (1995) *Wik-Ngathan Dictionary*. Caitlin Press, Adelaide.

Sutton P (2003a) *Native Title in Australia: An Ethnographic Perspective*. Cambridge University Press, Cambridge.

Sutton P (2003b) Sacred images and political engagements: a brief history of Wik sculpture. In *Story Place: Indigenous Art of Cape York and the Rainforest*. (Ed. I Were) pp. 54–59, 215–216. Queensland Art Gallery, Brisbane.

Sutton P, Snow M (2015) Iridescence. In *Re-materializing Colour*. (Ed. D Young) pp. 121–143. Sea Kingston Publishing, Oxford.

Sutton P, Walshe K (2021) *Farmers or Hunter-gatherers? The Dark Emu Debate*. Melbourne University Publishing, Melbourne.

Swain T (1990) A new sky hero from a conquered land. *History of Religions* **29**, 195–232. doi:10.1086/463193

Swain T (1993) *A Place for Strangers: Towards a History of Australian Aboriginal Being*. Cambridge University Press, Cambridge.

Taçon PSC (1991) The power of stone: symbolic aspects of stone use and tool development in western Arnhem Land, Australia. *Antiquity* **65**, 192–207. doi:10.1017/S0003598X00079655

Taçon PSC, South B, Hooper SB (2003) Depicting cross-cultural interaction: figurative designs in wood, earth and stone from south-east Australia. *Archaeology in Oceania* **38**, 89–101. doi:10.1002/j.1834-4453.2003.tb00532.x

Taçon PSC, Langley M, May SK, Lamilami R, Brennan W *et al.* (2010) Ancient bird stencils discovered in Arnhem Land, Northern Territory, Australia. *Antiquity* **84**, 416–427. doi:10.1017/S0003598X00066679

Taplin G (1874) The Narrinyeri. In *The Native Tribes of South Australia*. (Ed. JD Woods, 1879) pp. 1–156. ES Wiggs, Adelaide.

Taplin G (1879) *Folklore, Manners, Customs and Languages of the South Australian Aborigines*. Government Printer, Adelaide.

Taplin F (1889) An Australian native fifty years ago. *Adelaide Observer*, South Australia. 6 April, p. 41.

Taplin G (1859–79) *Journals*. Mortlock Library, Adelaide.

Taylor I (1985) Canberra once a home for flocks of emus. *Canberra Times*, Australian Capital Territory. 17 March, p. 10.

Taylor L (1996) *Seeing the Inside: Bark Painting in Western Arnhem Land*. Oxford University Press, Oxford.

Teichelmann CG (1857) Dictionary of the Adelaide dialect. Manuscript. South African Public Library, Cape Town.

Teichelmann CG, Schürmann CW (1840) *Outlines of a Grammar, Vocabulary and Phraseology, of the Aboriginal Language of South Australia*. Thomas & Co., Adelaide.

Telamon (1930) Australiana. *The West Australian*, Western Australia. 27 December, p. 4.

Tellurian (1935) Nature and science: Science notes. *The Australasian*, Victoria. 9 November, p. 46.

Tench W (1789) *1788. Comprising a Narrative of the Expedition to Botany Bay and a Complete Account of the Settlement of Port Jackson*. Republished in 1996. Text Publishing, Melbourne.

TenHouten WD (1999) Text and temporality: patterned-cyclical and ordinary-linear forms of time-consciousness, inferred from a corpus of Australian Aboriginal and Euro-Australian life-historical interviews. *Symbolic Interaction* **22**(2), 121–137. doi:10.1525/si.1999.22.2.121

Tenison-Woods JE (1882) Physical structure and geology of Australia. *Proceedings of the Linnean Society of New South Wales* **7**, 371–392.

Tennant Kelly C (1935) Tribes on Cherburg settlement, Queensland. *Oceania* **5**(4), 461–473. doi:10.1002/j.1834-4461.1935.tb00165.x

Teulon GN (1886) Bourke, Darling River. In *The Australian Race: Its Origins, Languages, Customs, Place of Landing in Australia and the Routes by which it Spread Itself over the Continent*. Vol. 2. (Ed. EM Curr) pp. 186–187. Trubner, London.

Thieberger N, McGregor W (Eds) (1994) *Macquarie Aboriginal Words*. Macquarie Library, Macquarie University, Sydney.

Thomas NW (1906) *The Natives of Australia*. Archibald Constable, London.

Thomas K (1991) *Man and the Natural World: Changing Attitudes in England 1500–1800*. Penguin, London.

Thomas WH (2010) Everyone loves birds: using Indigenous knowledge of birds to facilitate conservation in Papua New Guinea. In *Ethno-ornithology: Birds, Indigenous Peoples, Culture and Society*. (Eds S Tidemann and A Gosler) pp. 265–278. Earthscan, London.

Thomas M (2011) *The Many Worlds of R.H. Mathews: In Search of an Australian Anthropologist*. Allen & Unwin, Sydney.

Thomson DF (1933) The hero cult, initiation and totemism on Cape York. *Journal of the Royal Anthropological Institute of Great Britain and Ireland* **63**, 453–537. doi:10.2307/2843801

Thomson DF (1935) *Birds of Cape York Peninsula: Ecological Notes, Field Observations and Catalogue of Specimens Collected on Three Expeditions to North Queensland*. H.J. Green, Government Printer, Melbourne.

Thomson DF (1936a) Fatherhood in the Wik Monkan tribe. *American Anthropologist* **38**, 374–393. doi:10.1525/aa.1936.38.3.02a00030

Thomson DF (1936b) Notes on some bone and stone implements from north Queensland. *Journal of the Royal Anthropological Institute of Great Britain and Ireland* **66**, 71–74. doi:10.2307/2844118

Thomson DF (1939) The seasonal factor in human culture illustrated from the life of a contemporary nomadic group. *Proceedings of the Prehistoric Society* **5**(2), 209–221. doi:10.1017/S0079497X00020545

Thomson DF (1946) Names and naming in the Wik Mongkan tribe. *Journal of the Royal Anthropological Institute* **76**, 157–167.

Thomson DF (1949) *Economic Structure and the Ceremonial Exchange Cycle in Arnhem Land*. Macmillan, London.

Thomson DF (1975) *Bindibu Country*. Nelson, Melbourne.

Thomson DF (1983a) *Donald Thomson in Arnhem Land*. (Ed. N Peterson). Currey O'Niel, Melbourne.

Thomson DF (1983b) *Children of the Wilderness*. Currey O'Niel, Melbourne.

Thomson DF (1996) *Thomson Time: Arnhem Land in the 1930s – A Photographic Essay*. (Ed. L Allen). Museum of Victoria, Melbourne.

Thornton A (1940) Bush remedies: old timer's pharmacopoeia. *Chronicle*, South Australia. 14 November, p. 34.

Thurman J (2014) Cave men, luminoids and dragons: monstrous creatures mediating relationships between people and Country in Aboriginal northern Australia. In *Monster Anthropology in Australasia and Beyond*. (Eds Y Musharbash and GH Presterudstuen) pp. 25–38. Springer, The Hague.

Tidemann S, Gosford R (2006) Capturing indigenous ornithology. *Journal of Ornithology* **147**(5), 283.

Tidemann S, Gosler A (Eds) (2010) *Ethno-ornithology: Birds, Indigenous Peoples, Culture and Society.* Earthscan, London.

Tidemann S, Whiteside T (2010) Aboriginal stories: the riches and colour of Australian birds. In *Ethno-ornithology: Birds, Indigenous Peoples, Culture and Society.* (Eds S Tidemann and A Gosler) pp. 153–179. Earthscan, London.

Tidemann S, Chirgwin S, Sinclair JR (2010) Indigenous knowledges, birds that have 'spoken' and science. In *Ethno-ornithology: Birds, Indigenous Peoples, Culture and Society.* (Eds S Tidemann and A Gosler) pp. 3–12. Earthscan, London.

Tindale NB (1924) *Rough Diary of Trip to Northern Flinders Ranges S. Australia Nov.14–Dec. 18th 1924, Incorporating Various Notes on the Aborigines.* AA338/1/3, South Australian Museum Archives, Adelaide.

Tindale NB (1925) Natives of Groote Eylandt and of the west coast of the Gulf of Carpentaria: Part I. *Records of the South Australian Museum* **3**(1), 61–102.

Tindale NB (1930–52) *Murray River Notes.* AA338/1/31/1, South Australian Museum Archives, Adelaide.

Tindale NB (1931–34) *Journal of Researches in the South East of South Australia.* Vol. I. AA338/1/33/1, South Australian Museum Archives, Adelaide.

Tindale NB (1931–62) *Aboriginal Vocabulary Cards.* AA338/7/2/23-24 and 7/1/16, South Australian Museum Archives, Adelaide.

Tindale NB (1934–37) *Journal of Researches in the South East of South Australia.* Vol. II. AA338/1/33/2, South Australian Museum Archives, Adelaide.

Tindale NB (1934a) Vanished tribal life of Coorong blacks: tragedy of supplanted race. Country's changed aspect. *The Advertiser*, South Australia. 7 April, p. 9.

Tindale NB (1934b) *Map of the Coorong and Southeastern South Australia.* AA338/16/27, South Australian Museum Archives, Adelaide.

Tindale NB (1935) Legend of Waijungari, Jaralde tribe, Lake Alexandrina, South Australia and the phonetic system employed in its transcription. *Records of the South Australian Museum* **5**(3), 261–274.

Tindale NB (1936a) *Map for Emu and Brolga Text.* AA338/16/19/1-2, South Australian Museum Archives, Adelaide.

Tindale NB (1936b) Wonders of the Coorong: effect of sea movements and volcanoes. Did Aborigines see eruptions? *The Advertiser*, South Australia. 14 May, pp. 18,23.

Tindale NB (1936c) Notes on the natives of the southern portion of Yorke Peninsula, South Australia. *Transactions of the Royal Society of South Australia* **60**, 55–70.

Tindale NB (1937a) Native songs of the south east of South Australia. *Transactions of the Royal Society of South Australia* **61**, 107–120.

Tindale NB (1937b) Two legends of the Ngadjuri tribe from the middle north of South Australia. *Transactions of the Royal Society of South Australia* **61**, 149–153.

Tindale NB (1938) Prupe and Koromarange: a legend of the Tanganekald Coorong, South Australia. *Transactions of the Royal Society of South Australia* **62**, 18–23.

Tindale NB (1938–56) *Journal of Researches in the South East of South Australia.* Vol. III. AA338/1/33/3, South Australian Museum Archives, Adelaide.

Tindale NB (1938–60) *Tja:pukai Grammar, Kuranda Queensland and Research Notes on Queensland Tribes.* AA338/1/38, South Australian Museum Archives, Adelaide.

Tindale NB (1939) Eagle and crow myths of the Maraura tribe, Lower Darling River NSW. *Records of the South Australian Museum* **6**, 243–261.

Tindale NB (1940–56) *Journal of Campsites and Stone Implements of the Australian Aborigines and Others.* AA338/1/40/1, South Australian Museum Archives, Adelaide.

Tindale NB (1941a) A list of plants collected in the Musgrave and Mann Ranges, South Australia, 1933. *South Australian Naturalist* **21**(1), 8–12.

Tindale NB (1941b) Native songs of the south east of South Australia. Part II. *Transactions of the Royal Society of South Australia* **65**(2), 233–243.

Tindale NB (1951) Comments on supposed representations of giant bird tracks at Pimba. *Records of the South Australian Museum* **9**(4), 381–382.

Tindale NB (1955) *Vocabularies of Antakarinja, Kokata, Mirning, Nukunu, Potaruwutji, Marditjali, Pintubi, Pitjandjara, Tanganekald and Wirangu.* AA338/7/1-2, South Australian Museum Archives, Adelaide.

Tindale NB (1959) Ecology of primitive Aboriginal man in Australia. In *Biography and Ecology in Australia.* (Ed. A Keast) pp. 36–51. Springer, The Hague.

Tindale NB (1960) *Visit to Bentinck and Mornington Islands, Queensland by Norman B. Tindale.* AA338/1/23, South Australian Museum Archives, Adelaide.

Tindale NB (1963) *Journal of Visit to the Rawlinson Range Area in the Great Western Desert by Norman B. Tindale 24 October–25 November 1963.* AA338/1/26, South Australian Museum Archives, Adelaide.

Tindale NB (1964) *Murray River Notes.* AA338/1/31/2, South Australian Museum Archives, Adelaide.

Tindale NB (1964–65) *Entomology and General Field Trips by Norman B. Tindale.* AA338/1/43/3, South Australian Museum Archives, Adelaide.

Tindale NB (1966) Insects as food for the Australian Aborigines. *Australian Natural History* **15**(6), 179–183.

Tindale NB (1972) The Pitjandjara. In *Hunters and Gatherers Today: A Socioeconomic Study of Eleven such Cultures in the Twentieth Century.* (Ed. MG Bicchieri) pp. 217–268. Holt, Rinehart & Winston, New York.

Tindale NB (1974) *Aboriginal Tribes of Australia: Their Terrain, Environmental Controls, Distribution, Limits and Proper Names.* ANU Press, Canberra.

Tindale NB (1976) Letter to RMW Dixon, 6 March 1976. Correspondence files. Anthropology Archives, South Australian Museum, Adelaide.

Tindale NB (1977) Adaptive significance of the Panara or grass seed culture of Australia. In *Stone Tools as Cultural Markers.* (Ed. RVS Wright) pp. 345–349. Australian Institute of Aboriginal Studies, Canberra.

Tindale NB (1978) Notes on a few Australian Aboriginal concepts. In *Australian Aboriginal Concepts.* (Ed. LR Hiatt) pp. 156–163. Australian Institute of Aboriginal Studies, Canberra.

Tindale NB (1981) Desert Aborigines and the southern coastal peoples: some comparisons. In *Ecological Biogeography of Australia.* (Ed. A Keast) pp. 1855–1884. Junk, The Hague.

Tindale NB (1986) Milerum (1869–1941). In *Australian Dictionary of Biography. Vol. 10: 1891–1939 Lat-Ner.* (Eds. B Nairn and G Searle) pp. 498–499. Melbourne University Press, Melbourne.

Tindale NB (1987a) The wanderings of Tjirbruki: a tale of the Kaurna people of Adelaide. *Records of the South Australian Museum* **20**, 5–13.

Tindale NB (1987b) *Yaralde Tribe Place Names, June 1987.* AAA338/8/17, South Australian Museum Archives, Adelaide.

Tindale NB (c.1924–91) *Place Names: S.E. of S. Australia for Geo Names Board Adelaide.* AA338/7/1/43, South Australian Museum Archives, Adelaide.

Tindale NB (c.1931–91a) *Place Names: N.B. Tindale Ms SE of S Australia.* AA338/7/1/44, South Australian Museum Archives, Adelaide.

Tindale NB (c.1931–91b) *Birds I–II.* AA338/7/1/27-28, South Australian Museum Archives, Adelaide.

Tindale NB (2005) Celestial lore of some Aboriginal tribes. *Archaeoastronomy* **12–13**, 358–379.

Tindale NB, Hackett CJ (1933) Preliminary report on field work among the Aborigines of the north-west of South Australia, May 31st to July 30th, 1933. *Oceania* **4**(1), 101–105.

Tindale NB, Pretty GL (1980) The surviving record. In *Preserving Indigenous Cultures: A New Role for Museums.* (Eds R Edwards and J Stewart) pp. 43–52, 228. Australian National Commission for UNESCO and Aboriginal Arts Board of the Australia Council, Canberra.

Tobler R, Rohrlach A, Soubrier J, Bover P, Llamas B *et al.* (2017) Aboriginal mitogenomes reveal 50,000 years of regionalism in Australia. *Nature* **544**(7649), 180–184. doi:10.1038/nature21416

Tonkinson M (1994) Healers. In *The Encyclopaedia of Aboriginal Australia*. (Ed. D Horton) pp. 454–456. Aboriginal Studies Press, Canberra.

Tost R (1909) Emu oil advertisement. *Evening News*, New South Wales. 25 September, p. 15.

Trezise PJ (1969) *Quinkan Country: Adventures in Search of Aboriginal Cave Paintings in Cape York*. AH & AW Reed, Sydney.

Troy J (1994) *The Sydney Language*. Aboriginal Studies Press, Canberra.

Tsunoda T (2005) *Language Endangerment and Language Revitalization: An Introduction*. Trends in Linguistics: Studies and Monographs No. 148. Mouton de Gruyter, The Hague.

Tuan Y (1974) *Topophilia: A Study of Environmental Perception, Attitudes and Values*. Prentice-Hall, Englewood Cliffs.

Tunbridge D (1988) *Flinders Ranges Dreamings*. Aboriginal Studies Press, Canberra.

Tunbridge D (1991) *The Story of the Flinders Ranges Mammals*. Kangaroo Press, Sydney.

Tunbridge D, Coulthard A (1985) *Artefacts of the Flinders Ranges*. Pipa Wangka for Nepabunna Aboriginal School, Port Augusta, South Australia.

Turner GW (1966) *The English Language in Australia and New Zealand*. Longmans, London.

Turpin M (2013) Semantic extension in Kaytetye flora and fauna terms. *Australian Journal of Linguistics* **33**(4), 488–518. doi:10.1080/07268602.2013.857571

Turpin M, Ross A (2012) *Kaytetye to English Dictionary*. IAD Press, Alice Springs.

Turpin M, Ross A, Dobson V, Turner MK (2013) The spotted nightjar calls when dingo pups are born: ecological and social indicators in central Australia. *Journal of Ethnobiology* **33**(1), 7–32. doi:10.2993/0278-0771-33.1.7

Tweedie P (1998) *Aboriginal Australians: Spirit of Arnhem Land*. New Holland, Sydney.

Urry J (1985) Savage sportsmen. In *Seeing the First Australians*. (Eds I Donaldson and T Donaldson) pp. 51–67. George Allen & Unwin, Sydney.

Vaarzon-Morel P, Edwards G (2012) Incorporating Aboriginal people's perceptions of introduced animals in resource management: insights from the feral camel project. *Ecological Management & Restoration* **13**(1), 65–71. doi:10.1111/j.1442-8903.2011.00619.x

Van der Leeden AC (1975) Thundering gecko and emu: mythological structuring of Nunggubuyu patrimoieties. In *Australian Aboriginal Mythology*. (Ed. LR Hiatt) pp. 46–101. Australian Institute of Aboriginal Studies, Canberra.

Van Egmond ME (2012) Enindhilyakwa phonology, morphosyntax and genetic position. PhD thesis. School of Letters, Arts and Media/Linguistics, University of Sydney, Sydney.

Vane-Millbank RA (1936) The battlefields of Australia. Part V: Aboriginal lore. *Proserpine Guardian*, Queensland. 15 February, p. 1.

Verburgt B (1999) *They Called Me Tjampu-Tjilpi (Old Left Hand)*. Hesperian Press, Perth.

Verstraete J (2020) *A Dictionary of Umpithamu with Notes on Middle Paman*. Aboriginal Studies Press, Canberra.

Von Brandenstein CG (1977) Aboriginal ecological order in the south-west of Australia: meaning and examples. *Oceania* **47**(3), 169–186. doi:10.1002/j.1834-4461.1977.tb01286.x

Von Brandenstein CG (1982) *Names and Substance of the Australian Subsection System*. University of Chicago Press, Chicago.

Von Sturmer JR (1978) The Wik region: economy, territoriality and totemism in western Cape York Peninsula, north Queensland. PhD thesis. Department of Anthropology and Sociology, University of Queensland, Brisbane.

Waddy JA (1979) Ethnobiology of Groote Eylandt. a progress report. *Newsletter – Australian Institute of Aboriginal Studies* **11**, 46–50.

Waddy JA (1982) Biological classification from a Groote Eylandt Aborigine's point of view. *Journal of Ethnobiology* **2**(1), 63–77.

Waddy JA (1988) *Classification of Plants and Animals from a Groote Eylandt Aboriginal Point of View*. North Australia Research Unit, Australian National University, Darwin.

Walsh J (1951) Paradise for native huntsmen: west Kimberley. *Western Mail*, Western Australia. 12 April, p. 16.

Walsh M (1993) Classifying the world in an Aboriginal language. In *Language and Culture in Aboriginal Australia*. (Eds M Walsh and C Yallop) pp. 107–122. Aboriginal Studies Press, Canberra.

Walsh FJ (2008) To hunt and to hold: Martu Aboriginal people's uses and knowledge of their Country, with implications for co-management in Karlamilyi (Rudall River) National Park and the Great Sandy Desert, Western Australia. PhD thesis. School of Social and Cultural Studies (Anthropology) and School of Plant Biology (Ecology), University of Western Australia, Perth.

Walsh FJ, Mitchell P (2002) *Planning for Country: Cross-cultural Approaches to Decision-making on Aboriginal Lands*. Jukurrpa Books, Alice Springs.

Walshe K (2008) Pointing bones and bone points in the Australian Aboriginal collection of the South Australian Museum. *Journal of the Anthropological Society of South Australia* **33**, 167–203.

Warlpiri Lexicography Group (1986) *Warlpiri–English Dictionary: Flora Section: Yirdikari-Yirdikari Warlpiri-Yingkiliji: watiyapinkikirli*. Lexicon Project Working Papers No.4. Warlpiri Lexicography Group, Massachusetts Institute of Technology, Cambridge.

Warne RM, Jones DN (2003) Evidence of target specificity in attacks by Australian magpies on humans. *Wildlife Research* **30**(3), 265–267. doi:10.1071/WR01108

Warner WL (1937) *A Black Civilization: A Study of an Australian Tribe*. Revised edn, 1958. Harper & Row, New York.

Waterman PP (1987) *A Tale-type Index of Australian Aboriginal Oral Narratives*. Folklore Fellows Communications No. 238. Suomalainen Tiedeakatemia, Helsinki.

Wathen GH (1855) *The Golden Colony: or Victoria in 1854 with Remarks on the Geology of the Australian Gold Fields*. Longman Brown, Green, & Longmans, London.

Watson FJ (1946) Vocabularies of four representative tribes of south eastern Queensland with grammatical notes thereof and some notes on manners and customs. Also a list of Aboriginal place names and their derivations. *Journal of the Royal Geographical Society of Australasia (Queensland)* **48**(Suppl.), 1–114.

Watson P (1983) *This Precious Foliage: A Study of the Aboriginal Psycho-active Drug Pituri*. Oceania Monograph No. 26. University of Sydney, Sydney.

Webb TT (1933) Aboriginal bird names in east Arnhem Land. *Emu* **33**(1), 18–22. doi:10.1071/MU933018

Webb LJ (1960) Some new records of medicinal plants used by the Aborigines of tropical Queensland and New Guinea. *Proceedings of the Royal Society of Queensland* **71**(6), 103–110.

Webb LJ (1969) The use of plant medicines and poisons by Australian Aborigines. *Mankind* **7**(2), 137–146.

Webb LJ (1973) 'Eat, die and learn': the botany of the Australian Aborigines. *Australian Natural History* **9**, 290–295.

Wehi PM, Cox MP, Roa T, Whaanga H (2018) Human perceptions of megafaunal extinction events revealed by linguistic analysis of Indigenous oral traditions. *Human Ecology* **46**, 461–470. doi:10.1007/s10745-018-0004-0

Weldon G (1936) A glimpse of Yankalilla eighty years ago. In *A Book of South Australia: Women in the First Hundred Years*. (Ed. L Brown and Women's Centenary Council) pp. 50–51. Rigby, Adelaide.

Wesson S (2001) *Aboriginal Flora and Fauna Names of Victoria: As Extracted from Early Surveyors' Reports*. Victorian Aboriginal Corporation for Languages, Melbourne.

West D, Sim R (1995) Aboriginal muttonbird exploitation: an analysis of muttonbird (yolla) bone from the Beaton Rockshelter site, Badger Island, Bass Strait. *Australian Aboriginal Studies* **2**, 15–33.

White I (1983) Aboriginal childhood. In *Children of the Wilderness* (Ed. DF Thomson) pp. 1–5. Currey O'Niel, Melbourne.

White ME (1994) *After the Greening: The Browning of Australia*. Kangaroo Press, Sydney.

Whitehouse MW, Turner AG, Davis CKC, Roberts MS (1998) Emu oil(s): a source of non-toxic transdermal anti-inflammatory agents in Aboriginal medicine. *Inflammopharmacology* **6**, 1–8. doi:10.1007/s10787-998-0001-9

Whitley GP (1940) Mystery animals of Australia. *Australian Museum Magazine* **7**(4), 132–139.

Whyte KP (2013) Justice forward: Tribes, climate adaptation and responsibility. *Climatic Change* **120**, 517–530. doi:10.1007/s10584-013-0743-2

Wilhelmi C (1861) Manners and customs of the Australian natives in particular of the Port Lincoln district. *Transactions and Proceedings of the Royal Society of Victoria* **5**, 164–203.

Wilkie B, Cahir F, Clark ID (2020) Volcanism in Aboriginal Australian oral traditions: ethnographic evidence from the Newer Volcanics Province. *Journal of Volcanology and Geothermal Research* **403**, 106999. doi:10.1016/j.jvolgeores.2020.106999

Williams NM (1986) *The Yolngu and Their Land: A System of Land Tenure and the Fight for its Recognition.* Australian Institute of Aboriginal Studies, Canberra.

Wilson M (1937) Emus and the brolgas: an Aboriginal legend. *Chronicle*, South Australia. 3 June, p. 49.

Wilson E (1950) *Churinga Tales: Stories of Alchuringa – The Dreamtime of the Australian Aborigines.* Australasian Publishing, Sydney.

Wilson D (1998) *The Cost of Crossing Bridges.* Small Poppies Publishing, Melbourne.

Wilson G, McNee A, Platts P (1992) *Wild Animal Resources: Their Use by Aboriginal Communities.* Bureau of Rural Resources. Australian Government Publishing Service, Canberra.

Wilson G, Knight A, Liddle L (2004) Increasing the numbers of wildlife preferred by Aboriginal communities in the Anangu Pitjantjatjara lands, Australia. *Game and Wildlife Science* **21**(4), 687–695.

Wiminydji, Peile AR (1978) A desert Aborigine's view of health and nutrition. *Journal of Anthropological Research* **34**(4), 497–523. doi:10.1086/jar.34.4.3629647

Wiynjorrotj P, Flora S, Brown ND, Jatbula P, Galmur J *et al.* (2005) *Jawoyn Plants and Animals.* Northern Territory Government, Darwin.

Woinarski JCZ (1999) Fire and Australian birds: a review. In *Australia's Biodiversity – Responses to Fire: Plants, Birds and Invertebrates.* (Eds AM Gill, A York and JCZ Woinarski) pp. 55–180. Biodiversity Technical Paper. Environment Australia, Canberra.

Woinarski JCZ, Legge S (2013) The impacts of fire on birds in Australia's tropical savannas. *Emu* **113**(4), 319–352. doi:10.1071/MU12109

Woods JD (1879) Some notes on the Aborigines of Australia. *South Australian Advertiser*, South Australia. 11 February, p. 4.

Woodward E, Jackson S, Finn M, McTaggart PM (2012) Utilising Indigenous seasonal knowledge to understand aquatic resource use and inform water resource management in northern Australia. *Ecological Management & Restoration* **13**, 58–64. doi:10.1111/j.1442-8903.2011.00622.x

Worms E (1938) Onomatopœia in some Kimberley tribes of north-western Australia. *Oceania* **8**(4), 453–457. doi:10.1002/j.1834-4461.1938.tb00435.x

Worsley PM (1955) Totemism in a changing society. *American Anthropologist* **57**(4), 851–861. doi:10.1525/aa.1955.57.4.02a00090

Worsnop T (1897) *The Prehistoric Arts, Manufactures, Works, Weapons, etc, of the Aborigines of Australia.* Government Printer, Adelaide.

Wyndham FS, Park KE (2018) 'Listen carefully to the voices of the birds': a comparative review of birds as signs. *Journal of Ethnobiology* **38**(4), 533–549. doi:10.2993/0278-0771-38.4.533

Wynjorroc P, Long H, Long P, Avalon N, Coleman M *et al.* (2001) *Plants, Animals and People: Ethnoecology of the Jawoyn People.* Batchelor Institute of Indigenous Tertiary Education, Batchelor, Northern Territory.

Yallop C (1982) *Australian Aboriginal Languages.* Andre Deutsch, London.

Yates AM, Worthy TH (2019) A diminutive species of emu (Casuariidae: Dromaiinae) from the late Miocene of the Northern Territory, Australia. *Journal of Vertebrate Paleontology* **39**(4), e1665057. doi:10.1080/02724634.2019.1665057

Yengoyan AA (1968) Demographic and ecological influences on Aboriginal Australian marriage systems. In *Man the Hunter.* (Eds RB Lee, I DeVore and J Nash-Mitchell) pp. 185–199. Aldine Publishing, New York.

Young E (1983) *The Aboriginal Component in the Australian Economy: Tribal Communities in Rural Areas*. Development Studies Centre, Australian National University, Canberra.

Young E, Ross H, Johnson J, Kesteven J (Eds) (1991) *Caring for Country: Aborigines and Land Management*. Australian National Parks and Wildlife Service, Canberra.

Yunupingu B, Yunupingu-Marika L, Marika D, Marika B, Marika B *et al.* (1995) *Rirratjinu Ethnobotany: Aboriginal Plant Use from Yirrkala, Arnhem Land, Australia*. Northern Territory Botanical Bulletin No. 21. Conservation Commission of the Northern Territory, Darwin.

Zander KK, Austin BJ, Garnett ST (2014) Indigenous peoples' interest in wildlife-based enterprises in the Northern Territory, Australia. *Human Ecology* **42**(1), 115–126. doi:10.1007/s10745-013-9627-3

Zeanah DW, Codding BF, Bird DW, Bird RB, Veth PM (2015) Diesel and damper: changes in seed use and mobility patterns following contact amongst the Martu of Western Australia. *Journal of Anthropological Archaeology* **39**, 51–62. doi:10.1016/j.jaa.2015.02.002

Ziembicki MR, Woinarski JCZ, Mackey B (2013) Evaluating the status of species using Indigenous knowledge: novel evidence for major native mammal declines in northern Australia. *Biological Conservation* **157**, 78–92. doi:10.1016/j.biocon.2012.07.004

Zorc RD (1996) *Yolngu-matha Dictionary*. Education Technology Unit, Batchelor College, Batchelor, Northern Territory.

Index